POLLUTED RAIN

Environmental Science Research

Recent Volumes in this Series

A Continuation Order Plan is available for this series. A continuation order will bring
delivery of each new volume immediately upon publication. Volumes are billed only upon
actual shipment. For further information please contact the publisher.

POLLUTED RAIN

Edited by
Taft Y. Toribara
Morton W. Miller
and
Paul E. Morrow

University of Rochester Medical Center
Rochester, New York

PLENUM PRESS · NEW YORK AND LONDON

Library of Congress Cataloging in Publication Data

Rochester International Conference on Environmental Toxicity, 12th, 1979.
  Polluted rain.

  (Environmental science research; v. 17)
  "Proceedings of the Twelfth Rochester International Conference on Environmental
Toxicity organized by the Department of Radiation Biology and Biophysics, the
School of Medicine and Dentistry of the University of Rochester, held at Rochester,
New York, May 21–23, 1979."
  Includes index.
  1. Acid precipitation – Environmental aspects – Congresses. I. Toribara, Taft
Yutaka, 1917-   II. Miller, Morton W., 1936-   III. Morrow, Paul E. IV. Ro-
chester, N. Y. University. Dept. of Radiation Biology and Biophysics. V. Title.
[DNLM: 1. Environmental pollution – Congresses. 2. Weather – Congresses. 3. Envi-
ronmental pollutants – Adverse effects – Congresses. W1 EN986F v. 17/WA670
R676 1979p]
QH545.A17R62   682.5   80-285

ISBN-13: 978-1-4613-3062-2   e-ISBN-13: 978-1-4613-3060-8
DOI: 10.1007/978-1-4613-3060-8

Proceedings of the Twelfth Rochester International Conference on
Environmental Toxicity: Polluted Rain, held in
Rochester, New York, May 21–23, 1979

© 1980 Plenum Press, New York

Softcover reprint of the hardcover 1st edition 1980

A Division of Plenum Publishing Corporation
227 West 17th Street, New York, N.Y. 10011

PREFACE

This is the twelfth in a series of conferences on environmental toxicity sponsored by the Department of Radiation Biology and Bio- physics. The topics selected are intended to be of timely interest with a focus somewhat different from those with with similar titles. I would like to share with you some of the thought which went into the development of this year's program.

Everyone here is aware of the great amount of publicity being given to some phase of this year's topic of "Polluted Rain", but I wonder how many of the younger generation realize how the entire world got into this kind of predicament. I can remember as a young- ster that precipitation from the sky in the form of rain or snow was considered one of the ultimates in purity. That was in the era when a small number of automobiles were in use and the airplane was more of a curiosity in the developmental stage than anything of commer- cial significance. Many homes used iceboxes for refrigeration, and summer cooling was accomplished by electric fans by the more afflu- ent families.

For one who has witnessed such an era, it is easy to see that the home gadgets accepted as necessary in every household, the large number of automobiles, the great amount of air travel and the huge industrial plants to produce all these have increased the energy consumption by almost astronomical proportions. A little thought will bring out the fact that this energy must be supplied from some source, and a little more thought concerning the chemical reactions involved in this energy production should have elicited the fact that the products of the reactions could be a source of serious problems.

In the philosophy of the earlier years, the oceans and the atmosphere were considered to be infinite reservoirs in which wastes of all types would be diluted to insignificance and ultimately elim- inated by nature. One of the early indicators that such was not the case was the demonstration that the oceans could not dilute out the effects of DDT which was used rather indiscriminately after World War II. Its use has been banned, and much other progress in

cleaning up the rivers and lakes through the elimination of the direct dumping of wastes has been made.

The problem of treating atmospheric pollution is still in its
infancy. As late as the '60s, the atmosphere was still considered
a limitless sump in which noxious fumes created by the combustion
of fuels would be diluted to harmless proportions. Consequently,
when large power plants first were found to contaminate the immediate surroundings, the designers solved the problem by building
very tall smokestacks. This solved the local problem, but, as the
information in this conference will show, the noxious products were
not diluted to harmless proportions.

In recent times the subject of atmospheric pollution has been
brought home forcefully to the public by the large amount of exposure in the news media given to the phenomenon of "acid rain." A
number of scientific conferences have also been devoted to this
topic. However, in our concerns with the pollution of the atmosphere, we find that the acid is involved in more ways than the
effects of the protons as would be suggested by such a title. We
have, therefore, chosen the broader title of "Polluted Rain" in
order to emphasize that the rain is principal depositor of the
atmospheric pollution. We have further used the term "rain" in its
broadest sense--that of anything that "rains down from the sky."

Obviously we could not cover everything in the time alloted
for the conference, so we have selected topics that would result in
a total picture or complete story. We have started with the sources
of pollution, described some of the mechanisms for the formation of
the offensive materials, detailed some of the damage to flora and
fauna, shown damage to stationary structures, outlined the present
systems for monitoring the pollutants, and finally injected what is
the current status of the legal implications. The final portion
was included to emphasize the fact that the pollution problems
transcend boundaries, both interstate and international. Legal
action has been hindered by the lack of unequivocal evidence pin-
pointing the causes of the problem. It is the hope that conferences
such as this will produce the necessary evidence.

A final thought has been brought on by the recent traumatic
demonstration of the undesirability of dependence on foreign sources
of oil for energy. The prospects of massive pollution looms in the
large scale switch to coal. Hopefully the technology to eliminate
the pollution will accompany this switch.

Taft Y. Toribara
Conference Chairman

ACKNOWLEDGMENTS

    We gratefully acknowledge the support for this conference by the Department of Energy, Empire State Electric Energy Research Corporation, Environmental Protection Agency, National Institute of Environmental Sciences and the University of Rochester. For their time-consuming performances in taking care of all the details for a successful conference, we are especially indebted to Jane Leadbeter and Judy Havalack.

The Conference Committee
Taft Y. Toribara, Chairman
George G. Berg
Thomas W. Clarkson
Morton W. Miller
Paul E. Morrow

CONTENTS

SESSION II: THE MERCURY PROBLEM
George H. Tomlinson, Chairman

SESSION III: EFFECTS ON PLANTS
Morton W. Miller, Chairman

SESSION IV: ANTICIPATED PROBLEMS
AS YET NOT QUANTITATED
Paul E. Morrow, Chairman

SESSION V: CONTROL PROBLEMS:
MONITORING SYSTEMS AND LEGAL ASPECTS
Dennis J. Sugumele, Chairman

WELCOMING REMARKS

It is my pleasure to welcome you to the Medical Center of the University of Rochester for the Twelfth Rochester International Conference on Environmental Toxicity sponsored by the Department of Radiation Biology and Biophysics. There is general, world-wide concern for the condition of the environment today, but such was not the case when these conferences were started 12 years ago. We in Rochester feel that these conferences and the subsequent dissemination of their information through the publication in book form help in bringing about an awareness and understanding of timely toxicological problems, current or potential.

Indeed, we were living in what seemed an endless era of consumptive plenty, with little regard for the environmental effects of producing or disposing of so much. Our first conference accurately predicted a potential mercury problem in this country, and, we believe, the proportions of the problem were greatly diminished by these efforts. A similar situation arose with persistent pesticides, and again our conference proceedings directly impacted on the New York State legislature, eventually banning persistent pesticides. Thus, there is direct evidence of the value of these conferences: to scientists and educators, and to organizations dealing with scientific frontiers and technology as they affect our environment.

For those of you who are unfamiliar with the history of the Department of Radiation Biology and Biophysics which sponsors these conferences, I would like to give you a brief sketch. During World War II, the Manhattan District of the Army Corps of Engineers, which was responsible for developing the atom bomb, established in Rochester a unit which had as its principal function the responsibility of seeing that all perconnel involved in the Manhattan Project remained healthy. To this end many toxicological studies to establish the limits of radiation and chemical exposure were done. After the war the project was continued under civilian control to study problems arising from peacetime use of atomic energy. The Project became a preclinical department of the Medical School in 1948 as the Department of Radiation Biology which subsequently in

1965 underwent a name change to the Department of Radiation Biology
and Biophysics to reflect more precisely the nature of the ongoing
studies. Not reflected in the name, however, is the department's
close association with the Medical School's Toxicology program and
the Environmental Health Sciences Center.

The problems with toxic chemicals discussed in the first conference entitled "Chemical Fallout" have been unfortunately and
ironically demonstrated close to Rochester in Buffalo's "Love Canal
Incident", which has received national attention. It is obvious
now that a gross ignorance of chemical-soil functions in the earth
portion of the environment led to the tragically ignorant disposal
of toxic chemicals. Another demonstration of this ignorance is
coming to light in the discovery of damages to the environment in
places where none of the noxious substances is being generated.
Again we in Rochester and the Northeastern portion of the United
States are conscious of the damage close by in discovering that
some of the lakes in the Adirondacks no longer support life.

Since the latter type of pollution is obviously atmospherically
borne with much of it brought down by rain, the topic of this year's
conference "Polluted Rain" is a very timely one. I note that in
addition to the important topics of the causes, effects, and mechanisms of this pollution, some time is also being devoted to the
legal aspects of attempts to control such pollution. That science
is becoming more interdisciplinary than before is quite obvious.
Generated pollution is often deposited in an area outside its
origin and often crosses borders--interstate and international.
Clearly there are scientific and legal problems. I can see that
this conference should cover some very complicated issues.

You may have noted that the number of participants in this
conference is rather small. This has been done purposely in order
that all can participate readily in the discussions. The dissemination of the valuable information generated will be published in
book form available to all. Past experience has shown that this
format is a successful one, and I urge your full participation in
it.

Again, let me extend to you a warm welcome, and I join the
committee in the hope that we in Rochester will make your stay
pleasant, stimulating, and enlightening.

J. Lowell Orbison
Dean, School of Medicine
and Dentistry
The University of Rochester
Rochester, New York

WELCOME FROM THE DEPARTMENT OF RADIATION BIOLOGY AND BIOPHYSICS

On behalf of the Department of Radiation Biology and Biophysics I am delighted to welcome you to this 12th in the series of International Conferences on Environmental Toxicology. As the Dean has noted the Department is heterogenous in interests and faculty, and has had the remarkable breadth to support the variety of topics represented in the last eleven conferences. We consider the topic of this conference to be particularly germaine and certainly a timely one in terms of both regional and national concerns.

In addition to the opportunity to welcome you, I am pleased to acknowledge the individuals of our Department who have made this Conference possible. I congratulate Drs. Toribara, Berg, Clarkson, Miller and Morrow for their accomplishments in organizing this very promising meeting, and acknowledge with gratitude the work of Mrs. Havalack, Leadbeter and Marsden.

Agencies which have provided financial support include the Department of Energy, the Empire State Electric Energy Research Corporation, National Institute of Environmental Health Sciences, and Environmental Protection Agency; in addition to support from the University of Rochester. I'd like to acknowledge all of these because I think it's farsighted on the part of these agencies to support such a conference as this which is, by no means, parochial in its interests but has the objective of collating and presenting this important information in a form that will profit not simply the people here but throughout the world.

Finally, I wish you success in the conference, and as the Dean suggested, I think the interchange, the dynamics of the conference itself will be important, perhaps even more so than the publication. Thank you very much.

Paul L. LaCelle, M.D.

SESSION I:
THE CHEMISTRY OF POLLUTED RAIN

THE ACIDIC PRECIPITATION PHENOMENON. A STUDY OF THIS
PHENOMENON AND OF A RELATIONSHIP BETWEEN THE ACID
CONTENT OF PRECIPITATION AND THE EMISSION OF SULFUR
DIOXIDE AND NITROGEN OXIDES IN THE NETHERLANDS

Arend J. Vermeulen

Department of Environmental Control

The Netherlands

ABSTRACT

    The Netherlands is a nation which is highly industrialized
and densely populated. Along with other European countries
noting the increase in acidity of the precipitation, The
Netherlands became part of the extensive European network of
atmospheric measuring stations which are coordinated by the
International Meteorological Institute (IMI) in Stockholm. Three
stations are part of this network, and in addition the province
of North Holland has installed 22 sampling stations. The
extensive information available over a period of years has made
possible the correlation between the acidification of
precipitation and the acidic gas content of the atmosphere. A
pronounced relationship between the total interior emission of
SO_2 and the acidification was found, but no such relationship
could be shown for the nitrogen oxides. Factors which have an
influence such as the wind direction and the dust content have
been investigated.

    The total acidity in the years around 1966 was the highest in
the world for this country. Increased energy requirements
combined with a change from pit coal to oil resulted in the
increased sulfur dioxide emissions. The reduction after 1967 was
then credited to a large conversion to gas when the largest
coherent natural gas field in the world was found in the province
of Groningen. Because of a prior commitment for the exportation
of much of this gas, the domestic use will be reduced until the
year 2000. In the meantime a reconversion to oil and coal will
be necessary to satisfy the energy requirements, and the SO_2
emissions accompanied by increased acid deposition are expected
to rise.

INTRODUCTION

Various tests of the chemical composition of precipitation
indicate that acids are present in larger quantities than would
occur naturally. This phenomenon is attributed to ever
increasing sulfur dioxide (SO_2) and nitrogen oxides
(NO_x = NO + NO_2) pollution of the air. In Europe, the
acidification of precipitation was found to increase gradually
in various countries. Because of this phenomenon and its
possible environmental consequences, the number of rain water
analyses has been increased.

In 1952, at the initiative of Sweden an extensive European
network was installed; a few years later this network consisted
of more than 150 measuring locations. The co-ordination is
effected by the International Meteorological Institute in
Stockholm (IMI). In The Netherlands, the De Bilt, Witteveen
and Den Helder stations are part of this network.

Around 1966, the highest acid precipitation measurements in
the world (on a yearly basis) were made in The Netherlands.
After 1967, the acid content of rain decreased, simultaneously
with a considerable reduction in SO_2 emissions. In 1970, the
Government of the province of North Holland started an
experimental rain water sampling network consisting of 7
sampling locations. A few years later it graduated from the
experimental stage, and in 1974, this network consisted of 22
sampling locations. The sampling was performed on a daily
basis and some of the data on the acid content of the
precipitation are given in this report.

As the coming years will again be characterized by a sharp
rise in SO_2 emissions in The Netherlands, it is to be
expected that the acidity of precipitation will increase as
well, possibly up to values exceeding the 1967 figure at De
Bilt when the pH was 3.78 on a yearly basis. In this paper an
attempt will be made to indicate a relationship between the
development of acid concentration of precipitation in The
Netherlands and total interior emissions of SO_2 and NO_x.

GENERAL

The degree to which a liquid is acidified, is expressed by
the pH values of the liquid. For the nonexpert a short
description of this term is given in the following:

Pure water is to a small extent split into ions according
to the following reaction:

$$H_2O \rightleftharpoons H^+ + OH^-$$

water   hydrogen   hydroxyl
                  ions   ions

The product of H^+ OH^- is constant which according to very precise measurements equals 10^{-14}. The pH is defined as follows:

$$pH = -\log [H^+]$$

In pure water the H^+ concentration will equal the OH^- concentration; according to the above definition pure water therefore has a pH = 7.

If the water (rain water for example) becomes polluted with SO_2 and NO_x, a transformation into sulfuric and nitric acids takes place. As acids are characterized by their property to split off H^+ ions (Table I), the number of H^+ ions increases.

According to the definition the pH will become less than 7 and the liquid is called acid. As the pH is a logarithm, a decrease by 1 pH unit means an increase of the acidity by a factor of 10. A liquid having a pH 6 will therefore be 10 times as acid as one with pH 7. There are strong and weak acids. Strong acids are characterized by the complete splitting off of H^+ ions (Table I), whereas weak acids only partially split off H^+ ions (see equation for carbon dioxide). Rain water, normally, contains hardly any acids. In theory its pH should therefore lie near the neutral value 7, i.e. between 6 and 8.

The carbon dioxide (CO_2) contained in the atmosphere will, to a certain extent, dissolve in rain water and form carbonic acid (a very weak acid):

$$CO_2 + H_2O \rightleftharpoons HCO_3 + H^+$$

The minimum pH of rain water saturated with carbonic acid will be 5.5 to 5.6. In areas where the soil is calcareous, in case of a pollution of precipitation by dust raised by the winds, the pH will reach values somewhat above 7. From the ion balances of rain water samples preserved prior to 1930 (Tennessee), Cogbill has calculated that at that time the pH values were about 7.5. In glacier ice from the Cascade Mountains, minimum pH values of about 5.6 were found, which corresponded to the pH of water saturated with carbonic acid. Collinson and Mensching, performing rain water analysis (1919 –

1929, in Geneva, N.Y.S.) found relatively large quantities of
bicarbonate. Since bicarbonate cannot exist in the presence of
strong acids, the pH values of their samples must have been
about 7. In The Netherlands, in the 1932 - 1937 period,
Leeflang found pH values between 5 and 7, which he described as
being "relatively low".

In the last ten years, however, it appears that over large
areas in the Western Hemisphere, pH values of single
precipitations fluctuated between 2 and 6. This translates to
an acid content 10 to 10,000 times greater than is possible
naturally. The averages on a yearly basis[*] fall between pH 4
and pH 5 for large areas, and even below pH 4 in small areas.

[*]Note:
     In this paper, average pH values or pH values on a yearly
basis are frequently mentioned. When calculating an average pH
for a certain period, many authors make a fundamental error by
averaging single pH values measured in a certain period
according to the following formula:

$$\overline{pH} = \frac{\sum_i pH_i}{n}$$

Logarithms are then just added arithmetically and divided by
the number of observations, and the quantity of the
precipitation is not included in the calculation. Such
averaged pH values cannot be used for comparative examinations.

The correct method for calculating the average is to
convert the pH values into H^+ concentrations and to multiply
these H^+ concentrations by the corresponding precipitation
quantity. The total quantity of H^+ ions is divided by the
total precipitation quantity, and the logarithm is taken
according to the formula:

$$\overline{pH} = \log_{10} \frac{\sum_i H_i^+ \cdot mm_i}{\sum_i mm_i}$$

The average pH values of the IMI network and all other
examinations mentioned in this paper have been calculated in
this way.

From past data it is clear that the acidification of precipitation is a phenomenon of the present day, particularly in densely populated and highly industrialized areas of the Western Hemisphere.

It has been proven that the acidification of precipitation is caused principally by the presence in the atmosphere of the gaseous components sulfur dioxide (SO_2) and nitrogen oxides (NO and NO_2, the mixture indicated as NO_x). These gases are emitted in large quantities in the air by human activities. SO_2 is formed by oxidation of the sulfur present in fossil fuels, while NO_x is formed in every combustion process in which high temperatures arise. SO_2 and NO_x can be transformed in the atmosphere into compounds that are soluble in water, such as cloud water and raindrops, forming at the same time sulfuric acid and nitric acid (Table 1).

However, according to some authors there exist apart from sulfuric and nitric acids still other acids in rain water. During a test, Frohliger and Kane found concentrations of weak acids that were from 2 to 200 times as high as the sulfuric and nitric acid concentrations. They assumed that rain water should be considered to be a solution of a weak acid, and a number of objections have been raised against this conclusion (Science, vol. 194, 5 Nov. 1976). On the basis of more accurate examinations carried out by others in the U.S.A. and Europe, such conclusions could not be drawn; it appears that the acid contained in rain water consisted of an average of 80 to 100% of a mixture of sulfuric and nitric acids. Only a small percentage of organic acids was found. Likens looked for the existence of 33 weak acids in rain water; he found only one (isocitric acid) in a very small concentration.

The acid content of precipitation can be measured in different ways.

A) with pH-paper (indicator paper). This method is very inaccurate and can only be used for a rough indication.

B) with a pH-electrode. With this method only free protons can be measured, bound protons cannot. This method is the most applied one for measuring the free proton concentration (pH) of a liquid. The accuracy can be within 0.05 pH-units.

C) by titration*.

---

*With this method one can determine the total acidity of a sample, i.e. the total quantity of components that consume alkali.

Table 1. Some Chemical Reactions for the Formation of
Acids in Precipitation

1. CO_2 + H_2O $\rightleftharpoons$ HCO_3^- + H^+

HCO_3^- $\rightleftharpoons$ $CO_3^=$ + H^+

2. $2\ SO_2$ + O_2 $\longrightarrow$ $2\ SO_3$

SO_3 + H_2O $\longrightarrow$ $SO_4^=$ + $2\ H^+$

3. $2\ NO$ + O_2 $\longrightarrow$ $2\ NO_2$

$3\ NO_2$ + H_2O $\longrightarrow$ $2\ NO_3^-$ + NO + $2\ H^+$

4. NO + O_3 $\longrightarrow$ NO_2 + O_2

$3\ NO_2$ + H_2O $\longrightarrow$ $2\ NO_3^-$ + NO + $2\ H^+$

5. $2\ NO_2$ + O_3 $\longrightarrow$ N_2O_5 + O_2

N_2O_5 + H_2O $\longrightarrow$ $2\ NO_3^-$ + $2\ H^+$

6. $4\ NO_2$ + $2\ H_2O$ + O_2 $\longrightarrow$ $4\ NO_3^-$ + $4\ H_+$

The velocity and preponderance of each reaction depends on
the degree and kind of pollution of the ambient air. It can
range from a few minutes to some hours in heavily polluted air
and even to several weeks or more in non-polluted air.

Explanation of above equations

1. The solution of CO_2 in water; the minimum pH that can be
attained this way is about 5.5.

2. The stoichiometric equation for the oxidation and transformation of SO_2 into sulfuric acid.

3. The formation of nitric acid starting from nitrogen monoxide
(the reaction velocity is low).

4. The formation of nitric acid from nitrogen dioxide formed in
the presence of ozone (the reaction velocity is high).

5. The formation of nitric acid when the ozone concentration is
higher than 0.1 ppm.

6. The catalytic oxidation of nitrogen dioxide in fog and water
droplets. Catalysts are trace metals and/or SO_2.

When titrating a rain water sample according to an acid/base-titration, one sometimes finds that the sample consumes more alkali than should be expected on the basis of the acid concentration calculated from the pH. The surplus of alkali consumption is caused, however, by the simultaneous titration of the components Al^{3+}, Fe^{2+}, Mn and NH_4^+, the so-called Bronsted acids, sometimes contained in precipitation, and the titration of (if present) weak acids whose contribution to the free acidity can be neglected.

In summary the following can be stated:

-- rain water samples sometimes contain Bronsted acids. Their influence on the pH (H^+ ion concentration) is small and can in case of pH values under 5.5 - as far as the percentage is concerned - be neglected.

-- A weak acid that can be found in rain water in relatively large quantities is a carbonic acid; the minimum pH value of a solution of carbonic acid in rain water is about 5.5. In a rain water sample having a pH 5.5 the carbonic acid equilibrium is reversed and CO_2 will disappear from the sample. In such samples the pH value is entirely determined by the free H^+ ions of strong mineral acids such as sulfuric acid and nitric acid.

THE PHENOMENON OF LOW pH VALUES IN THE WESTERN HEMISPHERE

The considerable increase in the acidity of precipitation in the years after World War II coincided with enormous industrial expansion and the nearly exponentially rising consumption of energy in Western Europe accompanied by the emissions of SO_2 and NO_x. During a period of 15 to 20 years, the pH of precipitation declined at many measuring locations of the IMI network, on a yearly basis, from 6 to 4, which means an increase in acidity by at least a factor of 100. At some locations the increase was even larger.

In 1956, the center of acid precipitation in Europe was situated over the southeast of England, the northwest of France and Belgium, Luxembourg and The Netherlands; average pH values on a yearly basis ranged from 4.5 - 4.0. After 1965, the acidification further increased.

Figure 1. The acidification pattern in Europe about 1966 with
           the center over The Netherlands.

     In The Netherlands, around 1966, pH values fell below 4 (on
a year basis). In 1967, the year's average at De Bilt was
3.78, and regions with pH values between 4.5 and 4 were
increasing in number. As The Netherlands is the most densely
populated country in Europe and is also highly industrialized,
it is easily understood that in those years, with oil as the
main fuel, acidification of precipitation over The Netherlands
was so considerable (see Table 2).

     In the United States of America - particularly in the
Northeast - an acidification pattern has developed that is
comparable to the European experience. On the West Coast,
Likens found that in 1955 - 1966, on a yearly basis, the pH
values ranged between 5.5 and 6.0 and for the northwestern part
of the U.S. values were measured nearly 4.5. In the years 1972
- 1973 the values on the West Coast were between 4.6 and 5.6 on
a yearly basis and in the Northeast near 4.0 and 4.2. In the
northeastern part of the U.S. there is also the phenomenon that
the highest degree of acidification is measured in a densely
populated and highly industrialized area. Figure 2 shows the
pH distribution over the U.S.A. in a randomly selected month
(June, 1966).

Table 2. The SO_2 Emission Densities (tons-km^{-2}) in some European Countries around 1968 and 1974

	1968	1974
The Netherlands	3]	12
Belgium[+], Luxembourg, Britain, Ireland	20--24	23--33[+]
Western Germany	13	16
Denmark[+], France, Austria, Switzerland	3--6	3--14[+]
Southern Sweden	2	3
Finland, Norway, Iceland, Northern Sweden	< 1	< 3

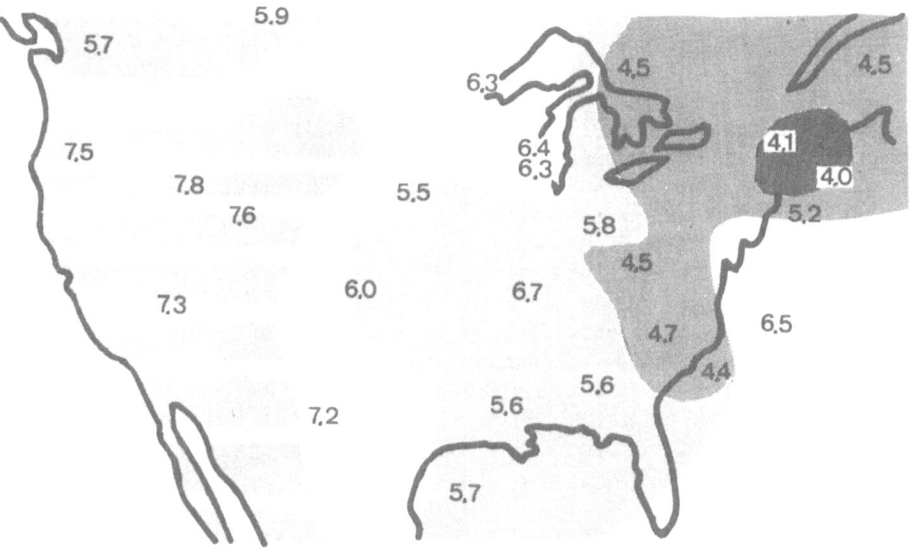

Figure 2. The acidification pattern over the U. S. A. for a random month (June 1966).

A STUDY OF THE ACID CONTENT OF PRECIPITATION IN THE NETHERLANDS

In the Netherlands, since 1956, in the framework of the
IMI-programme (see introduction and Figure 12) of three
locations, the precipitation has been collected on a monthly
basis and analyzed at a central point for a great number of
components, as well as the acidity, which is measured with a pH
meter.

In this paper, the pH figures found in precipitation from
samples at these three locations will be investigated more
closely, but first a close look at an experimental rain water
sampling network set up in 1970 by the Provincial Government of
North Holland with the intention to study the quality of the
precipitation over this province is in order.

In the beginning this network consisted of seven sampling
locations spread through out the province. This study was
different from other rain water studies, because samples were
taken on a daily as well as on a monthly basis. Every rainy
day, at all locations, samples were taken together with data on
wind direction, wind velocity and characteristics of the
precipitation. Originally, only sulfate, ammonium, fluoride
and the pH were determined. In this experimental state, a lot
of experience was gained with analyzing technics, sampling, the
height for the rain sampling rig, the size of the sampling
surface, evaporation of the sample, contamination of the
samplers by bird droppings, contamination of the samples by
insects, algae growth, the influence of bacteria on the
composition of the sample and the influence of dry deposition
on the composition of and the acidity of the samples. An
important factor also seemed to be the choice of materials. In
this study, all materials were of polyethylene which was free
of catalysts (polyethylene manufactured under high pressure).
The results of the study were encouraging, and in 1974 the
sampling network was extended to 22 sampling stations (see
Figure 3). The number of components to be analyzed was greatly
enlarged and the pH was determined with a pH meter with an
accuracy of 0.05 pH unit. From the results of this network we
could, concerning the acid content of precipitation, draw some
interesting conclusions.

A)   We found that drizzles are much more acidic than rains
        with drops of a larger diameter. This is obviously due to
        the relatively larger surface of the drizzle drop through
        which the wash-out efficiency of the air parcel between
        ground and clouds is high. Over 80% of the drizzles in
        North Holland had a pH of between 3.0 and 3.5. In some
        cases the pH was even between 2.5 and 3.0.

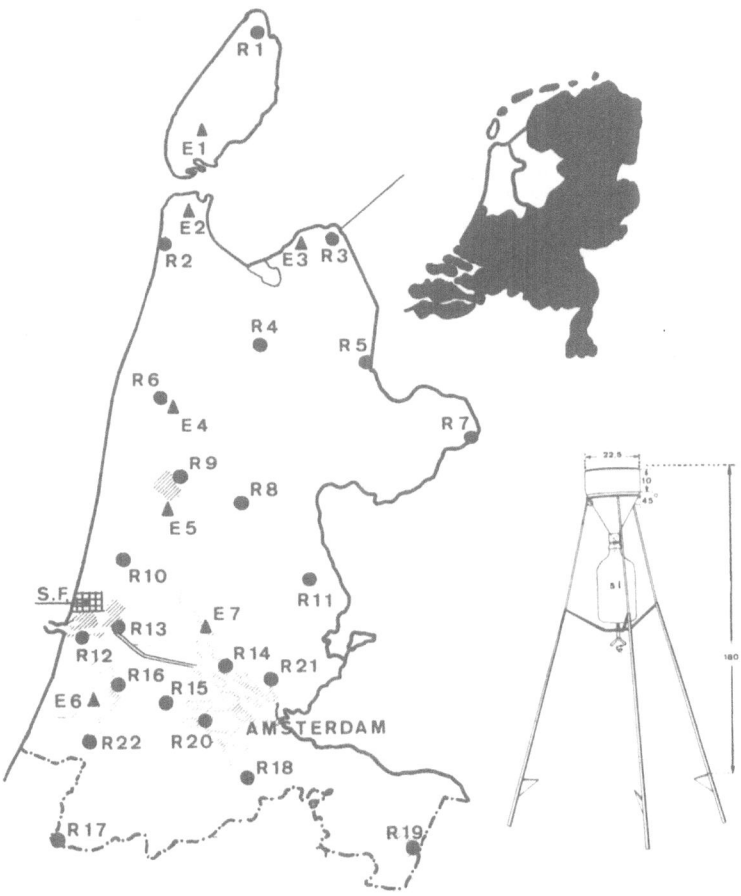

Figure 3. The rain sampling network in the province of North
           Holland.
           Inset: Top--The Netherlands with North Holland
                   Bottom--Rain sampling rig, measures in cm.

                  $E_1 \cdots \cdots E_7$ the experimental network
                  $R_1 \cdots \cdots R_{22}$ the provincial network
           S.F. = steel factory

           Hatched area is heavily industrialized.

B)   In general, the pH of precipitation in North Holland,
compared with other areas in Europe or areas in the Western
Hemisphere, is at a lower level. In other words, the acid
concentrations were and are rather high. Remarkably, the
acid concentration in the rural areas is higher than the
acid concentration of precipitation sampled in the vicinity
of the industrial areas which have a higher dust content in
the atmosphere. On rainy days this dust is caught by the
raindrops, while on dry days the dust is caught by the rain
sampler itself and finally collected into the sample. In
both cases, this dust partially neutralizes the initially
acid rain water.

          The industrial area from which the greatest influence
on the composition of the samples can be expected is given
in Figure 3, marked with S.F. In general, the closer the
sampling locations are situated near industrial areas, the
higher the pH level is, as we can see in Table 3.

Table 3.   pH Levels at Different Locations in Figure 3.

Site	pH < 5	pH < 4	N
E 1	75%	17%	88
E 2	·70	12	204
E 3	65	10	163
E 4	54	9	90
E 5	45	6	112
E 6	54	5	123
E 7*	85	31	114

*Strongly influenced by local emissions

          To get a picture of the pH level of the precipitation
of the sampling locations, some frequency distributions are
given over the period 1974 - 1976 in Figure 4.

          Remarkable is that some five years earlier (around
1970) a small part of the samples taken at that time had a
pH between 2.5 and 3.0. After 1974 however, except for an
incidental case, we did not find such levels of pH
anymore. The magnitude of the interior SO_2 emission is
strongly related to this.

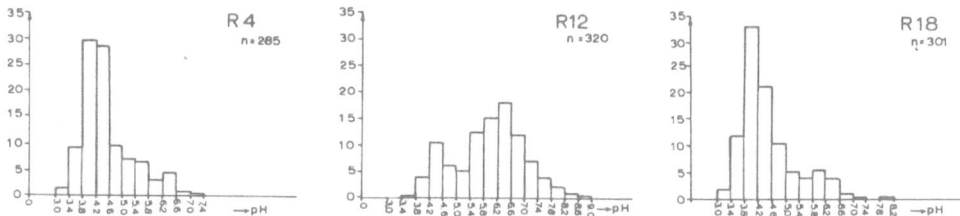

Figure 4. The frequency distribution of the pH for locations
in rural areas (R4 and R18) and an industrial area (R12) for
the period 1974-1976. The y-axis is in percent.

C)   For all sampling locations there was a relationship
between the pH and the prevailing wind direction during the
rainstorm or rain period. When the air passed over (large)
land areas, higher acid concentrations were found than when
the wind came more or less from the ocean. This means, for
the sampling network in North Holland, higher acid
concentrations result from directions of wind from East to
West (90°-270°) rather than wind directions from West to
East (270°-90°). An example is shown in Figure 5.

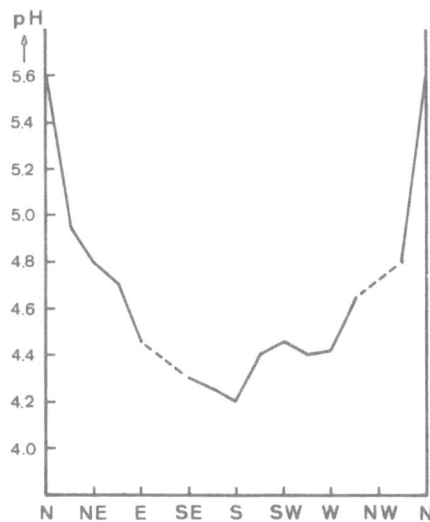

Figure 5. Annual average pH as a function of wind direction.
A similar picture was noted for all locations.

Furthermore, it was shown that, in general during rainy periods with an long lasting supply of air from northerly directions (except for northeast), because of the supply of relatively unpolluted arctic air, the pH of precipitation kept going up. With a continuing supply of air masses out of southerly directions however, because of the supply of relatively polluted air from the European continent, in general, the pH on days with rain gradually went down.

D)   No significant differences were shown between the pH of snow and rain. This is to the contrary of findings of other researchers. Lavrinenko (Russia) and Likens (U.S.A.) found higher acid contents in summer rains than in winter snows. Landsberg (U.S.A.) and Oden (Sweden), however, found lower acid concentrations in summer than in winter. Mrose (DDR) found, even as I did, no significant difference between the average value of winter snows and summer rains. Comissot (Lebanon) found snow to be more acidic than rain.

E)   It was shown that the calculation of an average pH of daily samples over a certain period, in almost all cases, lead to a lower pH value than the pH value of rain collected in one big sampling bottle over the same period. This can be explained by the fact that during the period (a month for example) the sampling bottle is in the field, chemical reactions in the collected sample take place between the acid present in this sample and the sampler collected dry deposits, such as dust particles, mostly originating from industrial processes. Furthermore, temperature, sunlight and bacteria can change the composition of the sample.

This all means that the pH should be measured on a daily basis, or better, directly after the rainstorm or rain period. It is also recommended when more components have to be determined, that samples should be brought to the lab and stored at low temperatures (below freezing points) as soon as possible after collection. It may be assumed that the pH figures of the last 20 years, because they have been measured by taking monthly samples, present an overly optimistic picture. Tn Table 4 some of the noticed differences between daily and monthly sampling are shown just in order to give an idea of the possible magnitude of those differences.

F)   It was remarkable that for each sampling location there was a significant relationship between the SO_4 and NO_3 concentrations in precipitation. The correlation

Table 4. A Comparison between Results of Monthly and Daily
          samples

A pH in a random monthly sample	B Average pH calculated from daily samples for the same month as in A	C H^+ ratio B/A in per cent
4.76	4.50	81
4.76	4.40	128
4.60	4.29	104
4.29	4.11	55

coefficient varied between 0.71 and 0.96 (n = 15), monthly
averages). In almost all cases, in single precipitation
samples, the concentration of SO_4 was about the double
that of the NO_3 concentration. Sometimes, very high
concentrations of SO_4 were found (200 mg/1). Although
one could expect that with such high sulfate concentrations
the pH of the sample would be low, in many cases, however,
this pH seemed to be on the high side.

A correlation test beteen the proton concentrations
and some relevant minerals in precipitation lead to the
following results:

H^+/NH_4^+ no significance (not a single case)

H^+/NO_3^+ a significant correlation in 4 cases

$H^+/SO_4^=$ a significant correlation in just one case

These calculations have been performed for 12 sampling
locations with monthly averages (n = 15).

With a multiple regression analysis, using the model

$$H^+ = \beta_0 + \beta_1 \cdot NH_4 + \beta_2 \cdot NO_3 + \beta_3 \cdot SO_4,$$

for 60% of the sampling locations there was a significant
relationship between the H^+ concentration on the one side
and NH_4 + NO_3 + SO_4 on the other. If there was such
a relationship, NO_3 in 45% was the dominant variable,
SO_4 in 20% and in 35% of the cases there was no dominant
variable at all. In all case, β_1 when tested, did not
differ significantly from zero, showing that the influence
of the concentration of ammonium on the pH of precipitation
was negligible.

Other investigators studying the influence of SO_2
emissions of big power plants on the surroundings,
sometimes found a significant relationship between the
SO_4 and H^+ in the collected precipitation samples. I
studied the figures and came to the conclusion that in most
cases the authors did their experiments in areas relatively
unpolluted by other components. The concentrations of the
major ions seemed to be much lower than in the
precipitation in Holland, where the composition of the
precipitation is rather complex with very high
concentrations of the major ions. For example, the
concentrations of NO_3 and SO_4 are probably the highest
in Europe and probably in the world (the yearly average
concentrations are about 5 and 9 $mg.l^{-1}$, respectively;
the total mineral load, including other components, varied
among the sampling locations from 200 to 500
$kg-ha^{-1}-year^{-1}$).

The SO_4 present in precipitation in our country not
only consists of sulfuric acid sulfate, but also of a
certain background quantity of sulfates of which the
greater part originates from other countries and an
important quantity is from seaspray sulfate (Na_2SO_4).
Such a complexity in most cases excludes the finding of a
significant relationship.

G)   No significant difference between the average pH in
the summer and that in the winter (heating season) was
indicated. However, an inspection of the 50 percentiles as
shown in Figure 6a does show that there is a seasonal
variation in pH. Despite these differences, I could not
find an expression which would indicate a difference in the
pH between the summer and winter seasons. We found that
the single pH values in the summer were higher, but because
pH is a logarithmic number, some low values (especially in
the dry summer season) had a large influence on the average
value. The H^+ concentration of the precipitation in the
summer was higher in the countryside than in urban areas,
and the situation was reversed in the winter.

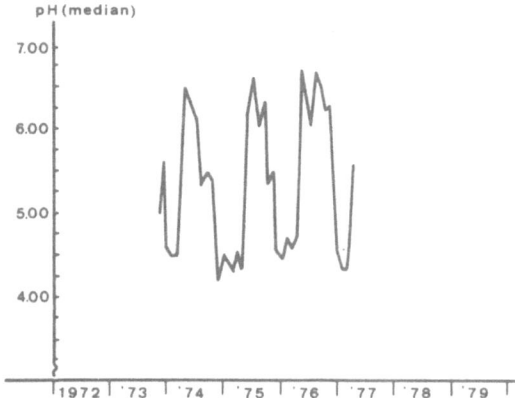

Figure 6a.   50 percentile pH in winter
              and summer seasons. All
              locations were similar.

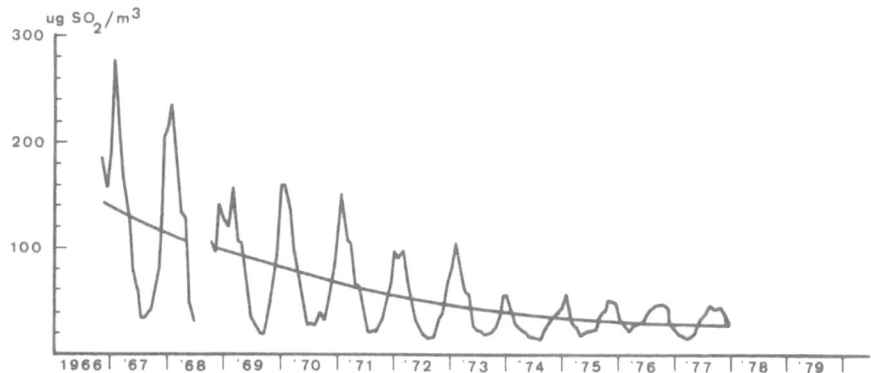

Figure 6b. SO_2 concentrations in winter and summer
              seasons. All North Holland locations
              were similar.

The course of the median pH corresponded remarkably
well with the SO_4 concentration of the ambient air as
measured at 31 locations in the province of North Holland.
As shown in Figures 6a and 6b, a high pH corresponds well
with a low SO_2 concentration and vice versa. This
correspondence encouraged me to determine whether a causal
relationship between the SO_2 emission, on the one hand,
and acid content of precipitation, on the other, could be
proven statistically. Such a relationship has been a
difficult one to establish by other investigators because
of the lack of the necessary information to cover all the
factors involved. Information such as a long series of pH
observations together with an accurate measure of the SO_2
emissions over the same period is necessary. It is also
possible that the emissions of NO_x may contribute to the
acidity, and this information should also be available. In
The Netherlands, a long series of pH observations from the
IMI sampling network (the values from three locations will
be used in this study: Den Helder, Witteveen and De Bilt)
are available, and these will be correlated with the course
of the interior SO_2 and NO_x emissions as calculated by
our National Bureau of Statistics.

THE CHANGES IN THE DUTCH FUEL RANGE AND THE CONSEQUENCES

WITH REGARD TO THE ACIDIFICATION OF PRECIPITATION

In the following an attempt will be made to explain the
consequences of the changes in the Dutch fuel range as far as
the acidification of precipitation is concerned.

After World War II, the greater part of the Dutch fuel
range consisted of coal, i.e. about 85%. Gradually, however,
oil obtained a competitive position. Especially between 1950
and 1960 a quick transition from coal to oil took place. In
the beginning of 1970, the share of coal was only 10% of the
total fuel range, and by 1977 it was only 4%. When things were
looking as though oil had won the battle, in 1959 in the
province of Groningen, The Netherlands, the largest coherent
natural gas field in the world was discovered having a total
proven gas reserve of about 2,100 billion cubic meters. In
1963 the exploitation of this field was started. Oil soon was
superseded by natural gas. Whilst the share of natural gas in
the total Dutch fuel range only amounted to 5% in 1965, in the
year 1977 this share had increased up to about 60% (Fig. 7).

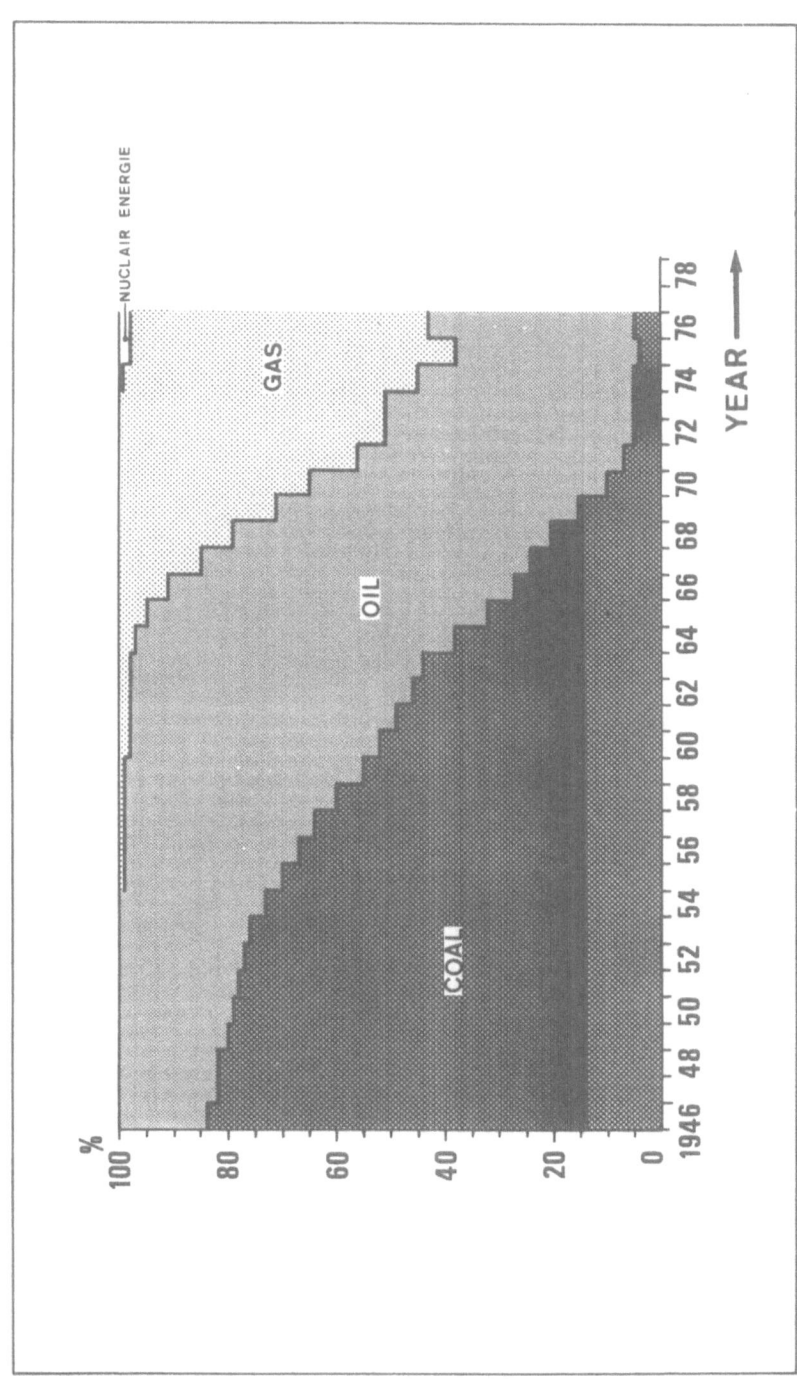

Figure 7. The share of coal, oil and natural gas in the Dutch fuel consumption.

The Interior Emission of Sulfur Dioxide

In 1946, when in The Netherlands mainly coal was burned, SO_2 emission amounted to roughly 200,000 tons a year. In 1956, the total SO_2 emission from burning coal and oil rose to about 400,000 - 500,000 tons. Owing to the expensive exploitation of pit coal, the growing energy needs in The

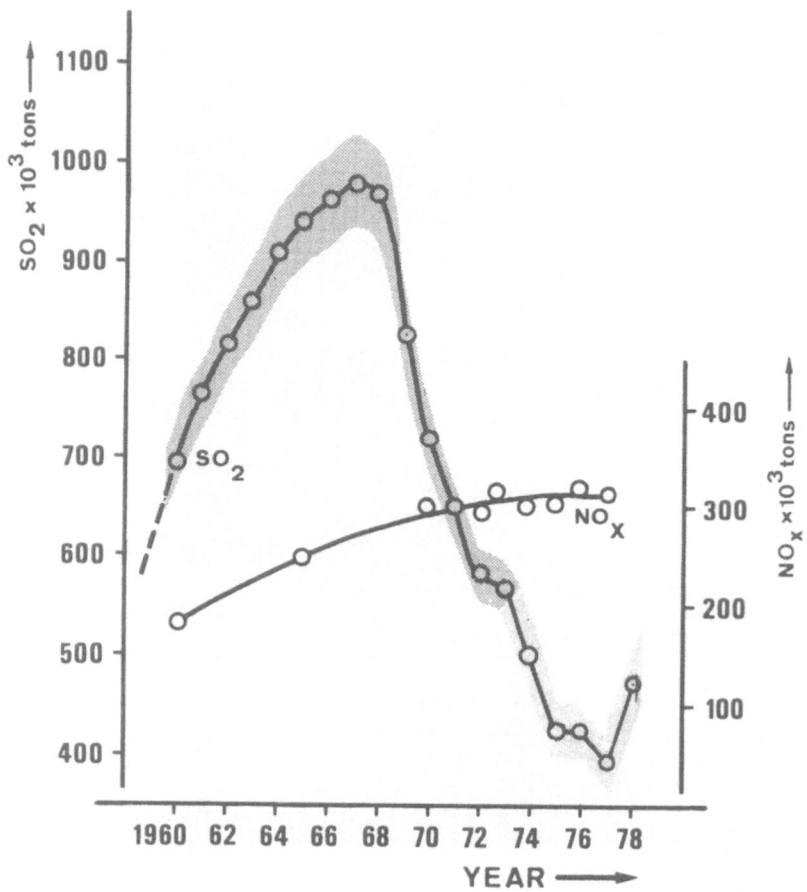

Figure 8. The total interior emissions of SO_2 and NO_x in The Netherlands. Accuracy for $SO_2 \pm 5\%$.

Netherlands were met by importing cheaper oil. However,
because of the ever increasing consumption of oil with its
higher sulfur content, SO_2 emissions rapidly increased.
Around 1967, the SO_2 emission reached a maximum value between
900 and 1,000 x 10^3 tons per year in other words an emission
pressure of about 30 tons per sq. km. (see Figure 8 and
Table 2).

The consequences of the high emission density were clearly
perceptible. In many places in The Netherlands, especially
under stagnating weather conditions, frequently ambient
concentrations between 500 - 2,000 g SO_2 per m^3 and even
higher were measured. Besides high levels, the acid content of
precipitation also seemed to have increased considerably.

In 1967 it was easily foreseen that - assuming a constant
rise in energy consumption, with oil as the main energy source
- within a period of 5 years the SO_2 emission density would
increase to 60 tons per sq. km. per year!

In order to keep The Netherlands habitable, drastic
measures were required immediately. In 1968 the Dutch Clean
Air Act was proclaimed. By means of this Act, the Dutch
Government among other things wanted to halt the increasing
SO_2 emissions. The Act became effective on December 29th,
1970. In anticipation of this, in the course of some years,
some industries began to clean up their emissions;
desulfurization installations were built, the use of heavy fuel
oil with a sulfur content of over 2.5% was prohibited, and
other types of oil with a lower S-content were imported; power
stations switched over to oil having a S-content below 1.7% and
sometimes even below 1%. The graph of SO_2 emissions (Fig. 8)
depicts decreasing SO_2 emissions after 1967 - 1968. However,
this absolute decrease was not brought about mainly by the
measures mandated by the Clean Air Act but principally by the
very lucky discovery in The Netherlands of - as would later
appear to be - the largest coherent natural gas field in the
world. In 1967 this natural gas accounted for about 18% of The
Netherlands energy consumption, five years later for about 50%
and in 1976 for about 60%. In 1975, natural gas supplied 85%
of the fuel needs of Dutch power stations.

From later calculations it became obvious that, if natural
gas had not been discovered, on account of the ever increasing
energy consumption, with the measures taken in compliance with
the Clean Air Act it would have been difficult to realize and
to maintain an SO_2 emmission of about 700,000 tons a year
from 1970 to the present (see Figure 9).

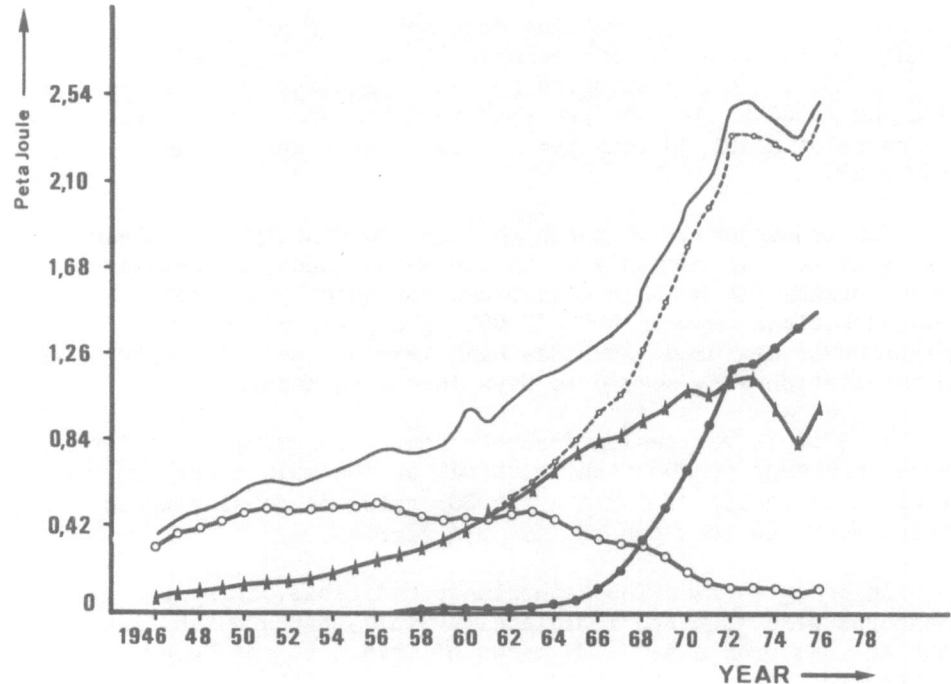

Figure 9.   Energy demand in The Netherlands.

────────── total energy demand
-o-o-o- consumption of coal
-●-●-●- consumption of natural gas
-▲-▲-▲- consumption of oil
-o--o··· consumption of oil if gas had
          not been found

In 1976 the Dutch Government issued a note stating that the
government's policy for the coming years would aim at lowering
and stabilizing the total SO_2 emission at about 500,000 tons
a year. This is an Utopian scheme! As a consequence of the
current exportation policy, natural gas for internal
consumption will be endangered in the long term. Moreover the
buildup of a strategic reserve (against oil boycotts) has
proven to be very important. Government policy for the years
to come, therefore, aims at terminating the export contracts
and reserving the natural gas for domestic heating only, until
after the year 2000. This means that power stations will have
to reduce their natural gas consumption from 85% in 1975 to 13%
in 1985, a loss they will have to make up with a proportionally

growing share of coal and oil. The discrepancy in the
government's aims is clear! In spite of the measures taken to
stimulate a further desulfurization of raw fuels and flue gases
for the power stations alone, SO_2 emissions of 400,000 up to
500,000 tons - equal to the total value aimed at by the Dutch
Government - can be expected in 1985. By 1985, the SO_2
emissions from the power stations will be equal to 25-30% of
the total interior SO_2 emissions. A simple exercise in
arithmetic shows that a total SO_2 emission of 1,200 - 1,500 x
10^3 tons can be expected, which is equal to a SO_2 emission
density of 38 - 47 tons per sq. km.

In view of the expected rise of energy consumption after
1985, the SO_2 density - when based on the most favorable
situation of an average S-content for all fuels of about 1%* -
can still reach 70 - 80 tons per sq. km in the year 2000. This
corresponds to an interior SO_2 emission of about 2,600 x
10^3 tons. This expectation is also based upon the assumption
that until 2000 the share of nuclear energy in the Dutch energy
production will not rise (in 1976: 1.65%, in 1985: max. 1.50%).

*Note:

To reach "over-all" percentages for the S-content of all
fuels for the years 1985 and 2000, respectively, the following
measures will be absolutely necessary:

-- a further development of the desulfurization
    installations. At present, for technical reasons all kinds
    of oil cannot yet be desulfurized.

-- the development of effective flue gas installations is an
    urgent necessity. In the U.S.A., good results are being
    obtained with coal fired power plants (fluidized bed
    system), while in Japan experiments with oil fired power
    plants give good results.

-- the development of methods allowing desulfurization during
    the combustion process: before the main combustion takes
    place, a partial combustion is effected at an increased
    pressure, whereby the SO_2 formed can be washed out. Even
    oils having a S-content of about 6 - 7% will, after having
    been treated, contain less than 1% sulfur (Shell
    Gasification Process).

Nuclear energy may contribute substantially to reduce the
SO_2 emissions. On account of the problems arising with this
form of energy generating (storage of active nuclear waste,
danger aspects), there is a strong opposition in The
Netherlands against the building of nuclear power plants. At
the moment, two nuclear power plants are in operation. The
announced construction of three new plants has been postponed.
In the scope of this report, the pros and cons of nuclear
energy will not be discussed any further. Solar energy and
wind energy are unlikely to play a major role in the total
generation of energy in The Netherlands during the coming years.

In order to combat air pollution by SO_2 - caused by the
reduction in natural gas consumption and the consequent
increase in oil usage - as much as possible and to counter the
threatening shortage of oil and gas during the next decades,
the general opinion in The Netherlands is that coal must be
restored to a place of honor among the fuels; not the direct
combusion of coal but its gasification to a useful fuel.

In The Netherlands, Shell is operating a coal gasification
pilot plant that has achieved considerable success. During the
process no - or hardly any - air pollution occurs. The sulfur
contained in the coal is released as elementary sulfur, while
the residues are transformed into a vitreous, pearl-like
material, suitable for several purposes in the construction
industry. At present the cost of this gas, produced in The
Netherlands, is approximately three times as high as that for
natural gas. Only when applied on an extensive scale (probably
not before 1986) will the price become competitive with those
of the fuels in use today.

As the world's coal reserve are so large that the entire
world population can draw energy from this source for hundreds
of years to come, it is obvious that a shortage of oil and
natural gas within a number of years will not necessarily mean
complete dependence on nuclear energy but will allow ample time
to seek other sources of energy.

Nitrogen Oxides

In contrast to the emission of SO_2, the emission of NO_x
shows only an upward trend although the rate of growth seems to
decrease somewhat during recent years (see Figure 8). In the
years 1974 and 1975 there was no increase at all; this
phenomenon can be attributed to the oil-crisis. The difference
in shapes of the curves results from the fact that the cause of
the NO_x emission is entirely different from that of SO_2;
NO_x is formed during combustion processes (at high

temperatures) from the nitrogen that forms part of the combustion air. This emission is pretty much independent of the kind of fuel used. With growing energy consumption, the NO_x emissions in The Netherlands will continue to rise. If we differentiate the NO_x according to their originators, the substantial part by traffic is at once obvious. From about 40,000 tons in 1960, the NO_x emission by traffic rose to 130,000 tons in 1976, accounting for more than 40% of the total interior NO_x emission. In other countries the percentage is of about the same order of magnitude. The statistical calculations will show that in The Netherlands, the interior emission of NO_x has not a great influence on the acidification of precipitation.

THE EFFECTS OF THE EMISSION OF SO_2 AND NO_x ON PRECIPITATION:

STATISTICAL ANALYSIS

In general, a direct connection between the emission of SO_2 and NO_x on the one side and the acidification of precipitation on the other will be difficult to prove.

As already stated the finding of such a relationship depends on a lot of factors, sometimes even accidental factors. The most important ones are:

-- the availability of data for a long period,

-- big variations in the SO_2 emission pattern to make it possible to prove a causal relationship, and

-- the precipitation sampling points have to satisfy certain rules.

All circumstances mentioned above were met in The Netherlands. The locations of the sampling stations are especially important. If the sampling locations are situated in the vicinity of a source or a source area, there is a great chance that the precipitation samples are strongly influenced by the dust in the emissions originating from the area considered. This dust is collected by the rain sampler (it forms part of the so called dry deposition) and finally collected together with the precipitation in the collecting bottle. The dust partially dissolves in the initially acid rain water. Part of the acid is neutralized in this way. As a consequence, normally occurring differences in pH values will be considerably reduced.

A well know example in The Netherlands is the industrial
area in the vicinity of Rotterdam; in spite of high SO_2
emissions in this area, relatively high pH values are being
found here, mostly more than pH 5 on a yearly basis, and the
fluctuations in pH values are not large. Another example is
shown in Figure 10. Here we see a distribution in the pH
pattern over the province of North Holland, caused by dust
emissions from a big steel factory. Close to the factory, the
pH is relatively high.

A sampling location also should not be situated at too far
a distance from the source or source area to be considered,
because the diluting and spreading effects will diminish the
SO_2 and NO_x concentrations of the air. A further
complication as a consequence of variations in wind direction
may result in a mixing with air masses supplied from
elsewhere. With increasing distances, it will become
increasingly difficult to indicate a connection between the
emissions of SO_2 and NO_x in a certain area and the acidity
of precipitation at the sampling locations. Therefore, in
situating sampling locations, some middle path may have to be
taken. Perhaps, transmission models may be utilized. Another
important factor that can strongly influence the average pH of
the precipitation sampled, even for a long period, is found in
certain weather conditions. For instance, at sampling
locations in The Netherlands, it was clearly ascertained that
for a rain period during a long lasting supply of air from
northern directions (Arctic air, relatively unpolluted), the
acid content of the daily precipitation gradually went down
(Fig. 11), whereas in the case of a prolonged supply of air
masses with a higher load of pollution because of the transport
across vast land areas (the European continent), the acid
content of the daily precipitation samples generally went up.

Since 1956, at the De Bilt, Den Helder and Witteveen
sampling locations the pH of precipitation has been measured
within the scope of the IMI measuring programme (see Figure 12).

From an evaluation of the measured values it may be seen
that from about the 1950s there is a tendency towards a growing
acidification of precipitation until around 1965, followed by a
rapid decrease. The H^+ concentrations calculated from the pH
values for these locations are indicated in Figure 13. Compare
this figure especially with Figure 8 (SO_2 emission). The
broken line in this figure indicates the moving average.

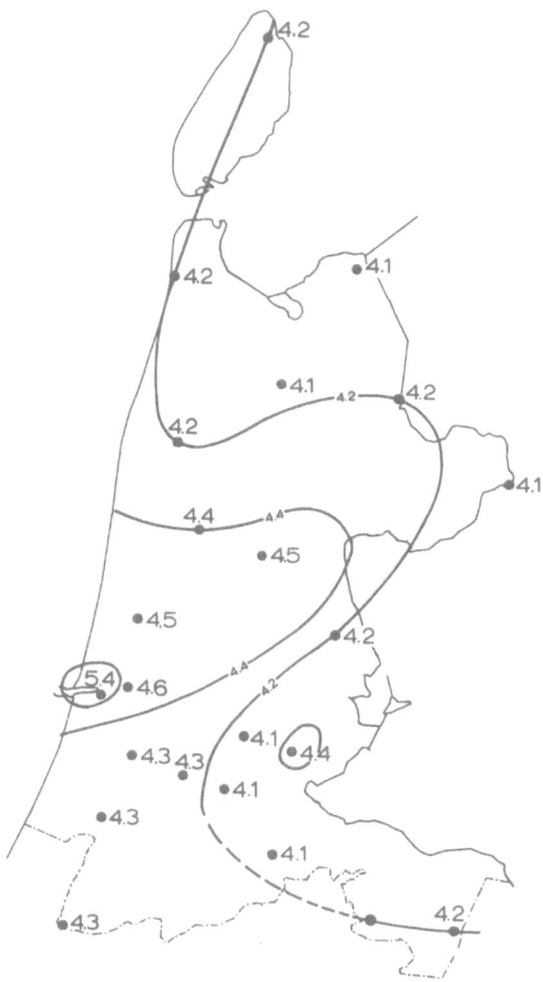

Figure 10.   The pH pattern over the province of North Holland
             resulting from dust emissions from a big steel
             factory. (See Figure 3 for location of factory).

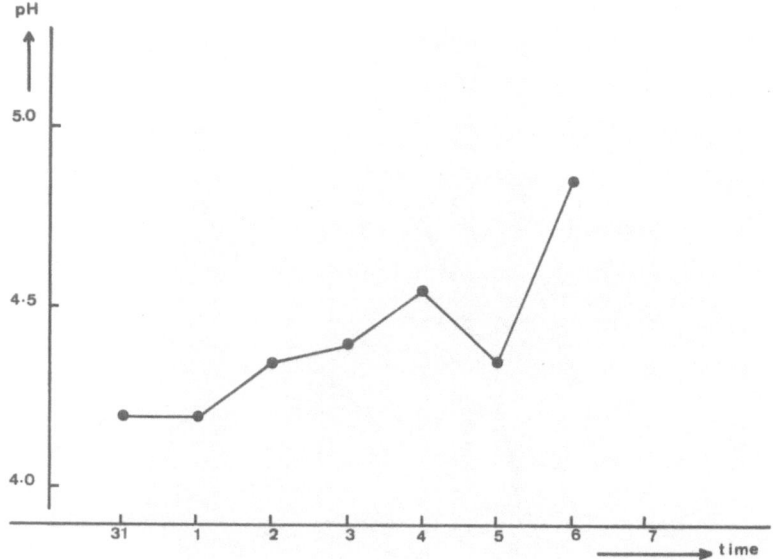

Figure 11.   The pH of rain samples taken October 31 – November 7.
The wind direction on all days was between WNW and N.

Figure 12.   The European rain sampling network coordinated by the
International Meterological Organization in Stockhold (IMI).

The curves of Figure 13 clearly show an increase and
decrease in H^+ concentration; from year to year there are
considerable fluctuations in H^+ concentration about the
moving average. With regard to these fluctuations, in my
opinion the variations in the general weather pattern that may
appear from year to year play an important role.

To indicate a relationship between the average acidity on a
yearly basis and the changes occurring in the total interior
emissions of SO_2 and NO_x, by means of a regression analysis
the best estimates have been determined for the parameters in
the following model:

$$y = \beta_0 + \beta_1 \cdot x_1 + \beta_2 \cdot x_2 + \varepsilon \qquad (1)$$

in which y = the average acidity on a yearly basis in
               rain water in $g. l^{-1}$, expressed as H^+
     x_1 = the yearly emission of SO_2 in 1000 tons
     x_2 = the yearly emission of NO_x in 1000 tons
      ε = a random variable indicating the variability
             of y around the basic relation

$$\beta_0 + \beta_1 \cdot x_1 + \beta_2 \cdot x_2$$

In this analysis it is assumed that ε is normally distributed
with an average of 0 and a variance of σ^2.

In view of the non-central position of the Den Helder
measuring site with its relatively larger distance from the
emission sources than Witteveen and De Bilt, and because eight
out of ten times air passing over Den Helder is transported
from the sea, we assume that the average acidity on a yearly
basis measured in Den Helder is not related to the total
interior yearly emission of SO_2 and NO_x. We therefore test
the hypothesis $\beta_1 = 0$ and $\beta_2 = 0$, and as alternatives the
hypothesis $\beta_1 \neq 0$ and $\beta_2 \neq 0$. With the aid of the F-test
we find a value $F = 2.48$ at 2 and 15 degrees of freedom, so
that the hypothesis β_1 and β_2 are significantly different
from zero can be rejected for the measuring location Den Helder.

For the 15 -20 years for which figures are available
(Table 5), the following estimates, b_i, were found for the
parameters β_i as indicated in Table 6. In this table the
standard deviations, s_i, of those estimates are also given.

For the sampling locations Witteveen and De Bilt, with the
same F-test the values $F = 23.39$ and $F = 15.07$ were found,
implying that for these locations the hypothesis $\beta_1 = 0$ and
$\beta_2 = 0$ with an unreliability <1% can be rejected. For both

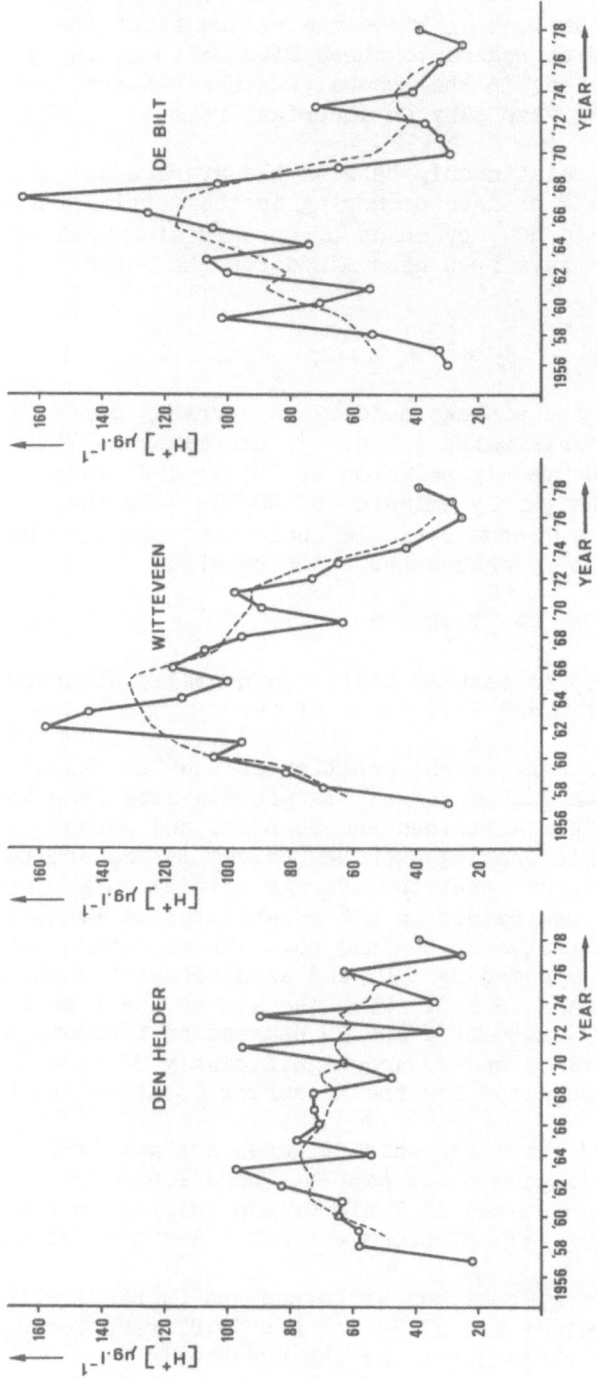

Figure 13. The acid content of precipitation (annual weighted averages) with moving average at the three sampling locations of the IMI ntework. The natural background value is between 1 and 5 µg H$^+$ IONS per liter.

Table 5. Data from De Bilt, Den Helder and Witteveen

Year	Total interior emissions kilotons SO_2	NO_x	Acid content of precipitation annual average $\mu g\ H^+/1$ D. B.	D. H.	Witt.
1956			29.5	-	-
57			31.6	21.9	28.8
58			53.7	57.5	69.2
59			102.3	57.5	81.3
60	694	182	69.3	64.6	104.7
61	756	205	53.7	63.1	95.5
62	816	215	100.0	83.2	158.5
63	858	230	107.2	97.7	144.5
64	908	240	74.1	53.7	114.8
65	940	249	104.7	77.6	100.0
66	960	260	125.9	70.8	117.5
67	980	270	166.0	72.4	107.2
68	970	280	102.3	72.4	95.5
69	844	294	64.6	47.9	63.1
70	723	294	28.8	63.1	89.1
71	650	300	31.6	95.5	97.7
72	580	301	37.2	32.4	72.7
73	572	314	72.4	89.1	64.6
74	530	301	40.7	33.9	42.7
75	419	303	-	-	-
76	463	315	32	63	25
77	393	305	25	25	25
78	475	315	40*	43*	37*

*New type of rain sampler, see Figure 3.

Table 6. Estimators b_i with Standard Deviations s_i for the
Parameters in the Model $y = \beta_0 + \beta_1 x_1 + \beta_2 x_2 + \varepsilon$

Location	b_0	b_1	b_2	s_0	s_1	s_2
De Bilt	-53.1	0.165	0.0164	63.0	0.035	0.160
Witteveen	123.5	0.103	-0.412	51.1	0.0284	0.136
Den Helder	43.0	0.047	-0.050	51.5	0.029	0.137

sampling locations there is a relationship between the H^+ concentrations of precipitation on the one hand, and the emission of SO_2 and NO_x on the other.

The De Bilt sampling location, because of its central position, can be considered as being the most representative location with respect to the total interior yearly emission of SO_2 and NO_x. We shall examine more closely the data with respect to this location.

For the sampling location De Bilt it may be seen (see Table 6) that the estimate for β_2 when tested will not differ significantly from zero, and thus for the center of The Netherlands, NO_x is not a parameter for the prediction of the concentration of acid in precipitation. Therefore, the model can be simplified to

$$y = \beta_0 + \beta_1 x_1 + \varepsilon \qquad\qquad (2)$$

For this model, the estimates for β_0 and β_1 are -47.3 and 0.163 respectively. The correlation coefficient between SO_2 and H^+ on this location is 0.82 (n = 18) and can be qualified as being very high (see Figure 4). On the basis of equation (2) an estimate can be made of the acid concentration of the De Bilt location with changing amounts of the total interior yearly emission of SO_2.

From a recent study, carried out by the Metra Consulting Group Limited in London, Great Britain, on behalf of our Ministry of Public Health and Environment, it was calculated that in The Netherlands, with no additional abatement, the quantities of SO_2, because of the switch over from natural gas to coal and oil, could possibly rise from 1000 to 1300 x 10^3 tons in 1985 and about 3000 x 10^3 tons in the year 2000.

Another recent calculation drawn up by the European Economic Community (EEC) gives the prognosis as shown in Table 7.

Table 7. Possible Future Emissions of SO_2

	Emissions of SO_2 in 10^6 kg/year		
	1976	1985	2000
The Netherlands	450	1,000--2,000	1,900--3,100
EEC	16,000	19,000--20,000	18,000--26,000

If we assume that, according to these predictions the emission of SO_2 will be about 1300 x 10^3 tons in 1985, the average concentration of acid (expressed as protons) will be 165 g-1^{-1}, with a standard deviation of 17.4. The 99% reliability interval gives pH values on a yearly basis of between 3.60 and 3.95. At an expected emission of about 2600 x 10^3 tons around the year 2000, on a yearly basis, pH values between 3.27 and 3.66 can be expected (at a 99% reliability interval), equal to proton concentrations of 220 to 530 μg-1^{-1} (375 μg as an average). As a comparison: the natural background concentration of protons would be less than 10 μg-1^{-1}. Table 8 gives the proton concentrations and pH values to be expected for different quantities of emitted SO_2.

Table 8. Proton Concentrations and pH Values Related to SO_2 Emissions for the Model $y = -47.304 + 0.1627 x$

x	y	St. Dev.	99% interval	pH interval
300,000	1.51	13.44	-37.74--40.75	4.39
400,000	17.78	10.89	-14.02--49.57	4.30--6.00
500,000	34.05	8.54	9.10--58.99	4.23--5.04
600,000	50.32	6.63	30.96--69.67	4.16--4.51
700,000	66.57	5.60	50.24--82.93	4.08--4.30
800,000	82.86	5.94	65.52--100.19	4.00--4.18
900,000	99.13	7.46	77.34--120.91	3.92--4.11
1000,000	115.40	9.63	87.30--143.49	3.84--4.06
1100,000	131.67	12.08	96.38--166.95	3.78--4.02
1200,000	147.94	14.69	105.03--190.84	3.72--3.98
1300,000	164.21	17.38	113.43--214.98	3.67--3.95
1400,000	180.48	20.12	121.96--239.26	3.62--3.91
1500,000	196.75	22.90	129.86--263.63	3.58--3.89
1600,000	213.04	25.69	137.97--288.06	3.54--3.86
1700,000	229.29	28.50	146.03--312.54	3.51--3.84
1800,000	245.56	31.32	154.06--337.05	3.47--3.81
1900,000	261.83	34.15	162.06--361.59	3.44--3.79
2000,000	278.10	36.99	170.05--386.14	3.41--3.77
2100,000	294.37	39.83	178.03--410.71	3.39--3.75
2200,000	310.64	42.68	185.98--435.29	3.36--3.73
2300,000	326.91	45.52	193.93--459.88	3.34--3.71
2400,000	343.18	48.38	201.87--484.48	3.31--3.69
2500,000	359.45	51.23	209.81--509.09	3.29--3.68
2600,000	375.72	54.08	217.74--533.70	3.27--3.66
2700,000	391.99	56.94	225.66--558.31	3.25--3.65
2800,000	408.26	59.80	233.58--582.93	3.23--3.63
2900,000	424.53	62.66	241.50--607.55	3.22--3.62
3000,000	440.80	65.52	249.41--632.18	3.20--3.60

x = yearly interior emission of SO_2 in tons

y = yearly average proton concentration in precipitation μg/l

The model used to calculate this table is assumed to be applicable for emissions greater than 1,300 x 10^3 tons.

From Figure 14 can be seen that the line
$y = \beta_0 + \beta_1 x_1$ intersects the x-axis at a total interior
SO_2 emission of about 300,000 tons. Below this figure the
rain water at this sampling location contains hardly any free
protons. A logical explanation seems to be that the rain water
has a certain buffering capacity but of course the diminishing
and spreading effects of the emitted SO_2 also play an
important role.

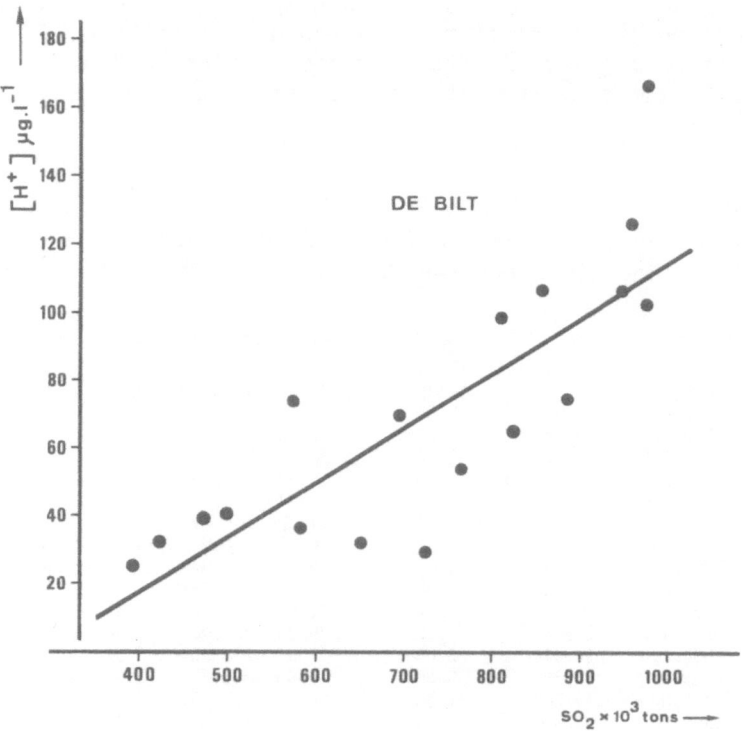

Figure 14.   Relationship between annual interior SO_2 emission
             and average annual H^+ concentration. The regression
             line $y = -47.6 + 0.163 x$ is clearly significant,
             $(r = 0.82, \quad n = 18)$.

From the statistical analysis we can draw the following
conclusions:

For the Den Helder measuring site, a relationship between
the average annual H^+ concentration on the one hand and
the total interior emission of SO_2 and NO_x on the other

cannot be found. Probably this is due to the non-central
position of this location, the high percentage of sea wind
and other factors already mentioned.

When comparing Figure 13 (Den Helder measuring
location) to Figure 8 (SO_2 emission), we find a certain
similarity (see moving average) although it cannot be
statistically determined. In Figure 13 we are not
confronted with a periodical phenomenon, because before
1956 in The Netherlands almost always H^+ concentrations
below about 30 g.-l^{-1} were measured.

--   For the De Bilt location a pronounced relationship
between the total interior SO_2 emission and the H^+
concentration in precipitation at this location was found.
No relationship was found between the annual NO_x emission
and the annual average H^+ concentration.

--   For the Witteveen sampling location for both SO_2 and
NO_x a relationship with the H^+ was found. Here the
aspect of a decreasing acid concentration at an increasing
NO_x emission is striking. It is assumed that there is no
causal relationship between NO_x and H^+.

The influence of ammonia on the pH

As mentioned previously, we could not find a correlation
between the proton and ammonium concentrations in
precipitation. The ammonium in rain water is a mixture of
ammonium from different orgins. We looked, therefore, for any
relationship between the proton concentraton in precipitation
on the one hand, and the quantity of ammonia emitted in The
Netherlands on the other. In The Netherlands, there are hardly
any ammonia emitters; the quantity emitted by industry is very
small. The major source of ammonia has to be found in our
fertilizers. Of course, only a few percent of the fertilizer
used for agricultural purposes will disappear into the
atmosphere. The quantities of fertilizer used in The
Netherlands amounted, expressed as nitrogen, from 307 x 10^3
tons in 1960 to 590 x 10^3 tons in 1978 (see Table 9).

When we tested the model

$$H^+ = \beta_0 + \beta_1 SO_2 + \beta_2 NH_3 \qquad (3)$$

for the sampling location De Bilt, we found that β_2 did not
differ significantly from zero, so for the central part of The
Netherlands NH_3 does not seem to be an explanatory variable
for the magnitude of the proton concentration.

Table 9. Quantity of Fertilizer Used for Agricultural
          Purposes in The Netherlands in kilotons of N
          per Year.

1960--365	1966--451	1972--545
1961--379	1967--467	1973--564
1962--397	1968--477	1974--594
1963--412	1069--496	1975--616
1964--420	1970--526	1976--598
1965--433	1971--534	1977--574

Just as with NO_x, the model (3) can be simplified to
model (2):

$$H^+ = \beta_0 + \beta_1 SO_2$$

The Deposition of Acid

The acidification of precipitation would perhaps not have
such consequences, if this precipitation was not deposited on
the soil. The deposition of acid - the quantity of acid that
is deposited by precipitation on the soil per unit of time and
per surface unit - depends on 2 factors, the acidity and the
quantity of precipitation. The yearly acid deposition is found
by multiplying the annual precipitation by the yearly average
concentration of acid. The natural yearly background
deposition is about 1 - 3 mg H^+ per m^2. The measured
values for De Bilt, Witteveen and Den Helder are much higher
(Fig. 15).

For the De Bilt measuring location for the year 1985 the
average deposition of acid expressed as H^+ can be rated at
about 74 to 91 mg per m^2 in case of a minimal annual rainfall
(about 500 mm), at about 118 to 145 mg per m^2 at normal
annual rainfall (about 800 mm) and at about 176 to 217 mg per
m^2 at maximum annual rainfall (about 1200 mm).

Under the same conditions, for the year 2000 the calculated
depositions amount to 161 to 215 mg per m^2, 250 to 344 mg per

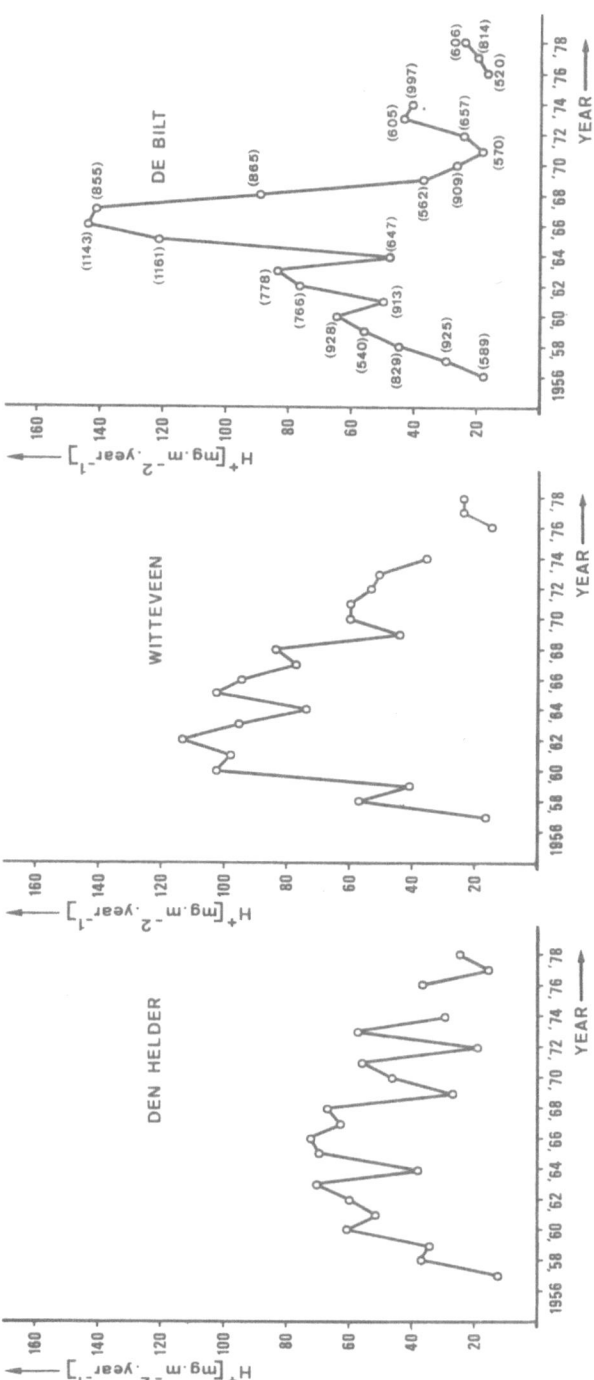

Figure 15. Acid deposition for the Den Helder, Witteveen and De Bilt sampling locations. Numbers in parentheses for De Bilt are annual amounts.

m^2 and 386 to 516 mg per m^2 for minimal, normal and maximal
annual rainfalls, respectively. These figures are based on an
SO_2 emission of 1300 x 10^3 tons in 1985 and 2600 x 10^3
tons in the year 2000, and calculated as mg H^+ ions per m^2.

The ecological effects of acid precipitation for the
greater part are still unknown but undoubtedly they will be
numerous and have a very complicated character.

The rough consequences, which can be pretty well quantified
on the basis of extensive studies, are

--   acidification of natural water sources and fresh water
     systems
--   leaching of the soil
--   damage to vegetation
--   damage to materials (corrosion, old monuments deterioration)

Because of the type of soil and the composition of surface
waters in The Netherlands, these ecosystems show a large
buffering capacity to acid deposition, although ecological
effects can still be demonstrated.

For example, a survey on diatoms and macrophytes in some
Dutch moorland pools shows that the biocenoses are seriously
changed by human activities, especially by agriculture,
recreation and acid precipitation. By inversion of the thropic
pattern of the landscape, thropic gradients cannot be
maintained and the biological differentiation within and
between the pools seriously decreases. Most probably, in the
future, only two types of moorland pools will survive: the
hyperthropic type (enriched by run-off and drainage water) and
the extremely acid type (by acid precipitation).

From many investigations in The Netherlands, it has already
become obvious that the abundance of species of the epiphytic
flora has considerably decreased. In the surroundings of
Bois-le-Duc (province of North Brabant), Wakker collected, in
the years 1900 - 1925, 117 lichene species. In the seventies,
of these 117 species only 47 were recovered!

Hovenkamp and Van Schaik made an investigation of the
bryophyte vegetation on ash tree trunks in the Fazentenbos
(near De Steeg, province of Gelderland). A comparison was made
with results of other investigations during the past 25 years
(see Fig. 16).

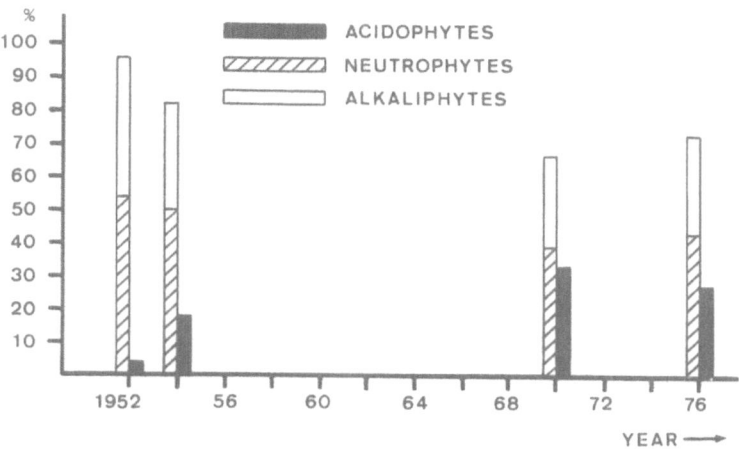

Figure 16.   The changing moss flora near De Steeg,
center of The Netherlands.

This figure clearly shows that before 1969 the acidophytes
were strongly increasing and the neutrophytes and basophytes
decreasing, whereas after 1969 the acidophytes decreased and
the basophytes and neutrophytes increased. See especially Fig.
13 and 15 for a comparison with Fig. 16.

Recent studies have shown that in other parts of The
Netherlands areas with acidophytic mosses and liverworts have
considerably extended.

The last example I want to show is given in Fig. 17 and
demonstrates the acidification of the soil in our country. In
this figure we see the average pH values of grassland in one of
our provinces. From about the 1950s we see a decreasing pH but
this decrease is not effected by the use of acidic
fertilizers. From about 1950, the major fertilizer in The
Netherlands (about 80% of the interior use) was a mixture of
ammonium nitrate and calcium carbonate with an N content of
23%. This fertilizer reacted neutral to grassland. Only after
1972, the N content was changed from 23 to 26% and the reaction
to grassland became slightly acidic. The acidification of the
grassland in the 1950-1952 period, however, has to be
attributed for the most part only by leaching of the soil by
acid precipitation.

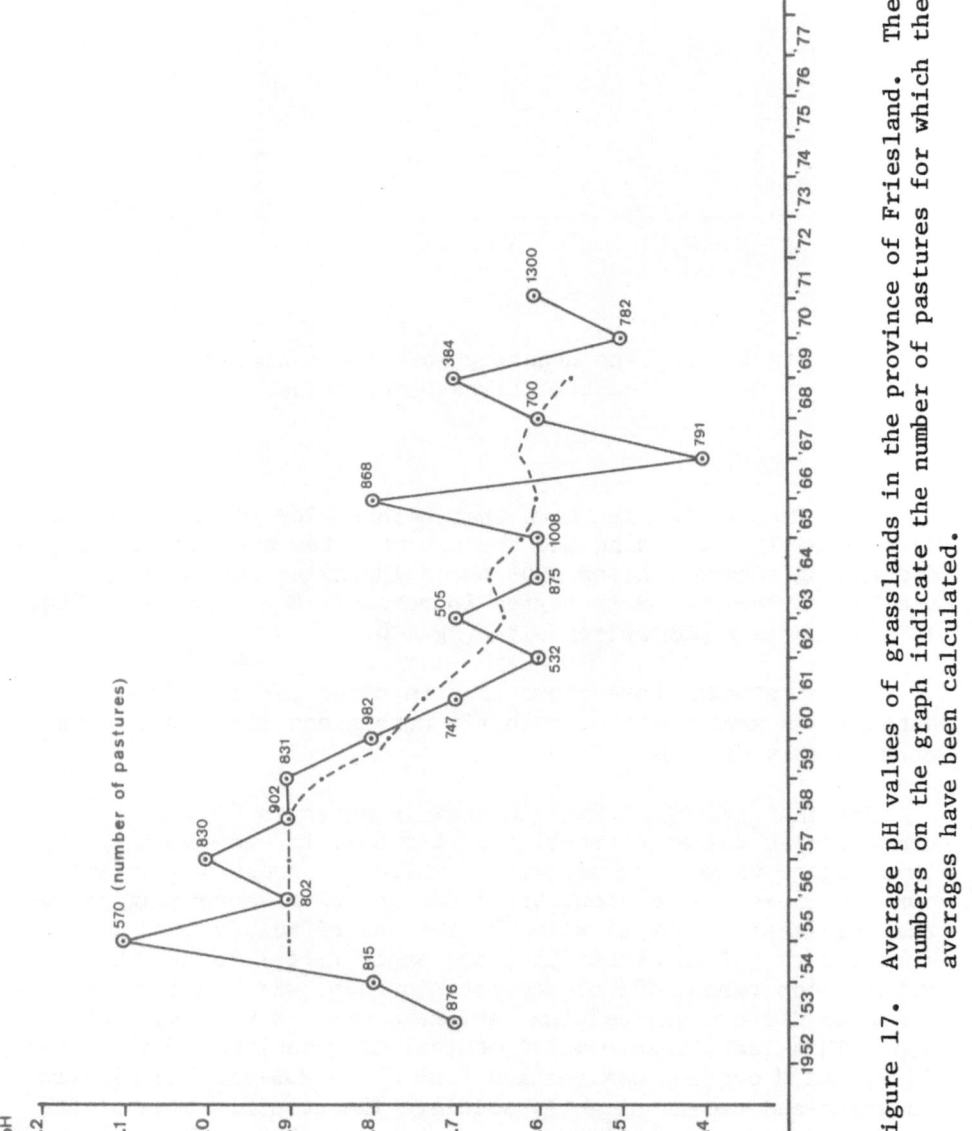

Figure 17. Average pH values of grasslands in the province of Friesland. The numbers on the graph indicate the number of pastures for which the averages have been calculated.

It is accepted as being normal at the moment that the
farmers in our country have to add enormous quantities of lime
to the soil to lower the acid content and to compensate for the
loss of calcium. It has been calculated that more than 20% of
the total yearly interior use of calcium for agricultural
purposes is used in compensating for the loss of calcium by
leaching of the soil by acid precipitation, and it appears that
the absolute quantities will increase very much in the coming
years.

A Danish report, I know, has suggested that lime be added
to the acidified Danish lakes to make possible fish life and
other aquatic life again.

The damage caused by acid precipitation in Scandinavia is
widespread and increasing at a fast tempo. A major part of the
air pollution produced in Western Europe is transported in the
northeasterly direction toward Scandinavia. At an average wind
velocity, distances of 1000 km and more can be covered in two
to four days. Because much of Scandinavia is underlain by
highly resistant granitic rocks with only a thin cover of
unconsolidated glacial till and soil, the inland waters are
characterized by low conductivities, low concentrations of
major ions and extremely low buffering capacities.

For the above reasons, large areas of Scandinavia are quite
vulnerable to pollutants deposited from the atmosphere.
Acidification of freshwaters with the disappearance of fish
population, frogs and other water organisms from hundreds of
lakes have been particularly rapid over the last decades. In
the Tovdal River catchment area (in southern Norway) for
instance, an area with 266 lakes, 48 lakes were without fish in
1950. In 1960 the number of fishless lakes increased to 75
and, in 1975, even to 175! Failure in body salt regulation is
suggested as the physiological mechanism behind fish death in
acid water. In addition to the acidification of surface
waters, leaching of the soil takes place, producing changes in
vegetation.

A situation comparable to that in Scandinavia is found in
the area of the Adirondack Mountains in the State of New York.

SUMMARY

In The Netherlands, a highly densely populated and
industrialized country, there is a pronounced relationship
between the total interior emission of SO_2 and the
acidification of precipitation. No such relationship could be
demonstrated for NO_x, probably because more than 40% of NO_x

is emitted at a height of 30 cm above road level, which makes it assumable that this part will be absorbed for the greater part by the direct surroundings. The remaining 60% (170,000 tons at this moment) will hardly play a role in acidification, also because NO_x reacts slowly to nitric acid so that the effects are probably not demonstrable in our country. Moreover, the quantities of emitted NO_x in our country not only are less than those of SO_2, but chemically seen, NO_2 reacting to acid only gives one proton where SO_2 gives two. In my opinion, the outlook for The Netherlands with respect to the acidification of precipitation is pretty grim. At this moment the Government's policy is to reserve the natural gas for domestic heating only. The power stations must, as they are already doing, switch over to oil and coal again. This means that in addition to the existing SO_2 and NO_x abatement program, an extra abatement program has to be carried out. If not, the SO_2 emission will rise from about 450,000 of tons now, to 1,500,000 of tons around 1985 and about 3 millions of tons in the year 2000. This means for The Netherlands acid depositions of about 8 grams of sulfuric acid per square meter per year in 1985 and more than 20 grams in the year 2000. This will bring enormous consequences for the environment. Because the emission will take place from high stacks, the pollution at the life level (the intake values) will hardly rise.

It can be postulated that from now till 1985, with a decreasing share of natural gas, oil will become the main fuel in The Netherlands. In the 1985 - 2000 period, it is most probably that oil will remain the main fuel for the first decade, but slowly, the gasification of coal (perhaps together with nuclear energy) will become competitive with oil. Solar, wind and nuclear energy are unlikely to play a major role in the total generation of energy before 1985, and after 1985 their contribution will hardly be more than about 5% in the year 2000, unless, within a few years, the Dutch Government begins building a number of nuclear power plants; but it appears that this is very unlikely.

With coal and oil as the most important fuels in the years to come it is obvious that a large and costly effort will have to take place in the field of desulfurization in order to prevent the acidification described in this paper. Recently, however, our Government issued a note of more than 500 pages about a very detailed study on the economic consequences of NO_x and SO_2 abatement in The Netherlands. In this note the expectations about the SO_2 emission in the next 20 years are more optimistic. At different categories of control they expect an SO_2 emission between 300,000 and 700,000 of tons for 1985, hardly increasing in the years after 1985. I hope it will come true, but I doubt it very much. Because most

desulfurization methods have not yet been technically perfected
to a degree that the efficiency is sufficiently high and the
costs are acceptable for large-scale application, in my opinion
hardly any positive effect can be expected before 1985.   After
1985, the acidification of precipitation will strongly depend
of the level of technology of desulfurization and the velocity
with which these installations will be built.   Except for the
costs, there must be the political will to handle the problem
vigorously, and that is a separate problem.

Additional Reading

-- Barrot, E. and G. Bodin, Tellus VII (1955).

-- "Acid Precipitation" Proceedings of a Conference on
    Emerging Environmental Problems, held on May 19-20, 1975.
    The Institute on Man and Science, Rensselaerville, New
    York. Report No. EPA-902/9-75-001 (1975).

-- Likens, E. and F.H. Bormann, "Acid Rain, a Serious Regional
    Environmental Problem'", Science, vol. 184, June 14 (1974).

-- Reynolds, R.C. and N.M. Johnson "Chemical Weathering in the
    Temperate Glacial Environment in the Northern Cascade
    Mountains", Geochim. Cosmochim. Acta (1972).

-- Collison, R.C. and J.E. Mensching. N.Y. Exp. Stn. Geneva,
    Tech. Bull. No. 193 (1932).

-- Granat, L.A., Report AC-18 (1972) Institute of
    Meteorology, University of Stockholm (M.I.S.U.).

-- Leeflang, K.W.H. "De chemische samenstelling van de
    neerslag in Nederland". Chemisch Weekblad 35 (1938).

-- Junge, C.E., "Air Chemistry and Radioactivity" Academic
    Press, New York and London.

-- Galloway, J.N., G.E. Likens and E.S. Edgerton "Acid
    Precipitation in the Northeastern United States: pH and
    Acidity". Science vol. 194, Nov. 12 (1976).

-- Persson, G., "The Acidity and the Concentration of Sulfate
    in Precipitation over Europe". Statens Naturvardsverk, the
    Swedish National Nature Conservancy Office, Air Quality
    Dept. Fack, 17120, Solna 1, Sweden.

-- Moss, M.R. "Spatial Patterns of Precipitation Reactions"
    Environment Pollution, 8 (1975).

-- Zeedijk, H. and C. Velds, "The Transport of Sulfur Dioxide
    over a Long Distance". Atmospheric Environment vol. 7, pp.
    849-862 (1973).

-- Scriven, R.A. and B.E.A. Fisher "The Long Range Transport
    of Airborne Material and its Removal by Deposition and
    Wash-out: II.-The Effect of Turbulent Diffusion".
    Atmospheric Environment vol. 9, pp. 59-68 (1975).

-- Førland, E.J., "A Study of the Acidity in the
    Precipitation in Southwestern-Norway" Tellus XXV No. 3
    (1973).

-- Emmelin, L., "Air Pollution Across National Boundaries",
    Environmental Planning in Sweden No. 23, Oct (1971). Royal
    Ministry of Foreign Affairs, Sweden.

-- Schofield, C.L. Jr. "Water Quality in Relation to Survival
    of Brook Trout 'salvelinum fontinalis' (Mitchill)" Trans.
    Amer. Fisheries Soc. 94, No. 3, pp. 227-235 (1965).

-- Schofield, C.L. Jr. "Lake Acidification in the Adirondack
    Mountains of New York, Causes and Consequences". First
    Intl. Symp. on Acid Precipitation and Forest Ecosystems.

-- Dochinger, L.S. and T.A. Seliga "Acid Precipitation and the
    Forest Ecosystems" Report from the First Intl. Symp.
    Journal of the Air Pollution Control Association, vol. 25,
    No. 11, Nov. (1975).

-- Oden, S. and R. Andersson "The Long Term Changes in the
    Chemistry of Soils in Scandinavia due to Acid
    Precipitation". Paper, Dept. of Soil and Science,
    Agricultural College, 750 07 Uppsala, Sweden.

-- Oden, S. and R. Andersson "The Long Term Changes in the pH
    of Lakes and Rivers in Sweden" Paper, address: Dept. of
    Soil and Science, Agricultural College, 750 07 Uppsala,
    Sweden.

-- "Impact of Acid Precipitation on Forest and Freshwater
    Ecosystems in Norway. Summary report on the research from
    the phase I (1972 - 1975) of the SNSF-project. Fagrapport
    FR 6 (1976).

The joint research project of the Agricultural Research
council of Norway, The Norwegian Council for Science and
Industrial Research, was started in 1972. The objective of
the report is to:

1. establish as precisely as possible the effects of acid
   precipitation on forest and freshwater fish.

2. investigate the effects of air pollutants on soil,
   vegetation and water, which are required to satisfy
   point 1.

Reports with information of this subject are obtainable free
of charge from:

SNSF-project,
NISK
1432 Aas-NHL
Norway

-- Proceeding of the First International Symposium on Acid
   Precipitation and the Forest Ecosystems, U.S. Forest
   Service, General Technical Report N.E. (1074 pp).

-- Dam, H. van, and H. Kooyman-van Blokland "Man-made Changes
   in some Dutch Moorland Pools, as Reflected by Historical
   and Recent Data about Diatoms and Macrophytes". Int. Revue
   Ges. Hydrobiologie, in press (1978).

-- Energienota, problemen en perspectieven. Provinciaal
   Electriciteitsbedrijf van Noord-Holland, September 1977.
   Verkrijgbaar: Provinciale Griffie Noord-Holland.

-- Zeedijk, H. "Het brandstofinzetplan centrales en de
   luchtverontreiniging". Chemisch Weekblad No. 24, June 1978.

-- Granat, L. and Henning Rodhe "A Study of Fallout by
   Precipitation around an Oil-Fired Power Plant".
   Atmospheric Environment, Pergamon Press, 1973, Vol. 7, pp.
   781-792.

-- Ta-Yung Li, H.E. Landsberg "Rainwater pH close to a Major
   Power Plant", Atmospheric Environment, Vol. 9, pp. 81-88,
   Pergamon Press, 1975.

-- Hutcheson, M.R. and Hall, F.P. "Sulfate Washout from a
   Coal Fired Power Plant Plume". Atmospheric Environment,
   Vol. 8, pp. 23-28, Pergamon Press, 1974.

-- Högström, U., "Wet Fallout of Sulfurous Pollutants Emitted
   from a City During Rain or Snow". Atmospheric Environment,
   Vol. 8, pp. 1291-1303, Pergamon Press, 1974.

-- Larson, T.V., et al. "The Influence of a Sulfur Dioxide
   Point Source on the Rain Chemistry of a Single Storm in the
   Puget Sound Region", Water, Air and Soil Pollution, 4,
   1975, pp. 319-328.

-- Georgii, H.W., "Investigation on the Incorporation of
    Sulfur Dioxide into Fog and Rain Droplets". Paper No. 4,
    Dept. of Meteorology and Geophysics, University of
    Frankfurt, Germany.

-- Heuvel, A.P. van den and B.J. Mason "The Formation of
    Ammonium Sulfate in Water Droplets Exposed to Gaseous
    Sulfur Dioxide and Ammonia", Quart. J. Meteor. Soc. 89,
    1963.

-- Enger, L. and U. Högström "Dispersion and Wet Deposition
    of Sulfur from a Power Plant Plume", Atmospheric
    Environment Vol. 13, 1979, pp. 797-810.

-- Coesel, P.F.M., R. Kwakkestein and A. Verschoor",
    "Oligotrophication and Eutrophication Tendencies in some
    Dutch Moorland Pools, as Reflected in their Desmid Flora"
    Hydrobiologia, Vol. 61, pp. 21-31, 1978.

DISCUSSION

EVANS: Why do you think the correlation between sulfur oxides
and nitrogen oxides and protons is only 0.82? What do you
think are the major components that would not solidify the
nitrogen?

VERMEULEN: I assume that you mean the correlation between the
emitted quantity of SO_2 + NO_x on the one hand, and the
proton content of precipitation on the other, and not the
correlation between those components in the precipitation.
Well, it surprises me to hear you saying "only 0.82" because in
my opinion it is very high. Other investigators searched for
just such a correlation but never found one. The most
important reason they did not, in my opinion was, that they had
not at their disposal a long continued series of observations
of pH measurements as well as data on the quantities of SO_2
and NO_x emitted. Moreover, there are other factors, I
mentioned in my paper, which are very important for a positive
result of the calculations. The 0.82 I found, with n = 18 –
the correlated figures were on a yearly basis – can be
qualified as being highly significant.

Concerning the second part of your question, I suppose with
nitrogen you mean nitrogen oxides. I don't know if the NO_x
emitted will not be solidified by certain components in
precipitation. Anyway, I did not contend the opposite. I
think that under certain circumstances, NO_x in contact with
rain water or cloud water will be transformed into NO_3^-, for
example at high ozone concentrations. It is a fact that under
normal conditions the oxidation of NO_x to higher oxides needs
much more time than the oxidation of SO_2 into SO_3.

EVANS: Did you put an NO_x component into your calculations
or not?

VERMEULEN: Yes, I did. As I showed you on one of my slides, I
tested the model

$$H^+ = \beta_0 + \beta_1 \cdot x_1 + \beta_2 \cdot x_2 + \varepsilon$$

in which x_2 is the yearly interior quantity of NO_x emitted
in our country.

What I found with a multiple regression analysis was, that
SO_2 and not NO_x was, in our country, the dominant variable
in the model. For the De Bilt location, situated in the center
of The Netherlands, there was no influence of NO_x on the
proton concentration at all.

I think I can explain that somewhat. First, the emitted quantity of NO_x in our country is less than of SO_2. Next, a very great part, 40 to 60% of NO_x is emitted in the form of NO and rapidly oxidized to NO_2. The emission height is about 30 cm so I believe a great part of this NO_2 will be absorbed by the direct surroundings. As I told you in my first answer, the half-life time of NO_2 is much longer than that of SO_2, so it is possible that before the oxidation of the greater part of the NO_x emitted in our country has taken place, the NO_x has passed our borders. This makes it difficult to find a relationship between the emitted quantity of NO_x in our country and the acid content of precipitation over The Netherlands. This is to the contrary for SO_2. A lot of authors demonstrated a high wash-out efficiency of SO_2 from the ambient air during rain, while in the rain samples taken no or hardly any SO_2 was found, mostly less then 5%. The rest of the original SO_2 was present in the form of sulfate. This indicates that in the presence of rainwater or cloud water the reaction velocity from SO_2 to SO_3 - and of course the transformation into sulfuric acid - is high, and probably among other things due to the presence of catalysts in the precipitation.

There is still another thing. Chemically seen, the influence of NO_x on the acid content of precipitation is less than that of SO_2 as SO_2 transformed into sulfuric acid gives two protons, where NO_x, transformed into nitric acid, gives one.

I hope these are reasons enough to make it explicable that, in contrast to SO_2, NO_x, originating from emissions in our country, will not be a dominant variable in the tested model.

MATTESON: Did you find any correlation between the mineral salts and the acidity?

VERMEULEN: Indeed I did, but as you will read in my paper, just in a single case. I looked for a correlation between protons, ammonia, sulfate and nitrate in rain water and found some in a small number of cases. I think the composition of precipitation is too complex to find good correlations between protons and one or more of the minerals present. The level of major compounds in precipitation over The Netherlands is very high, and one component (for example SO_4) has different origins, which makes finding a good correlation with the proton content very difficult.

GAURI: Is there any data on the drainage waters in Holland that would indicate that the bicarbonate ion in water has decreased against the increment of sulfate ions?

VERMEULEN: What you mean is a common phenomenon in the inland waters in Scandanavia. Over there, however, many of the inland lakes are more or less isolated and this phenomenon can be studied very well. In Holland, all the surface waters are more or less connected with the rivers Rhine and the Meuse. Perhaps you don't know, but let me tell you than that the Rhine is the most polluted river in the world. So the level of the major ions and also the buffering capacity of our inland waters is very high and also varies very much with the supply of the mentioned rivers. This makes it impossible to do studies as you mention in your question. Only in some parts of Holland we have some isolated moorland pools. We found that the greater part of these pools have been strongly acidified by acid precipitation. For example, in the 1919-1925 period one found that most of the pools had a pH above or close to 7. In the fifties, the pH values had dropped to about 5.5 and in 1975 many of the pools had a pH below 4. We did not study what you meant with your question, but we did find very high sulfate concentrations in the most acidified pools up to over 40 mg/l which is very high for these kinds of water systems in Holland.

McLEAN: I should like to give you some indication of recent developments in acid precipitation in Canada. Over the last two years, an Environment Canada network of 49 precipitation sampling stations (Canadian Network for Sampling Precipitation or CANSAP by short) has taken monthly measurements of common ions in the precipitation. One can clearly see from this data that the region which is affected by acid precipitation stretches from the Ontario-Manitoba border to the Atlantic; the mean pH is around 4.5, with a gradual increase from the south to north. West of this area, precipitation pHs are generally high, with a mean around 5.5.

Over the last nine months, there has been a strike at the biggest single SO_2 emission source in the world at Sudbury, Ontario. During this period, there appears to be little change in the precipitation chemistry compared to the fifteen months before the strike. In fact, there is very little evidence of an effect on the pH and the sulphate content of the precipitation falling in the downwind direction of this emitter.

This beings me to the point of your discussion -- that SO_2 emissions determine the level of acid precipitation.

VERMEULEN: Yes, in our country it is.

McLEAN: In Canada anyway, the role of the nitrogen oxides is now being examined more closely. Anither area which is now receiving much attention is the conversion rate of sulfur dioxide to sulfate. For example, the concentration of ozone in the atmosphere greatly influences the rate of formation of sulfate from SO_2 and the ozone concentrations in the atmosphere over Northern Ontario and Quebec are considerably lower than those in the Northeastern U.S.

VERMEULEN: Well, you had a long forward run before you put your question and I think this needs a long answer.

At first I will tell you something about studies carried out by others on the influence of SO_2 emissions of power plants on the composition of precipitation in the surroundings.

Högström did in 1974 a study of the wet deposition of sulfur from a power plant plume in Uppsala, Sweden. He found that 40 - 70 percent of the emitted SO_2 amount was deposited within 60 km of the stack. This was a very detailed study. The SO_2 washed out by precipitation consisted of 98 percent of sulfate and only 2 percent of SO_2. In the plume, this relation was just the opposite. He also found a good correlation between sulfate and protons in the rain water samples. Hales and others collected, during a study of the Keystone generating station in Pennsylvania, highly acidic rain water samples within 4 miles of the 800 ft high stack. Just as Högström did, Hales found an increase in sulfate, but a very small amount of SO_2 in the rain water. He stated that the relative importance of sulfate aerosols and SO_2 scavenging in the very complex sulfur washout process is significantly influenced by the background acidity of the rain, the height of the pollutant source and a function of the rainfall rate. Moreover, the washout factor is highly dependent on the geometry of the plume and, in my opinion, also of the size of the droplets and the content of the impurities in the rainwater. For example, when clean rainwater is falling through a low elevation, concentrated plume, a concentrated washout results, while acid rainwater falling through either a low or high elevation, concentrated plume, will fail to record its presence, when we only watch the pH. Little differences in pH yet represent rather big differences in proton concentration in the lower pH range. During an investigation in South Maryland, in the direct surrounding of a 710 MW power plant, Landsberg found pH values between 3.0 and 5.7. The samples were clearly influenced by the emitted and washed out SO_2. Another investigation of Landsberg carried out with 13 rain samplers in the surroundings of a power plant confirmed his earlier results. In a study by Granat and Rhode, in which they made use of 78 rain samplers situated about a power plant, they found a significant influence of the emitted SO_2 on the proton concentration and the concentration of sulfate.

Hutcheson and Hall had the same experience during their study
in a similar situation.

Recently, Högström and others studied the wet fallout of
sulfate and the SO_2 to sulfate transformation rate at an oil
fired 1000 MW power plant in Sweden. This project comprised a
number of wet fallout tests with not less than 100 sampling
locations up to 60 km from the source. In their experiments,
they made use of an inert tracer, SF_6. They found that about
70 percent of the emitted amount was deposited within 80 - 120
km from the source. This confirmed the earlier results from
1974. In the dispersion experiments, they found that on a day
with close to 100 percent relative humidity, 70 percent of the
sulfur from the power plant occurred as sulfate at a distance
of 30 km from the source. On a much drier day, the
corresponding figure was only 10 percent at 30 km.

I don't know under which conditions the studies you
mentioned have been carried out, but I gave you enough examples
of studies whereby the influence of SO_2 emitters on
precipitation have been proven. I think that, next to factors
such as high or low stacks, high or low emissions, weather
conditions including the humidity of the ambient air, the
starting pH of the falling rainwater, the presence of catalysts
in the precipitation, etc., the density of the sampling network
is of importance in this kind of study.

Concerning your comment on the transformation rate from
SO_2 to sulfate, from the examples I mentioned, it is obvious
that on rainy days the washed out amount of SO_2 is rapidly
transformed into SO_3 and sulfuric acid, producing the acid
precipitation.

In The Netherlands, the transformation time can be
accelerated by the very high concentrations of other components
in precipitation among which are trace metals. Also the ozone
concentration in ambient air in our country is rather high, up
to 300 µg per m^3 (98 percentile).

Now I come back to the beginning of your question again:
"Is SO_2 the main bad actor"? In The Netherlands it is, but I
can imagine that this is different in other parts of the
world. Possibly, just as you mentioned, the nitrogen oxides
are more important in the precipition in the area you spoke
about. I remember that during the conference on emerging
environmental problems, held during May 19-20, 1975 in The
Institute on Man and Science, Rensselaerville, New York, Likens
showed a figure of the Hubbard Brook experiment, where he found
a good relationship between the annual hydrogen ion input and

the annual nitrate input during a period of 10 years.

Going back to the transformation from SO_2 into sulfate, I thought that next to ozone, ammonia also is a very important compound for controlling the reaction for the transformation of SO_2 to SO_4.

KRUPA: I can make a couple of comments, Number one, it is generally the opinion that the conversion rate of SO_2 to SO_4 in a plume is photochemical. The conversion rates even in a most liberal sense estimated to be anywhere from .5 to 4 percent per hour. At night the conversion is almost zero. This is in an unscrubbed plume. Now, concerning ammonia, you ae not really talking about ammonia as a catalyst for sulfate formation. The actual process is SO_2 to sulfuric acid, followed by ammonia neutralization. The theory about SO_2, water, ammonia catalysis is being questioned. The real process seems to be SO_2 to sulfuric acid, then reaction with ammonia to form ammonium sulfate or ammonium bisulfate. In fact, in scavenging the sulfate, as the particle size increases, the scavenging efficiency also increases. Sulfuric acid aerosols are submicron particulates for which scavenging is very inefficient (Marsh, Atmos. Environ. 12:401-406, 1978). So what you are finding in a rain droplet is perhaps what is below the cloud or before the condensation nucleus stabilizes.

VERMEULEN: Well, I think I did not express myself quite well. Concerning ammonia, what I really mean is, that ammonia controls the reaction from SO_2 to sulfate in a way that we can be thankful that, not all the SO_2 present in the ambient air and washed out partially by precipitation will be transformed into an equivalent quantity of sulfuric acid. It is just as you said that ammonia plays a neutralizing role by forming the salts you mentioned. So I agree with you that ammonia is not a real catalyst as a catalyst is defined, but I don't know if ammonia does not accelerate the transformation. I think it does, because from different recent studies at oil fired power plants it appeared that next to the fact that most of the emitted SO_2 is deposited in the direct surroundings of the plant during rain, several independent factors like a high deposition rate of ammonium ions related to power plants fallout patterns with short residence times let's say 60 km, and much less fallout of ammonium with a residence time of, let's say 200 km, also related to the power plant fallout pattern, favor a kind of catalytic process with the participation of ammonia to explain the rapid transformation rate and subsequent fallout rate. I can add to this that with respect to these kinds of studies, diffusion estimates indicated that the radial turbulent influx of background NH_3

is likely to be effective enough for the ammonium fallout rates
observed.

Concerning your first comment you are right about the
conversion rate when you are speaking about a plume just
leaving the stack under dry weather conditions and in the
absence of relatively high levels of ozone or trace metals and
the presence of a low relative humidity of the ambient air.
From studies with airplanes, as well as from other studies and
those I named in my answer to Dr. Mc Lean it is known that
under the conditions of relatively high levels of ozone or high
relative humidity, the transformation rate is much higher as
you indicate. Transformation rates of 70 percent within 30 km
of a plant are given.

Next to that, the transformation rate from SO_2 to sulfate
in the washout fraction in precipitation is close to
100 percent, because the content of SO_2 in precipitation was,
as far as I know, in no case more than a few percent of the total
quantity of sulfur compounds present in the sample.

Now I want to inject a comment just for information.

When ammonia is present in the atmosphere, the washout
SO_2, partially transformed into sulfuric acid can be
neutralized more or less by forming ammonium salts. Now in
some parts of The Netherlands I found that because of the
presence of peroxides and ozone and perhaps other oxidizing
agents, the content of ammonia in the ambient air and the
presence of ammonium in precipitation was very small to
negligible. When ammonia in the atmosphere is present, by the
formation of ammonium bisulfates in precipitation,
theoretically, the pH can reach a minimum value of about 2.6.
In certain parts of The Netherlands, due to lack of sufficient
ammonia in the atmosphere I suppose, in single precipitation
samples before 1974 I found pH values between 2 and 2.6.

MASS TRANSFER OF GASES TO GROWING

WATER DROPLETS

Warren W. Flack and Michael J. Matteson

School of Chemical Engineering
Georgia Institute of Technology
Atlanta, Georgia   30332

INTRODUCTION

The absorption of gases by clouds, fogs and water droplets is
a key step in the removal process for many trace gases in the atmosphere. The high acid content of rainfall in many industrial regions
is attributed to the absorption of sulfur and nitrogen oxides. Upon
the release of fossil fuel stack gases, containing water vapor, which
are rapidly cooled from 250°F to ambient temperatures, much of the
associated oxides of sulfur and nitrogen is dissolved as the water
is condensed upon suspended particles and the smoke plume is formed.

The object of the work reported here is to determine the rates
of absorption of SO_2, NO_2 and oxygen at stack gas concentration by
water droplets undergoing growth due to water vapor condensation.

PART I: SO_2 ABSORPTION

Gartrell et al. [1] observed that there is little conversion
of SO_2 to SO_3 downwind in the plume. However conversion may occur
during plume formation. Bogaevskii [2] showed that gas absorption
is enhanced during condensation as compared with a drop in the no-
growth condition. Wills [3] studied SO_2 absorption enhancement in
falling droplets. Oliver [4] and Herrmann [5] have shown that absorption of oxygen and nitrogen dioxide was enhanced during condensation of a humid gas stream on a cold suspended droplet. These
studies pointed out that, in the condensed water film, supersaturated
amounts of absorbed gas are accumulated during the growth phase.
When the droplets begin to evaporate, the excess gas is desorbed
until a saturated condition is attained.

The absorption of a gas during condensation of water vapor on a cold water droplet is a complex process characterized by unsteady state mass and heat transfer [6]. In the classical development of absorption of a gas in a liquid, three theoretical models have ensued: The film theory, the penetration theory, and the boundary layer theory. Each model invokes different assumptions which result in different conclusions.

Angelo [7] has shown that during periods of continual surface renewal, the actual mass transfer coefficient may be fifteen times as large as that predicted by boundary layer theory. Thus, the unsteady state absorption during surface renewal is a more complex situation not covered by these theories. In order to describe a fog or mist formation, it is necessary to study droplet growth by condensation with no internal turbulence. Bogaevskii [2] reported water droplets growing by water vapor condensation in a mine shaft to absorb about six times more sulfur dioxide than that predicted by steady state absorption.

There are several possible mechanisms to explain the enhancement of absorption during surface renewal. The Marangoni Effect results from the fact that dilute solutions of water (10^{-3}N) exhibit abnormalities in regard to surface tension. Jones and Ray [8] have observed that absorption of ions at the surface continues until a specific number of sites are occupied. The concentration of these sites is about 5 per 10^5 surface molecules. If liquid vapor is continually condensed on the drop surface, new surface for sites is being formed at a rate fixed by condensation. A second mechanism for enhancement, Stefan flow, is a trapping of the gas molecules into the liquid phase by the condensing vapor flux. The Stefan flow flux can be expressed as [9]:

$$F_{SF} = \frac{D_{AB}P}{RT} \ln[(P-P_{sat})/(P-P_w)] \qquad (1)$$

Once growth stops, desorption of the gas occurs until the equilibrium concentration is reached. However, if the gas can be complexed by a fast enough chemical reaction, then it would be possible to contain the gas. Additives which result in a chemical reaction are present in the stack gases themselves. Several additives for sulfur dioxide are vanadium pentoxide, manganese sulfate, and soot (carbon). Preliminary work in this area was done with manganese sulfate by Matteson et al. [10].

INSTRUMENTATION AND PROCEDURE

The flow scheme for the gas delivery system is shown in Figure 1. The purpose of the system is to combine the nitrogen, sulfur dioxide and water vapor in a manner such that the SO_2 and water

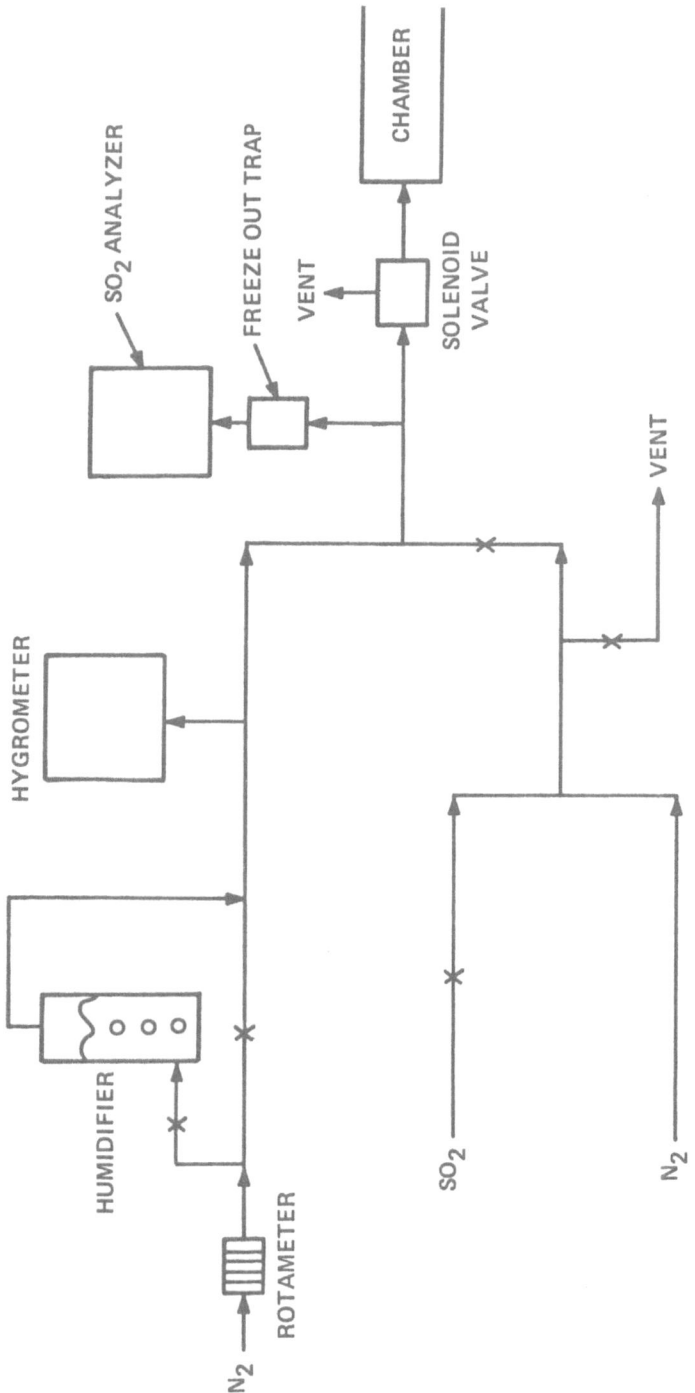

Figure 1. Experimental Arrangement

vapor concentrations can be accurately controlled. A Matheson
Rotameter, Type 605, was used to regulate the nitrogen flowrate.
The humidity of the stream was controlled by a water bubble column
and monitored by a Cambridge Systems Model 990 Thermoelectric Dew
Point Hygrometer. The sulfur dioxide concentration was regulated
by manipulating three precision needle valves. The gas was analyzed
with a Beckman Model 215A Infrared Sulfur Dioxide Analyzer equipped
with a Sargent Model MR Chart Recorder. The humid gas mixture was
then sent to a solenoid valve, which when activated permitted the
stream to flow into the contact chamber. The flowrates of all the
sidestreams are very small in comparison to the initial flowrate
of nitrogen. Thus, the initial flowrate can be considered as the
effective flowrate into the contact chamber.

The contact chamber, shown in Figure 2, consists of a glass
tube 60 cm in length and 3.5 cm in diameter. The contact chamber
included two drains and glass joints at both ends to connect to
tubing to allow the stream to enter and exit the chamber. The
sampling station was located approximately 40 cm from the entrance.
Above the sampling station was a 24-gauge hypodermic needle connected
to a plastic tube. Part of the tubing was coiled in the constant
temperature ice bath to allow the water in the coil to reach thermal
equilibrium. The other end of the tubing was connected to a Sage
Instruments Model 341 Syringe Pump with a timer mechanism. The
syringe pump/timer was controlled to give uniform size droplets on
the end of the hypodermic needle. A second 24-gauge hypodermic
needle, a YSI Model 524 Probe Thermistor, could also be inserted
into the chamber so that the temperature of the drop could be
monitored. The thermistor was connected to a YSI Model 43 TD Meter
which could be connected to the Sargent ME Recorder.

After an initial warmup period, the main nitrogen valve is
opened and the flowrate regulated to 21 liters per minute in order
to obtain a velocity (V) of 50 centimeters per second past the drop.
The humidity of the stream is then set in order to obtain the desired
supersaturation ratio. The supersaturation ratio (SSR) is defined
as the partial pressure of the water in the gas stream divided by
the partial pressure of the water at the initial temperature of the
droplet (5.5°C). The supersaturation ratios used were 1.0, 1.5,
2.0 and 2.5 which correspond to dew point values of 5.5, 11.5, 16.0,
and 19.5°C. These values were obtained by adjusting the valves
leading to the humidifier and bypassing it. The hygrometer was
used to measure the dew point values.

Next, the sulfur dioxide line was opened and the three needle
valves were used to regulate the sulfur dioxide to obtain 1000,
2000 or 3000 parts per million (ppm). The infrared analyzer was
used to measure the sulfur dioxide concentration.

The syringe mechanism of the syringe pump was then filled with

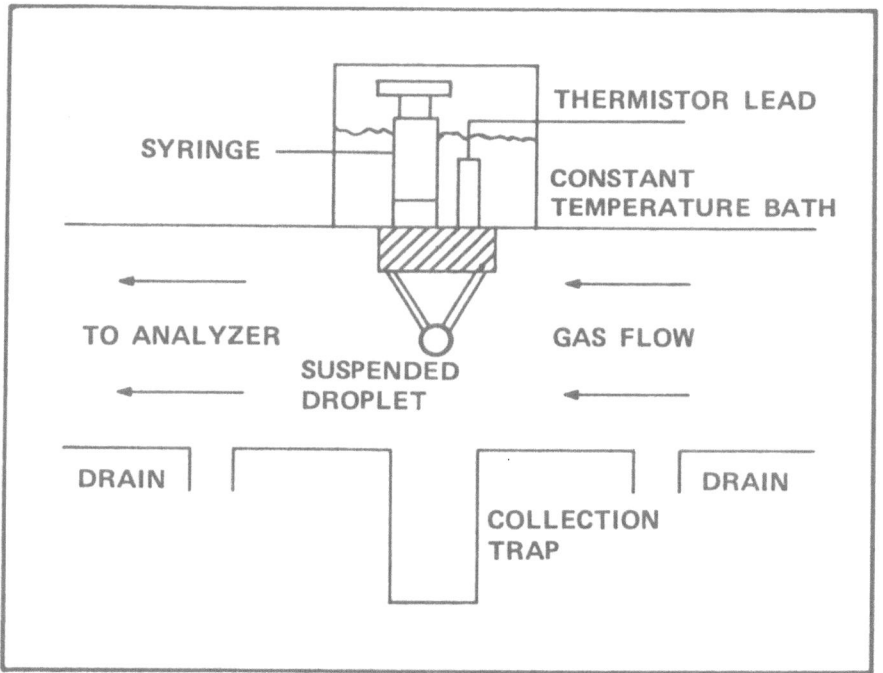

Figure 2. Contact Chamber

the desired solution to study. In the initial tests distilled, deionized water was used. In later tests the following solutions were studied: 10, 100, and 1000 μm vanadium pentoxide (V_2O_5); 10, 50, 100 and 200 mM manganese chloride ($MnCl_2$); and 0.344 mg/cm^3 of carbon black with a geometric mean diameter of 1.27 μm and a log normal standard deviation of 1.67. The syringe pump and timer mechanism were then activated to give uniform size drops of 0.23 cm in diameter. At the same time the drop was formed on the hypodermic needle, the solenoid valve was activated so that the humid gas stream would be in contact with the cold droplet. The desired exposure time was then measured. The exposure times used were 5, 10, 20, 30, 60 and 90 seconds. For drops exposed to gas at low supersaturation ratios it is not possible to collect the drops at the 90 second time because of evaporation. Also, it is not possible to obtain exposure times shorter than 5 seconds.

At the end of the exposure period, the exposed drop is released into a 10 ml aliquot of sodium tetrachloromercurate (TCM). This agent fixes the sulfur dioxide by forming a dichlorosulfitomercurate complex. Five exposed drops were collected per aliquot to bring the sulfur dioxide content into the middle of the test range. The sulfur dioxide content is then determined by the West-Gaeke Colorimetric Analysis adapted from ASTM D2914 [11]. This method is designed for determination of atmospheric sulfur dioxide and was used because of its sensitivity, 0.003-5.0 ppm. The Bausch and Lomb Model 20 spectrophotometer was used for determination of absorbance at 548 nm wavelength.

RESULTS

The experimental work was roughly divided into three divisions. The first part consisted of determination of the initial temperature of the drop and the resulting temperature profiles for different supersaturation ratios. The second part consisted of using sulfur dioxide concentrations of 1000, 2000 and 3000 ppm to test the concentration of sulfur dioxide in pure water drops at given supersaturation ratios and exposure times. The third part was a study of the effect of dissolved salts and suspended carbon particles on the concentration of the sulfur dioxide in the drops.

The temperature profiles for the different supersaturation ratios are indicated by Figure 3. These results are expressed in the reduced temperature, θ, versus time. This dimensionless temperature is defined as:

$$\theta = \frac{T_d - T_o}{T_\infty - T_o} \qquad (2)$$

where T_d is the drop temperature, T_o is the initial drop temperature

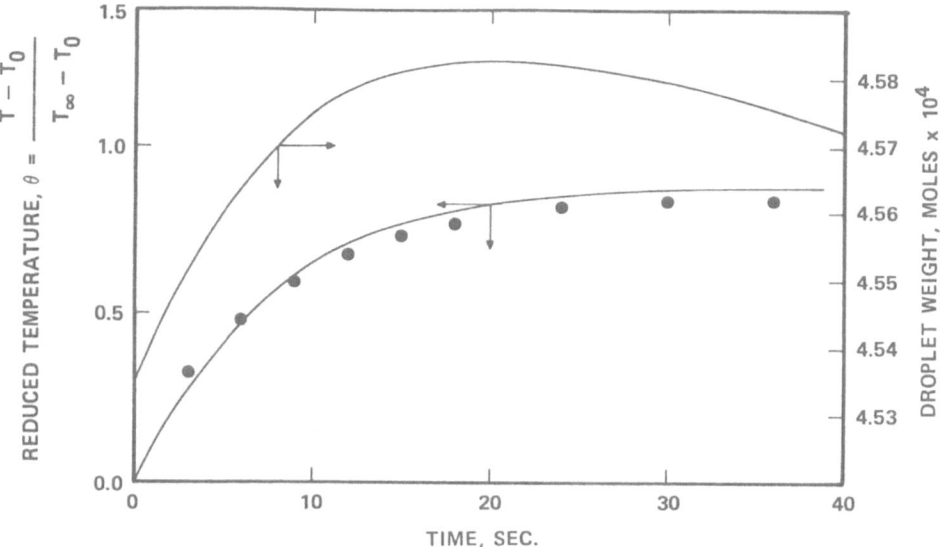

Figure 3. Droplet temperature and mass variation with time. Solid
lines represent Runge–Kutta approximations based on
energy-material balance for droplet with initial temperature of $T_0 = 5°C$ exposed to flowing air at $T_\infty = 25°C$
and water vapor pressure = 2.5 times that for water at
5°C. Points are experimentally obtained values.

and T_∞ is the ambient temperature. Using the thermistor probe over repeated measurements it was determined that $T_0 = 5.5°C$.

These tests show that the droplet temperature rises rapidly during the first 20 seconds and then slows down to an apparent equilibrium value lower than ambient gas temperature. The higher the supersaturation ratio, the closer the equilibrium temperature came to the ambient temperature.

The results presented in Figures 4, 5, and 6 show the sulfur dioxide absorbed by the pure water droplets as a function of time for various supersaturation ratios and gas phase sulfur dioxide concentrations. These results are expressed in a normalized sulfur dioxide concentration where the average concentration of the drop $(\bar{C})$ is divided by the saturation concentration (C_{sat}) at the temperature of the drop when sampled. The saturation concentration is determined using the Henry's Law Constant appropriate to that temperature.

Since the droplets tested were quite large, the core of the droplet did not become saturated with the sulfur dioxide. This is the reason that $\bar{C}/C_{sat}$ is found to be less than unity. For droplets in the aerosol size range, values greater than unity might well be observed.

These results indicate that as the supersaturation ratio is increased, an increase in absorption is experienced. This continues for approximately 30 seconds, and then droplet concentration begins to decline. The dashed curve has been inserted to show the concentration that would occur if absorption were controlled by diffusion in the liquid phase [6]:

$$\frac{\partial C}{\partial t} = D_{SO_2-H_2O} \nabla^2 C_{SO_2} \tag{3}$$

The effects of dissolved salts and suspended carbon particles are shown in Figures 7, 8, and 9. These results are expressed in the same format as before. Based on results from the second section, it was felt that it would be representative to conduct these tests at 2000 ppm sulfur dioxide and a supersaturation ratio of 2.5. Various concentrations of vanadium pentoxide and manganese chloride were investigated. However, due to the difficulties of working with a carbon particle suspension, only one concentration of the carbon was studied.

It is apparent that the addition of all three additives accelerates the absorption rate of sulfur dioxide. For the two salts, as the concentration is increased, the absorption is also increased. The vanadium pentoxide appears to be unique among the three additives

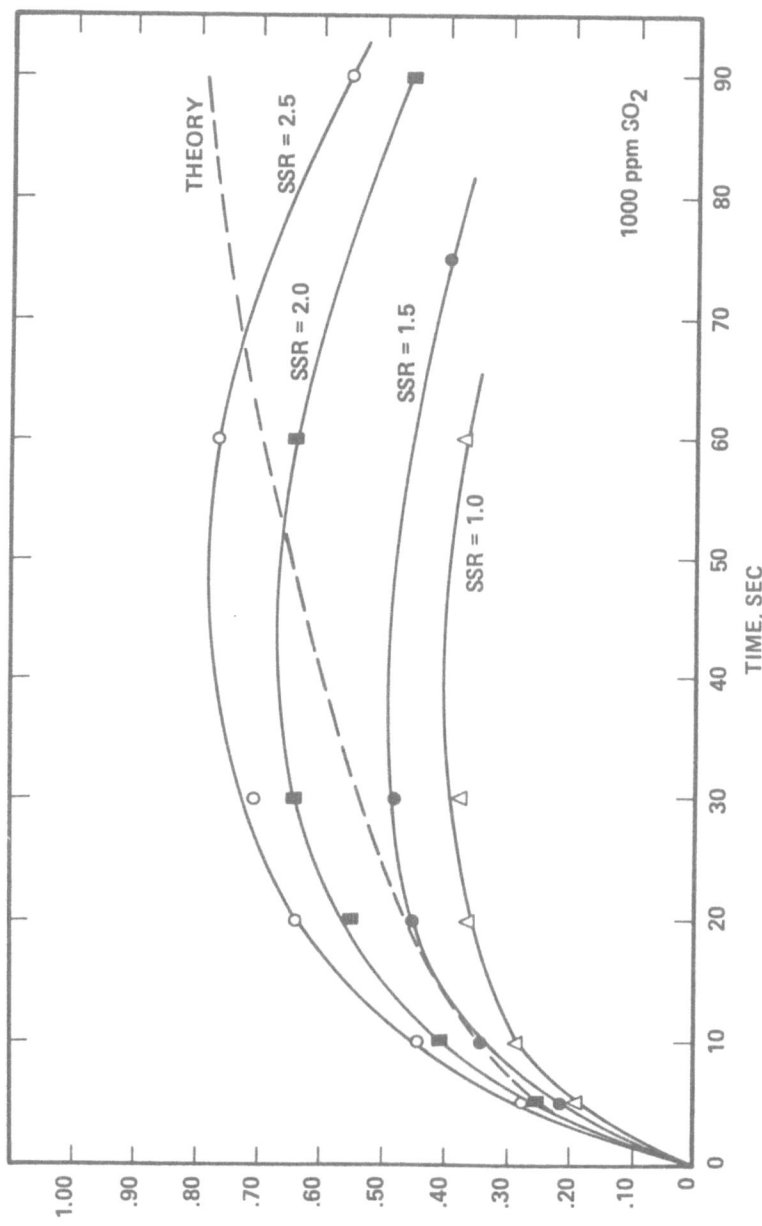

Figure 4. Absorption of 1000 ppm SO_2 from N_2 vs time at various supersaturation ratios (water vapor condensation rates). Sulfur dioxide concentrations are normalized to saturation value at the temperature of the droplet. Dashed line represents theoretical absorption rate based on aqueous phase diffusion.

WARREN W. FLACK AND MICHAEL J. MATTESON

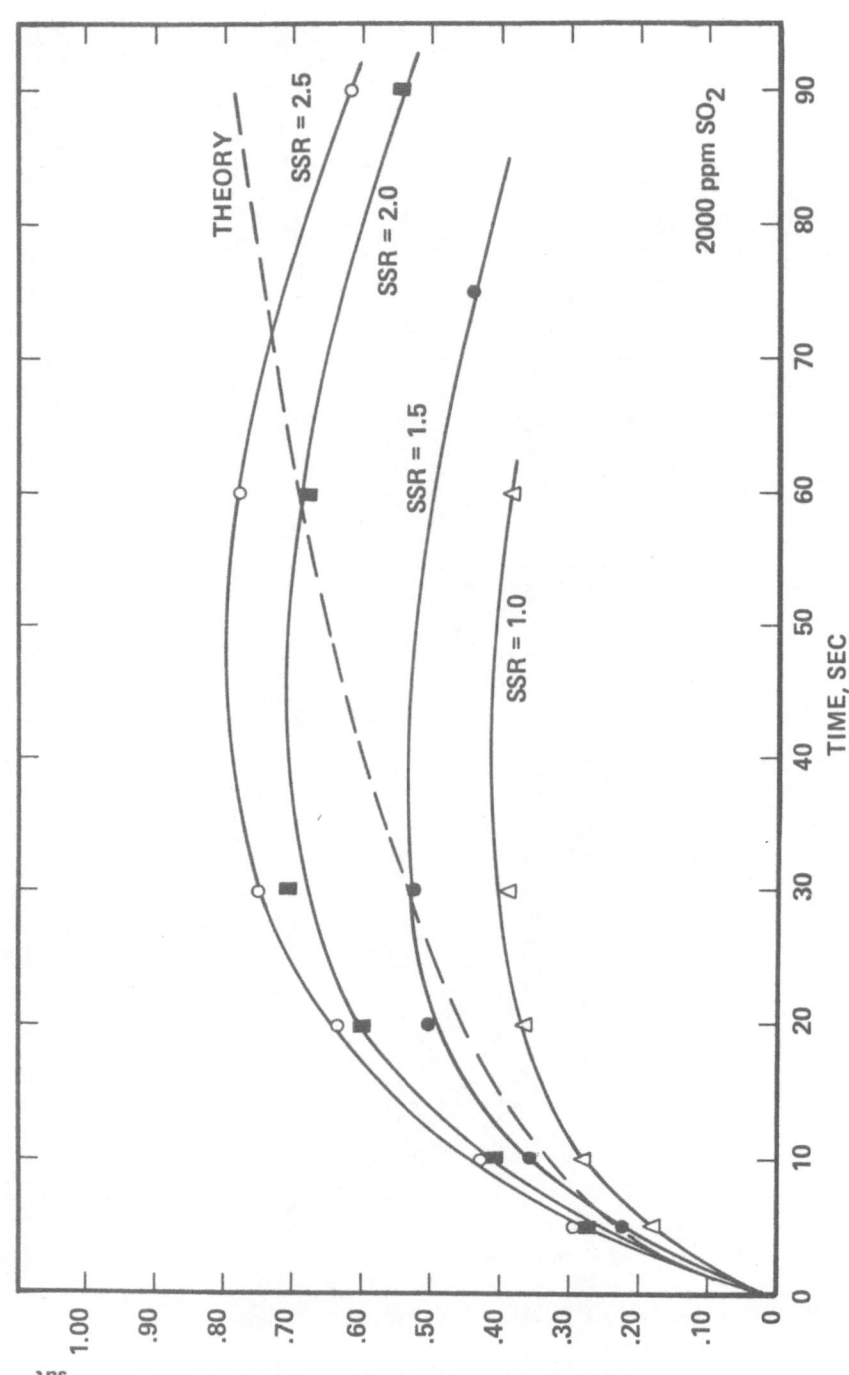

Figure 5. Relative concentration vs time for absorption of 2000 ppm SO$_2$.

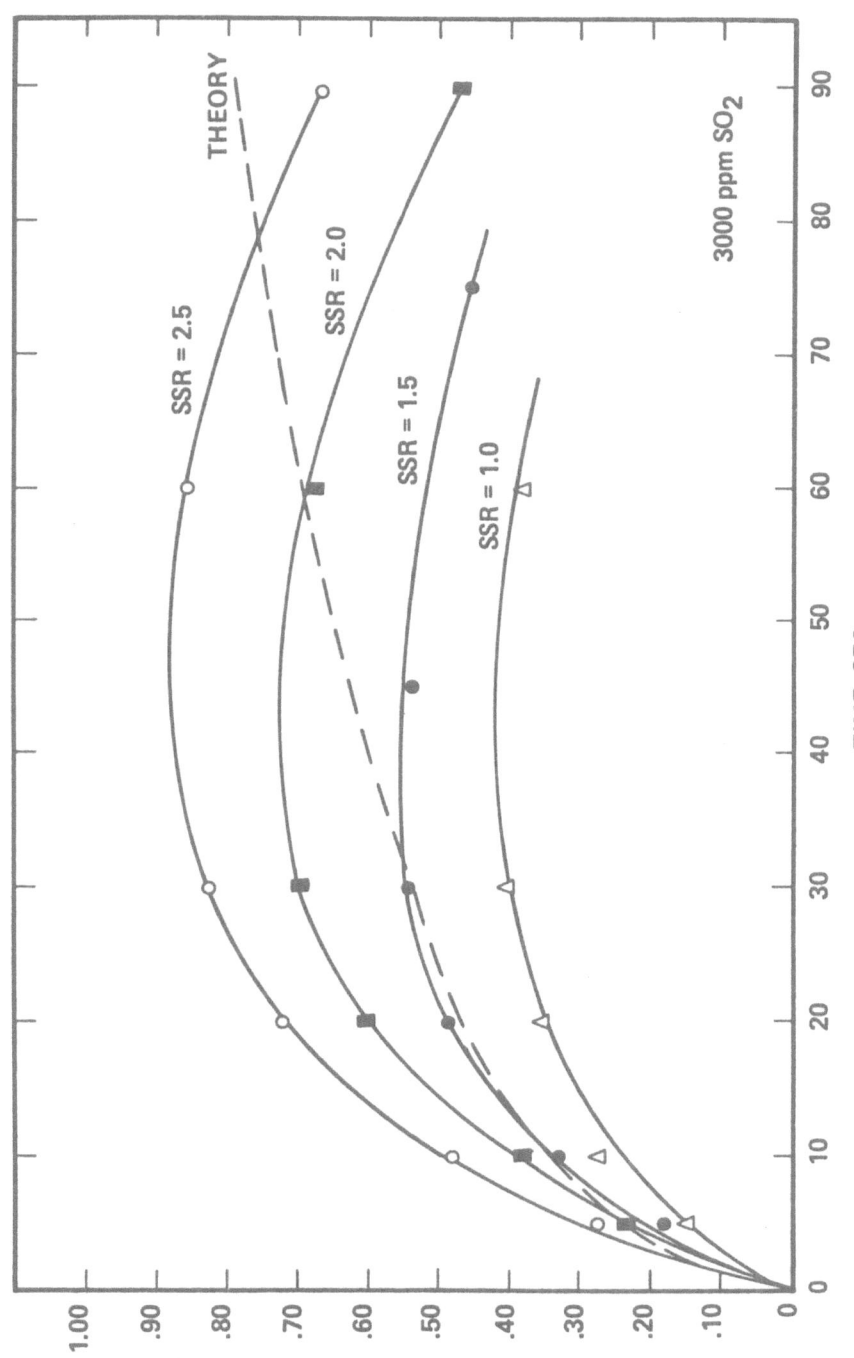

Figure 6. Relative concentration vs time for absorption of 3000 ppm SO$_2$.

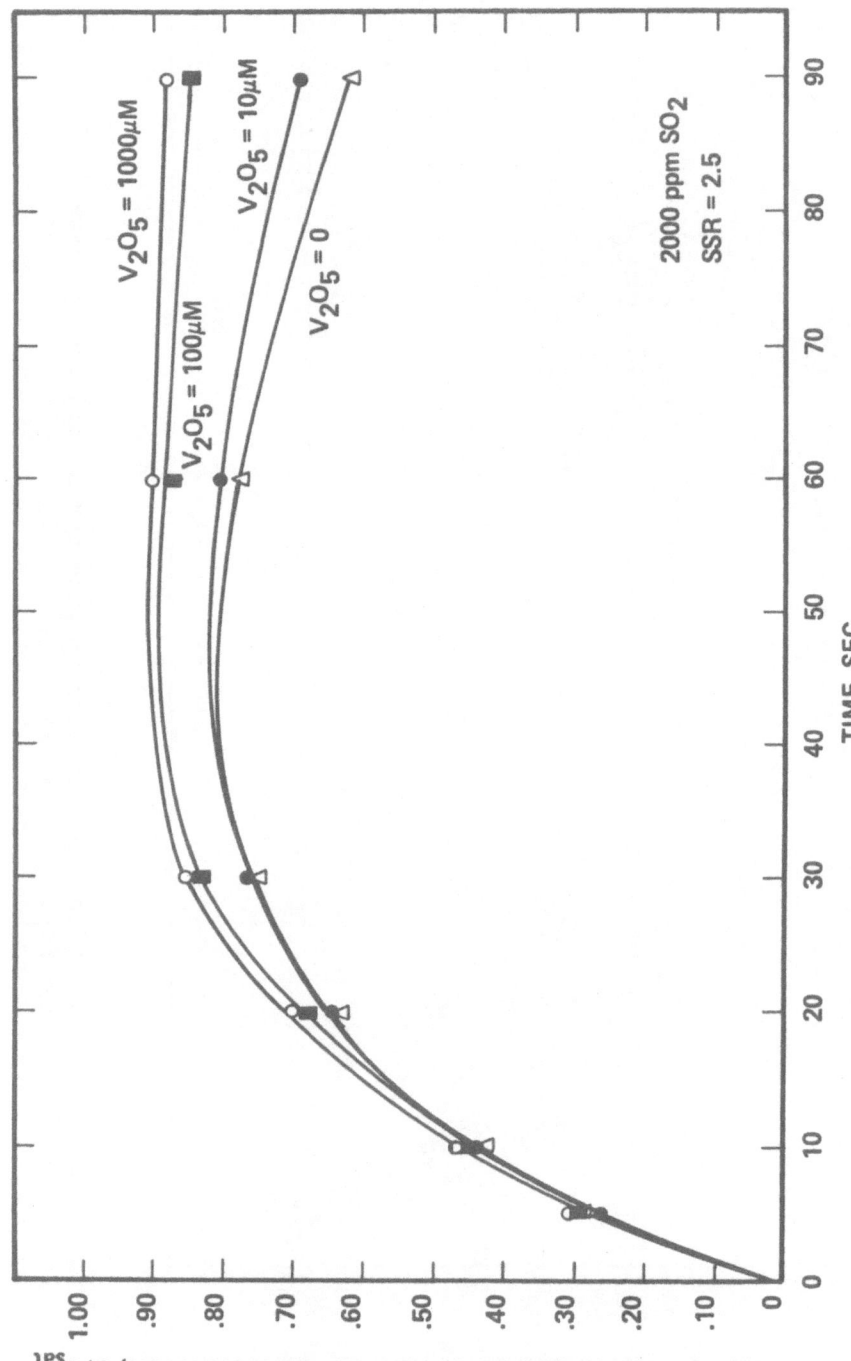

Figure 7. Relative concentration vs time for absorption of 2000 ppm SO_2 by droplet containing various concentrations of V_2O_5.

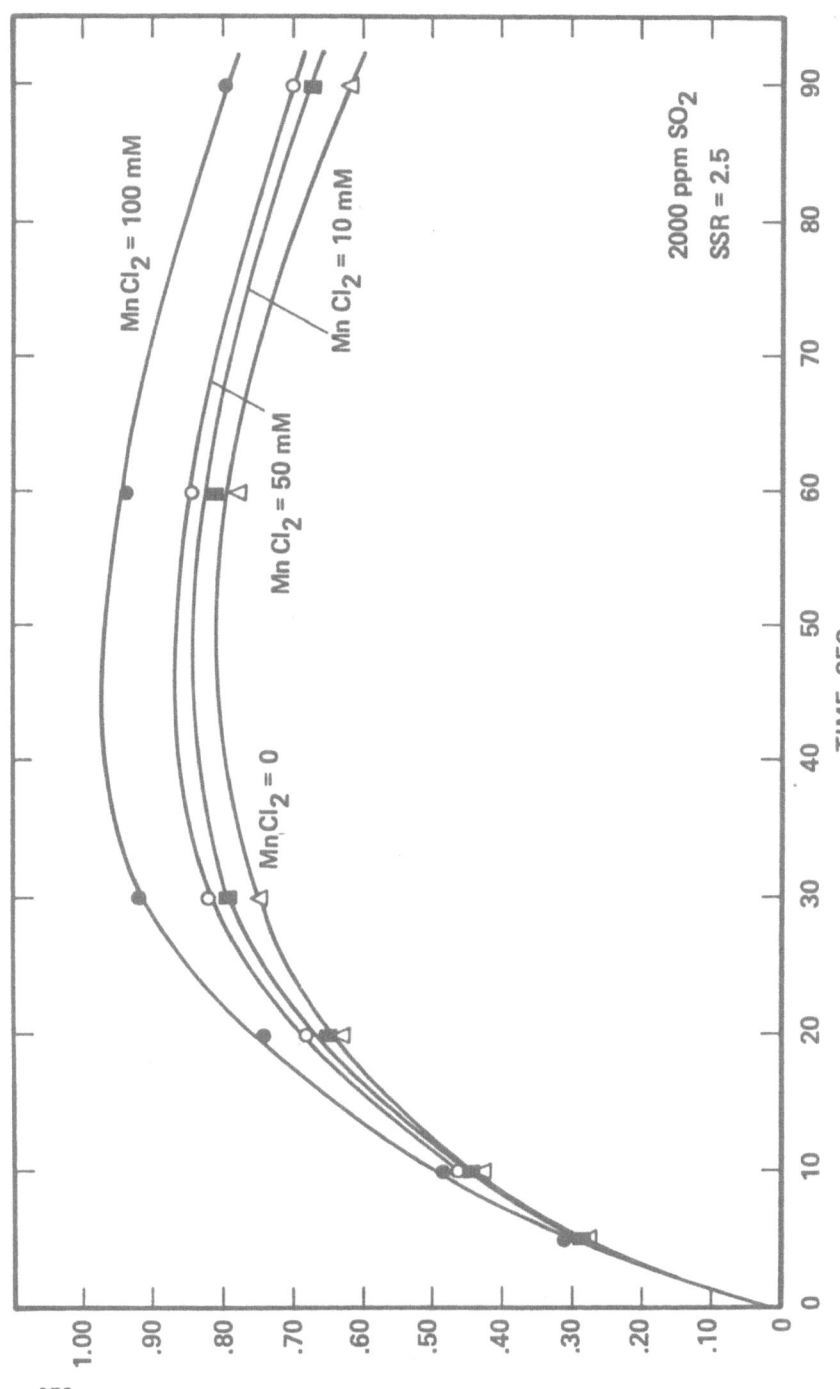

Figure 8. Relative concentration vs time for absorption of 2000 ppm SO_2 by droplet containing various concentrations of $MnCl_2$.

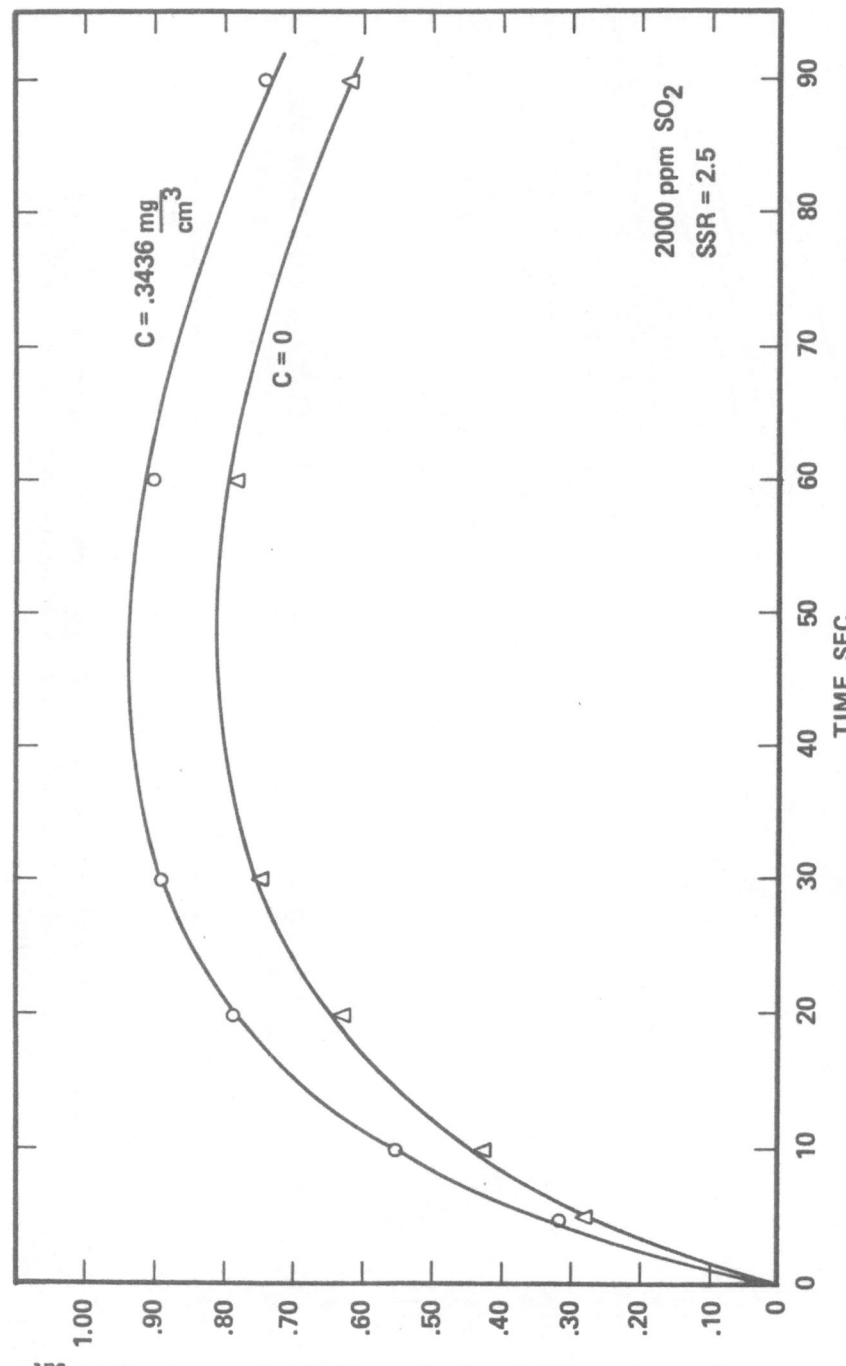

Figure 9.   Relative concentration vs time for absorption of 2000 ppm SO_2 by droplets containing suspension of carbon black.

in that if it is present in high enough concentrations, the droplet concentration remains approximately constant.

It should be noted that for the carbon suspension, it is possible that part of the sulfur dioxide was complexed on the carbon surface and not measured by the West-Gaeke analysis. Thus, the measured values might have an inherent source of error.

PART II: NO_2 ABSORPTION

Investigations into water absorption at low concentrations (ppm range) of nitrogen dioxide are relatively few. From Borok's [12] data, the Henry's Law constant for NO_2 in water is $H = 100$ atm^{-1}. Palmes et al. [13] and Crecelius and Forwerg [14] found that the NO_2 is present primarily as NO_2^- with only trace amounts of NO_3 when absorbed by water in the ppm range. A possible reaction sequence for the low concentration range may be:

$$3 \ NO_2 + H_2O \rightleftharpoons 2 \ HNO_3 + NO \qquad (4)$$

$$HNO_3 + 2 \ NO + H_2O \rightleftharpoons 3 \ HNO_2 \qquad (5)$$

$$HNO_2 \rightleftharpoons H^+ + NO_2^- \qquad (6)$$

with the ionization proceeding to 100 percent completion as the NO_2 concentration is decreased.

Dekker [15] concluded that the main resistance to mass transfer shifts from the gas phase to the liquid phase as the concentration is lowered. With this background it was decided to test NO_2 in the 100 - 300 ppm range for enhanced mass transfer via water vapor condensation.

The results presented in Figures 10 and 11 show the NO_2 absorbed by water as a function of time for various SSR and gas phase NO_2 concentrations. The results are normalized to the saturation concentration at the temperature of the droplet when sampled. For the tests at both 100 and 300 ppm NO_2 in the gas phase, the droplets were practically saturated with NO_2 within 5 seconds of exposure. The dashed lines are based on an assumed first order reaction during non-steady state diffusion into a semiinfinite medium (penetration theory).

There is a definite effect on supersaturation as the SSR is increased above 1.0. This continues for about 15 seconds and then droplet concentration returns to the saturation level. Tests at 60 seconds showed that the droplets remain at the saturation concentration. An explanation for the rapid decline in NO_2 supersaturation is indicated in Figure 3 which shows that condensation

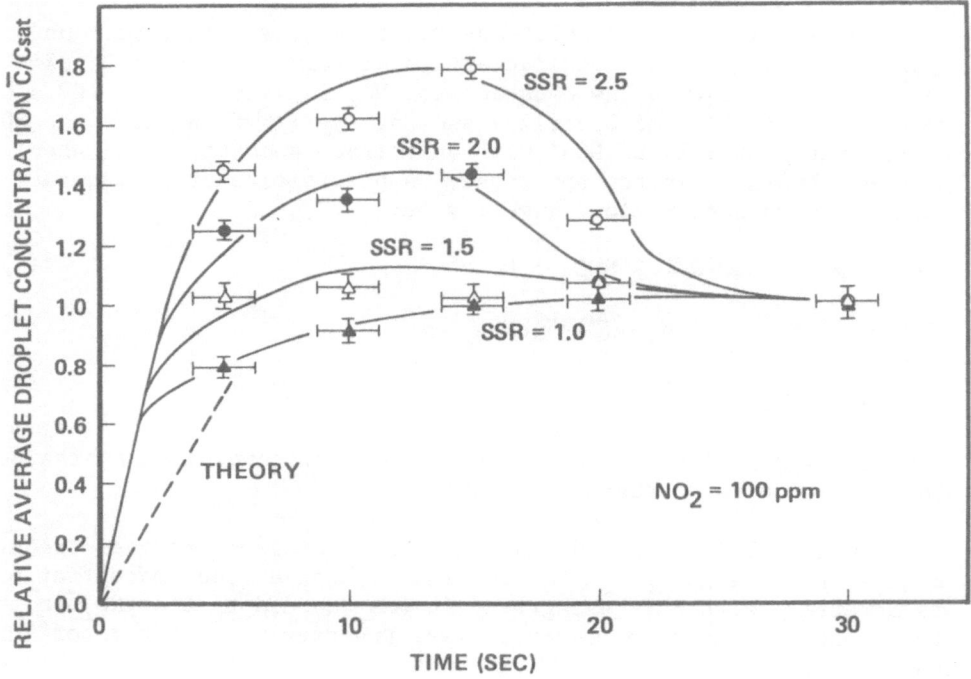

Figure 10.   Absorption of 100 ppm nitrogen dioxide from N_2 at various
             supersaturation ratios (water vapor condensation rates).
             Nitrogen dioxide concentrations are normalized to sat-
             uration value at the temperature of the droplet. Dashed
             line represents theoretical absorption rate based on
             N_2O_4 data.

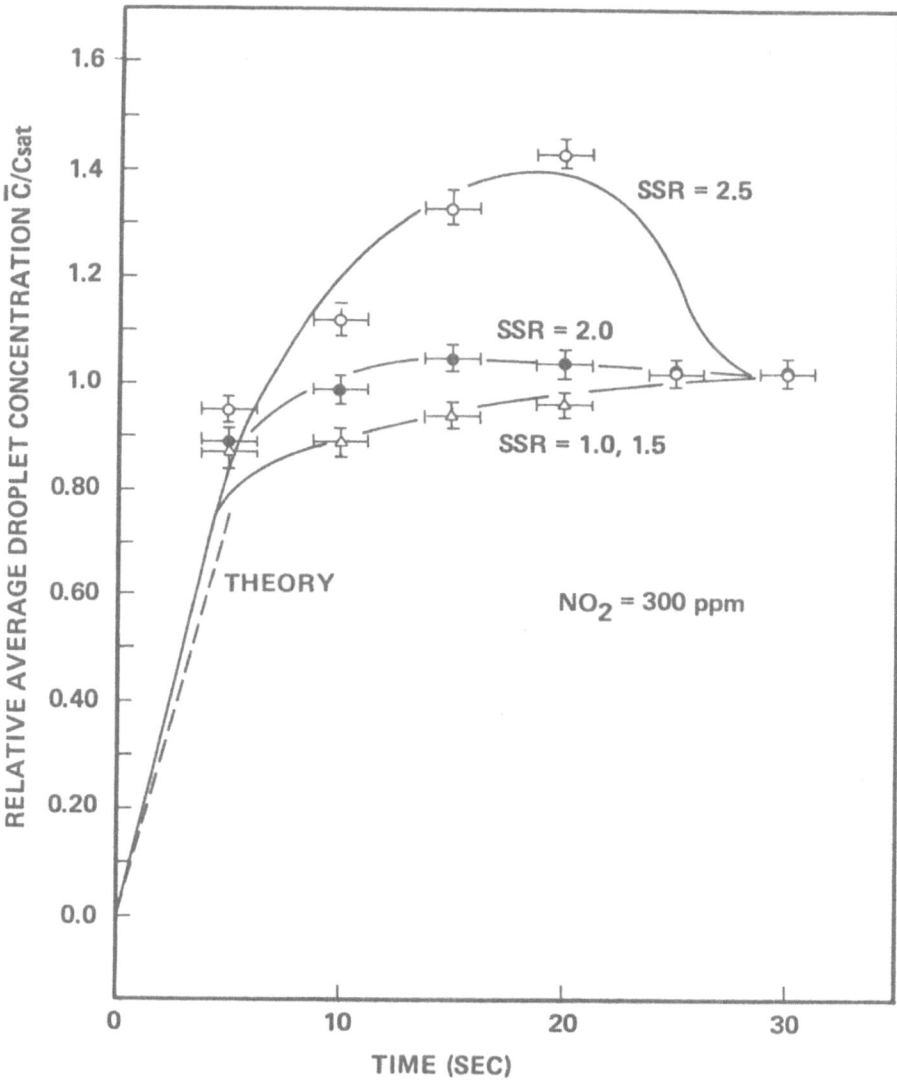

Figure 11. Absorption of 300 ppm NO_2 at various supersaturation
ratios.

ceases after 15-20 seconds and evaporation begins. Therefore, the supersaturation of NO_2 is strongly connected to the condensation of water vapor. For the case of SSR = 1.0, no condensation and only evaporation occurred. As expected there is no evidence of supersaturation of NO_2. Greater supersaturation was possible at the lower than at the higher NO_2 level, as was experienced in the investigations of SO_2.

PART III: O_2 ABSORPTION

Oxygen was tested to determine if a non-hydrolyzing, non-reacting gas would respond to the same enhancement effects of absorption during water vapor condensation as did SO_2 and NO_2. Furthermore, many oxidation processes in fogs depend on the concentration of dissolved O_2.

Since oxygen is so sparingly soluble in water we used the same contact cell as used with the SO_2 and NO_2 tests. To measure the amount of O_2 absorbed, we adapted a Natelson Microgasmeter, normally used for blood oxygen analysis. Concentrations of O_2 in N_2 tested were 21%, 40%, 60%, and 80%; and SSR were, again, 1.0, 1.5, 2.0 and 2.5.

Results are shown in Figures 12 and 13 for 21% and 80% oxygen, respectively. The results are normalized to the oxygen saturation value at that temperature. Also a curve has been inserted showing the concentration one would expect if absorption were controlled by diffusion in the liquid phase. As with the tests on SO_2 and NO_2, we noticed that oxygen absorption is definitely enhanced by water vapor condensation.

DISCUSSION

The rate of gas absorption during water vapor condensation or evaporation is a function of both the rate of diffusion in the liquid phase, and the rate at which the water is transported. In order to isolate the effect of the water vapor transport on the rate of absorption, the increase in the amount of gas absorbed and the corresponding increase in the amount of water condensed for different SSR was determined for fixed values of time.

Tests on all three gases indicate that the effect of water vapor transport on absorption follows the relation

$$\frac{d\bar{c}_A}{dt} = k(Y_A)^n \frac{d(H_2O)}{dt} \tag{7}$$

such that

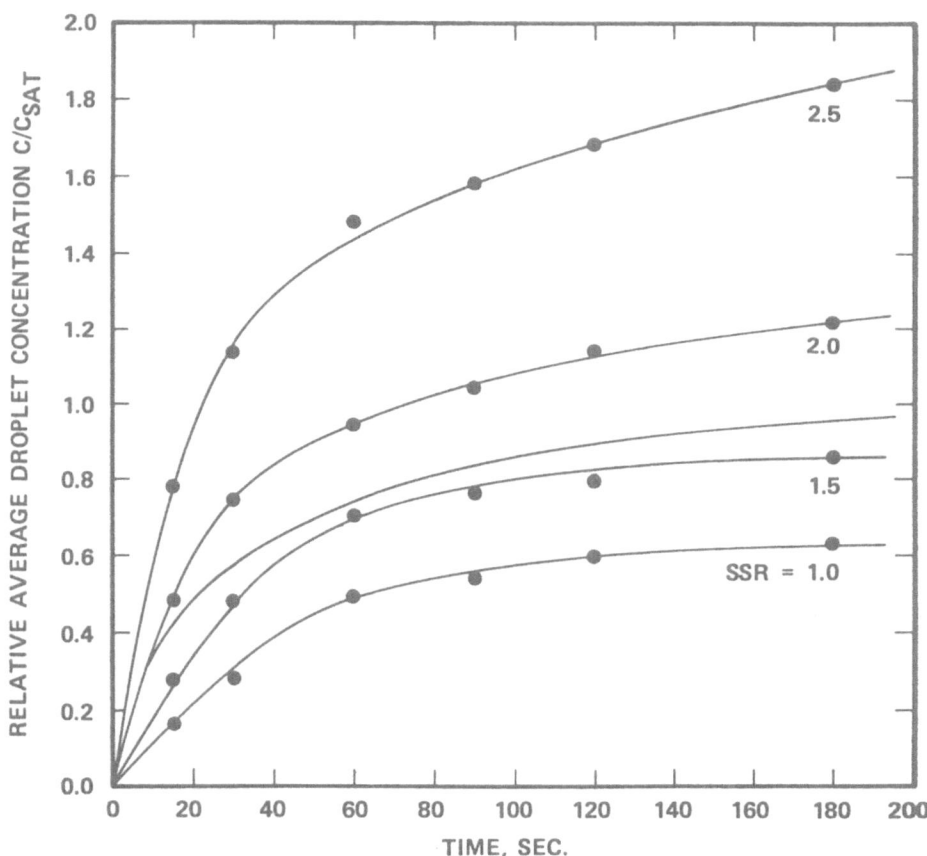

Figure 12. Absorption of oxygen from air (21% O_2) at various supersaturation ratios. Oxygen concentrations are normalized
to saturation value at the temperature of the droplet.
Solid line without experimental points represent theoretical variation of average concentration vs time.

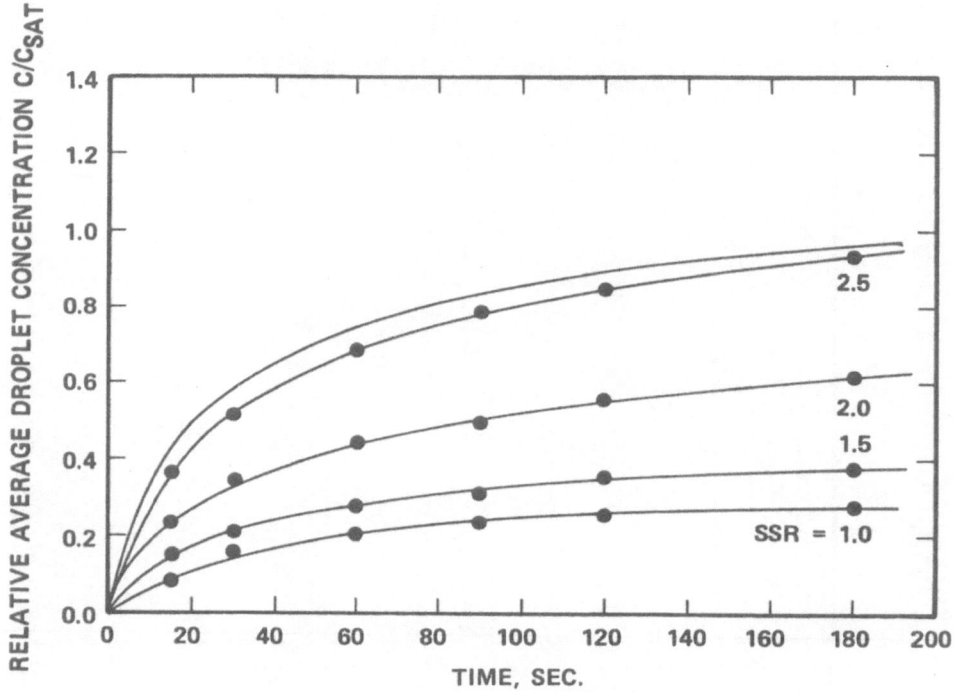

Figure 13.   Absorption of oxygen from mixture 80% O_2, 20% N_2, at
various supersaturation ratios.

$$\bar{C}_A = k(Y_A)^n \ (H_2O) \tag{8}$$

where Y_A = absorbed gas concentration in gas phase, mole fraction

(H_2O) = moles water condensed

Gas	k	n
SO_2	6.53	1.27
NO_2	3.68×10^{-4}	0.105
O_2	5.66×10^{-4}	0.20

A possible explanation for this absorption enhancement may be that under conditions of water condensation, there is a net movement of water vapor toward the water surface, tending to drag extra gas molecules along, whereas in the steady-state condition there exists an equimolar exchange of water molecules both to and from the surface. The gas is absorbed in both cases, but in the non-steady state the amount of gas entering the surface is enhanced somewhat by the condensing H_2O molecules. The resistance to absorption is primarily in the liquid phase which is governed by the dissolved gas concentration at the air-water interface. Therefore, if more gas molecules can be "packed" into the surface, this would tend to enhance the overall absorption process.

Another possible explanation for the enhancement effect is the generation of surface instabilities. The transfer of water vapor to the liquid phase may occur in localized amounts over the surface, causing adsorbed gas to be more concentrated at some points in the surface than in neighboring regions. Consequently, at the points of higher concentration the transfer of adsorbed gas to within the water bulk moved at a faster pace, dragging surface solution with it. The net result are many surface driven instabilities and an overall surface turbulence, which tends to reduce the liquid phase resistance to gas absorption.

We noted that the enhancement phenomenon is diminished at higher gas phase concentrations. There may be a limit to the number of extra molecules that the condensing water is dragging along.

It is generally assumed, in the steady-state case, that the surface of the droplet is saturated with oxygen and that the saturation concentration is determined by the gas phase concentration at the gas-liquid interface. Under evaporation conditions, this interfacial concentration may be less than that of the bulk gas phase concentration because of the migration of water molecules in the opposite direction. For example, the rate of evaporation at

SSR = 1.0 at 100 seconds is about 100 times faster than the rate of O_2 absorption over the range of oxygen concentrations tested. Therefore, the liquid phase saturation concentration would tend to be less than one would expect from steady-state conditions. However, acting to oppose this is the escape of water molecules from the surface solution, which would tend to concentrate the oxygen in the liquid at the interface. The net effect as shown in Figure 8 for SSR = 1.0, 1.5 is to produce an equilibrium concentration somewhat lower than the steady-state case for $t \to \infty$.

CONCLUSIONS

Droplets growing by water vapor condensation are able to concentrate gases at levels that far exceed saturation. The implication of these results when extended to growing fogs and clouds is that oxidation processes involving conversion of dissolved SO_2 or NO_2 to sulfates and nitrates may proceed much more rapidly than expected when reactant calculations are based on steady-state concentrations.

ACKNOWLEDGEMENT

This work was supported by Contract No. EE-77-S-05-5592 from the Department of Energy.

NOMENCLATURE

$\bar{C}$	average drop concentration (mg/cm^3)
C_{sat}	saturation drop concentration (mg/cm^3)
D	diameter of drop (cm)
D_{AB}	diffusivity of A into B (cm^2/sec)
F	flux ($gmole/cm^2/sec$)
H	Henry's Law constant (moles/liter atm)
K	absorption constant
P	vapor pressure (atm)
ppm	parts per million
R	gas law constant (cm^3 atm/gmole K)
SSR	supersaturation ratio
t	time (sec)
T	total drop concentration (moles/liter)
T_d	temperature of drop (°C)
T_o	initial temperature of drop (°C)
T_∞	ambient gas stream temperature (°C)
V	velocity of gas stream (cm/sec)
Y	mole fraction in gas phase
θ	reduced temperature

REFERENCES

1. Gartrell, F. E., F. W. Thomas, and S. B. Carpenter, "Atmospheric
    Oxidation of SO_2 in Coal-Burning Power Plant Plumes", Amer.
    Ind. Hyg. Assoc. J., 24, 113 (1963).
2. Bogaevskii, O. A., "Absorption of a Gas on a Growing Drop", Zh.
    Fiz. Khim., 43, 719 (1969).
3. Matteson, J. M. and T. L. Wills, "Colloid and Interface Science",
    vol. II, pp. 95-105, M. Kerker, ed., Academic Press, Inc.,
    New York, N.Y. (1976).
4. Matteson, M. J. and M. J. Oliver, "The Absorption of Oxygen by
    Condensing and Evaporating Water Droplets", Am. Ind. Hyg.
    Assoc. J., 39, 783 (1978).
5. Herrmann, J. P. and M. J. Matteson, "The Absorption of Nitrogen
    Dioxide by Condensing Water Droplets", AIChE 70th Annual Mtg.,
    New York, November 13-17, 1977.
6. Bird, R. B., W. E. Stewart and E. N. Lightfoot, "Transport
    Phenomena", John Wiley and Sons, Inc., New York (1960).
7. Angelo, J. B., E. N. Lightfoot, and D. W. Howard, "Generalization
    of the Penetration Theory for Surface Stretch", J. Am. Inst.
    Chem. Engr., 12, 751 (1966).
8. Jones, G. and W. H. Ray, "Surface Tension of Solutions of Elec-
    trolytes as a Function of Concentration", J. Amer. Chem.
    Soc., 63, 288 (1941).
9. Kamenetskii, F., "Diffusion and Heat Exchange in Chemical
    Kinetics", Princeton University Press, Princeton, N.J.
    (1955).
10. Matteson, M. J., W. Stober and H. Luther, "Kinetics of the
    Oxidation of Sulfur Dioxide by Aerosols of Manganese Sul-
    fate", Ind. and Eng. Chem., 8, 677 (1969).
11. West-Gaeke Method, ASTM, D2914 (1971).
12. Borok, M. T., "Dependence of the Degree of Absorption of
    Nitrogen Dioxide in Water on its Concentration in a Gaseous
    Mixture", Zh. Prik. Khim., 33, 8, 1761 (1960).
13. Palmes, E. D., A. F. Gunnison, J. Dimattio and C. Tomezyk,
    "Personal Sampler for Nitrogen Dioxide", Am. Ind. Hyg. Assoc.
    J., 37, 570 (1976).
14. Crecelius, H., and W. Forwerg, "Investigations of the Saltzman
    Factor", Staub., 30, 7, 23 (1970).
15. Dekker, W. A., E. Snoeck and H. Kermers, "The Rate of Absorption
    of NO_2 in Water", Chem. Eng. Sci., 11, 61 (1959).

DISCUSSION

KRUPA: I have a couple comments. Number one, you almost find an identical situation with CO_2.

MATTESON: Is that so? We haven't looked at CO_2.

KRUPA: We did a little study with CO_2 that almost identically parallels the SO_2 retainment in water droplets during condensation.

The other point I was going to make is that you did mention that these were large droplets. When you get down to the small particulates, I think the retainment ratio is going to increase with the increase in surface to volume ratio?

MATTESON: Yes. I hope so. I would expect to see much more retention.

KRUPA: Particularly I think you have very little re-evaporation because the sedimentation velocity is quite low with those particulates of almost a micron or sub-micron range.

MATTESON: With the particulates, you have the different dissolved species, and so we plan to generate aerosols from different materials that would also serve as condensation nuclei and reactants at the same time.

STENSLAND: Are the "biggies" that you just referred to, the size of raindrops?

MATTESON: Two and a half to three millimeters.

STENSLAND: Those are good sized raindrops. So, really you are considering the removal of SO_2 by falling raindrops, when they are condensing and increasing the absorption. Would you say that it is basically a stirring mechanism which is causing the increased mass transfer? The reason that I bring that up is that within a falling raindrop there is good mixing. That would give the upper limit of the curve that you showed.

MATTESON: We tried to avoid internal mixing. We wanted to look at surface effects due to condensation only. We saw internal mixing in the early experiments during droplet formation, where we had very high rates of mass transfer. We're not sure exactly what's accounting for this surface effect. It could be a Stefan-flow phenomenon where a lot of water vapor molecules push along the SO_2, or the third gas component, into the droplet involuntarily or at greater rates.

It could be some kind of Marangoni effect where you have surface-driven instabilities.

TOMLINSON: Back in the Thirties there was a major study that was carried out at McGill University on SO_2 absorption in water using various bases like lime or magnesia. They established that the reaction of dissolved molecular SO_2 with water to form sulfurous acid is somewhat slower than the rate of absorption. Formation of sulfurous acid is really the controlling reaction when you're reacting sulfur dioxide with calcium carbonate, lime or something of that nature.

MATTESON: Well, if you put some peroxide in there as we did, a conversion to sulfuric acid occurs, and that problem is eliminated.

TOMLINSON: In more recent experiments, we found, for instance, if you're picking up sulfur dioxide with a limestone slurry in an absorption tower, you really have to recycle the slurry and give it dwell time at the bottom of the tower for the reactions to take place before returning it to the top of the tower. Once through doesn't normally do.

PRECIPITATION CHEMISTRY TRENDS IN THE NORTHEASTERN UNITED STATES

Gary J. Stensland

Illinois State Water Survey
P. O. Box 232
Urbana, IL   61801

INTRODUCTION

Gases and particles injected into the atmosphere by natural or anthropogenic sources are returned to the earth's surface through wet and dry removal processes. Therefore historical trends of the concentrations of the various trace substances in precipitation can serve as an indication of the changes in the levels of these substances in the air.

This paper will examine three topics related to the precipitation chemistry data base for the United States. First, since the pH of precipitation is of major interest, techniques to calculate this quantity will be considered. This calculation is necessary when the pH was not reported, as is frequently the case for the older data sets. Next, the changes in the precipitation chemistry for a central Illinois site will be discussed. This site is at the western edge of present high atmospheric sulfate levels, and thus it is interesting to examine changes in the precipitation pH at this site and the reasons for them. The third and final topic will be to consider changes in the sulfate and nitrate at this central Illinois site and other sites further East, and to compare these changes to the time trends in the anthropogenic source emissions.

METHODS TO CALCULATE PRECIPITATION pH

The 1954 data set to be discussed in the next section of this paper did not include pH measurements so the following procedure has been used to calculate the pH.[1,2] In a rain or melted snow solution, a charge balance is maintained. If a term called Net Ions is defined as

$$(\text{Net Ions}) = (SO_4^=)+(NO_3^-)+(Cl^-)-(Ca^{++})-(Mg^{++})-(NH_4^+)$$
$$-(Na^+)-(K^+), \tag{1}$$

then the charge balance equation, consisting of the major ions, is

$$(H^+)-(HCO_3^-)=(\text{Net Ions}), \tag{2}$$

with each concentration in units of microequivalents per liter ($\mu eq/\ell$). Although Eq. 2 is not exactly correct, since some ions have not been included, the relation has been found to work reasonably well for rain and snow samples. With the appropriate chemical equilibrium constants, it can be shown that $(HCO_3^-) \simeq 490\ (OH^-)$ and for samples with pH<8, $(CO_3^=)<<(HCO_3^-)$. Thus, the two ions OH^- and $CO_3^=$ need not be considered in the charge balance equation. Assuming that falling raindrops are in equilibrium with atmospheric carbon dioxide ($P_{CO_2} = 320 \times 10^{-6}$ atm), then the chemical equilibria relationships can be used to give

$$(HCO_3^-)=K_H K_1 P_{CO_2}/(H^+) \tag{3}$$

where K_H is the constant in Henry's Law and K_1 is the first disassociation constant of CO_2 in water (for 25°C, $K_H = 0.034 \times 10^{+6}$ $\mu eq/\ell/atm$ and $K_1 = 4.5 \times 10^{-1}\ \mu eq/\ell$). Substituting Eq. 3 into Eq. 2 and solving the quadratic equation for (H^+) gives

$$(H^+)=\{(\text{Net Ions}) \pm [(\text{Net Ions})^2 + (4K_H K_1 P_{CO_2})]^{0.5}\}/2. \tag{4}$$

In Eq. 4, only the plus sign in front of the bracketed term gives positive and therefore physically realistic solutions. Eq. 4 is rewritten in terms of pH as

$$pH=+6-\log_{10}\{\{(\text{Net Ions})+[(\text{Net Ions})^2+(4K_H K_1 P_{CO_2})]^{0.5}\}/2\} \tag{5}$$

where the +6 factor results from the fact that the $\mu eq/\ell$ concentration unit is required for the terms in Eq. 5. Figure 1 illustrates Eq. 5 for a variety of temperature T and P_{CO_2} values. Curve E was used to calculate the pH for the 1954 data.

From 1959 to 1964 the U. S. Public Health Service collected monthly precipitation samples at 39 locations throughout the continental United States. Equation 5 was used to calculate the pH values for this data set of 1295 samples. Mg^{++} was not reported, so it was estimated by assuming it to be related to Ca^{++}, with the $(Ca^{++})/(Mg^{++})$ ratio being assigned regional values based on a very

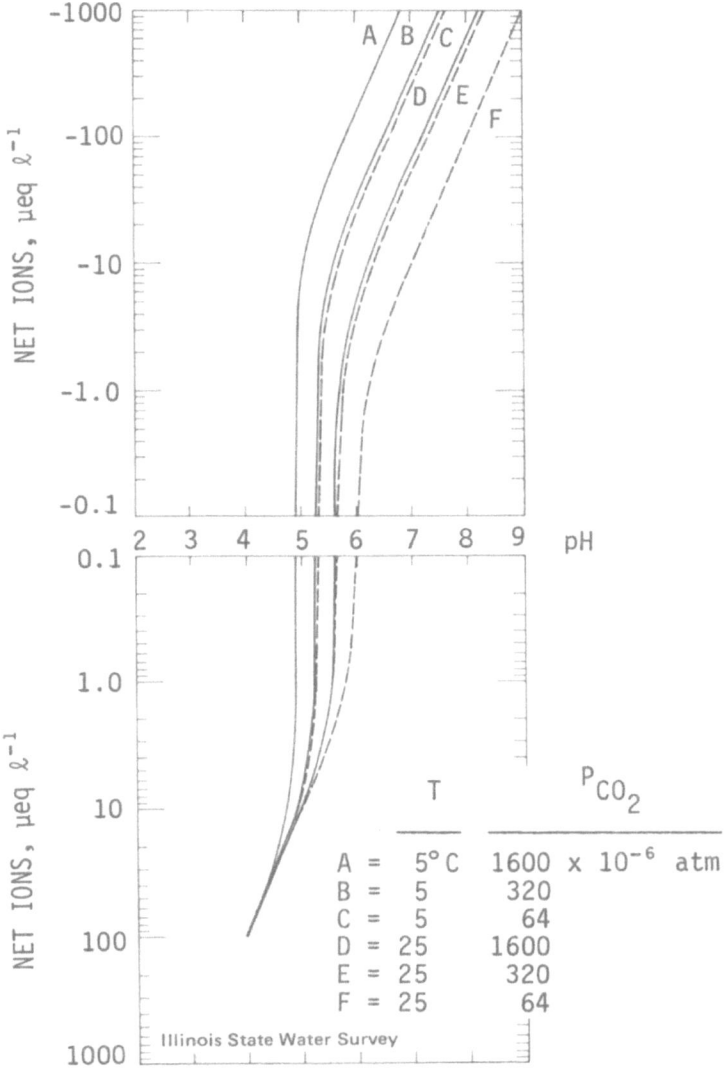

Figure 1.   The concentration of Net Ions versus pH for precipitation
samples with different values of T (temperature) and P_{CO_2}
(from ref. 2).

limited data base. The ratio was >>1 except along the sea coasts.
The results of these calculations are shown in Figure 2, where the
median of the calculated pH values for each measured pH class (0.1
unit in width) are plotted. The line segments which have been fitted to the data points can be used to correct the calculated pH

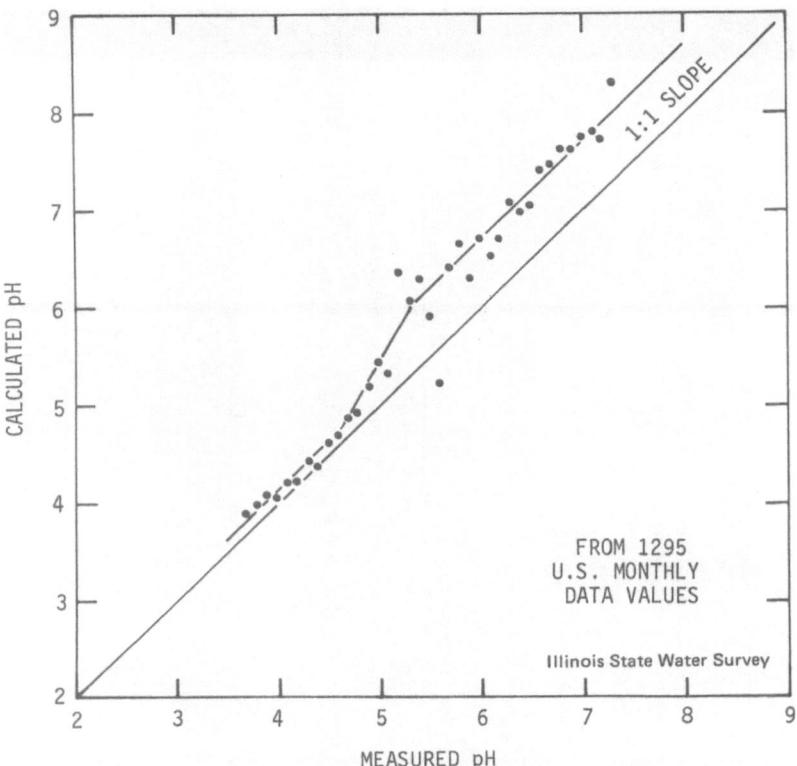

Figure 2.   Calculated pH versus measured pH for monthly precipitation
            samples (from ref. 2).

values such that they will agree with the measured pH values.   The
equations for the three line segments are

$$pH \text{ (corrected)} = pH \text{ (calculated)} - C, \qquad\qquad (6)$$

where C = 0.70 for pH (calculated) $\geq$ 6.0 and C = 0.15 for pH
(calculated) $\leq$ 4.8, and

$$pH \text{ (corrected)} = 0.54 \; pH \text{ (calculated)} + 2.06, \qquad\qquad (7)$$

for 4.8 < pH (calculated) < 6.0.   The data reported by Granat,[3] for
about 1500 monthly samples from the European chemistry network, were
also used to prepare a graph like Figure 2.   Equations 6 and 7 also
fit the Granat data very well.   A firm explanation as to why Eqs. 6
and 7 are needed for the precipitation data is not yet available.

        Finally, with respect to calculating pH, Eq. 1 can be compared
with two other somewhat different approaches in the literature.

Granat[3] assumed that all the Na^+ and Cl^- were due to sea salt, and thus neither ion was included in his equations. Also the other ions were corrected for sea salt. Granat's final equation, expressed in the form of Eq. 1, was

$$(\text{Net Ions})_{GRAN} = (SO_4^= - SSSO_4^=) + (NO_3^-) - (Ca^{++} - SSCa^{++}) - (Mg^{++}$$
$$- SSMg^{++}) - (NH_4^+) - (K^+ - SSK^+). \tag{8}$$

The SS prefix for an ion indicates that the quantity is the amount due to sea salt, Na^+ being used by Granat as the reference sea salt ion.

Cogbill and Likens[4] also corrected each ion for sea salt but included Na^+ or Cl^- in the final equation. For example, if $(Na^+)/(Cl^-)$ was greater than the ratio for sea water, then the excess Na^+ was included in the equation and the base ion for making the sea salt correction was Cl^-. In this case the final equation, expressed in the form of Eq. 1, was

$$(\text{Net Ions})_{COG} = (SO_4^= - SSSO_4^=) + (NO_3^-) - (Ca^{++} - SSCa^{++}) - (Mg^{++}$$
$$- SSMg^{++}) - (NH_4^+) - (Na^+ - SSNa^+) - (K^+ - SSK^+). \tag{9}$$

Equations 8 and 9 were used in place of Eq. 1 to produce equations analogous to Eq. 5. The three pH equations were then applied to the same data set, namely Junge's 1955-1956 data for 63 sites in the United States.[5] Mg^{++} was not reported by Junge, so regional ratios of Ca^{++} to Mg^{++} were used to estimate Mg^{++}, as was previously discussed. Also, Junge reported only quarterly values for NH_4^+ and NO_3^-. Thus, these values were weighted with the quarterly precipitation at the sites to produce annual averages for NO_3^- and NH_4^+.

The pH results are shown in Figures 3 and 4. In Figure 3 the pH values from Eq. 5 are plotted on the abscissa and the values with the assumptions of Cogbill and Likens are plotted on the ordinate. The agreement is very good, with the linear correlation coefficient r = 0.99. In Figure 4, the results from Eq. 5 are compared with the pH values using Granat's assumptions. The agreement is not as good, with r = 0.90 and the line of best fit being y = 0.96x. The pH values calculated with Granat's sea salt assumptions are generally lower than those calculated by Stensland with Eq. 5, a change which is in the same direction as that resulting from the empirical correction in Figure 2. Therefore, if the empirical corrections in Eq. 6 and 7 are not used, then the pH calculated with Granat's assumptions would probably agree better with measured pH values than if the assumptions of Cogbill and Likens or Stensland were used.

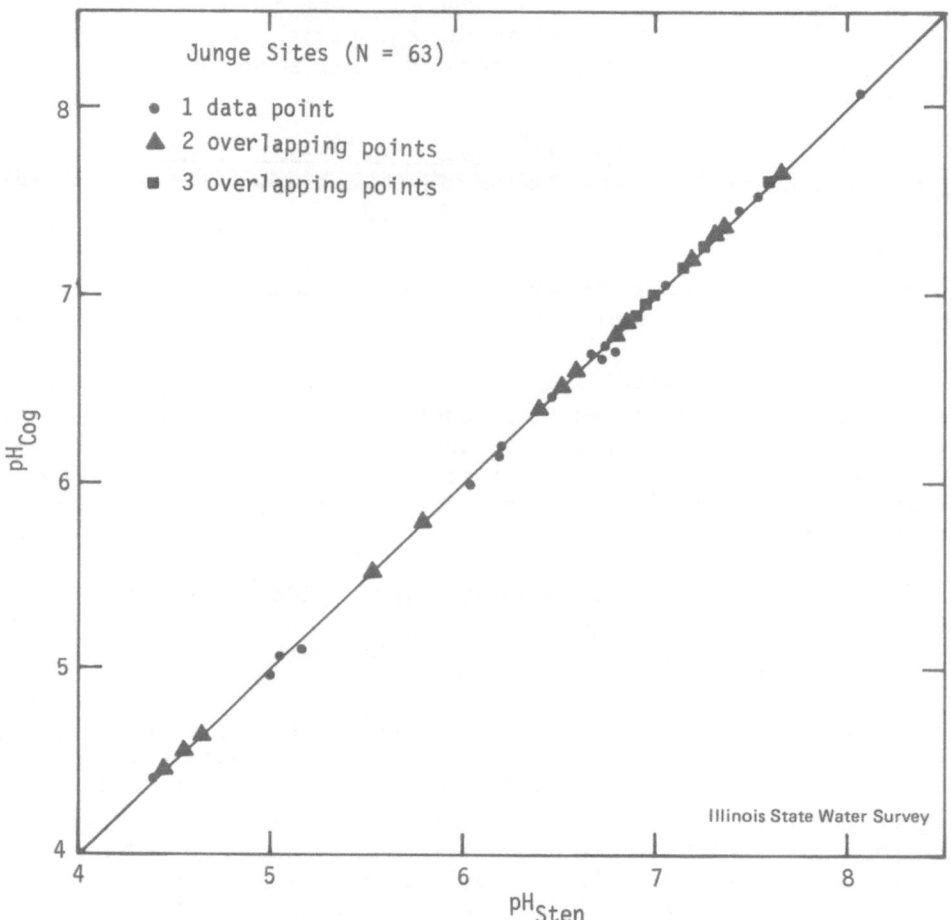

Figure 3.   The pH calculated with the assumptions made by Cogbill
            and Likens versus the pH calculated with the assumptions
            by Stensland, for the 1955-56 Junge data (from ref. 2).

PRECIPITATION CHEMISTRY IN CENTRAL ILLINOIS IN 1954 AND 1977

     The difference in the precipitation chemistry for a rural,
east-central Illinois site in 1954 as compared to 1977 was investigated to ascertain whether or not the precipitation became more
acidic and if so, why.* The 1954 data set[6] was unique for its time

---

*This topic is covered in more detail in reference 1 and in a paper
 that has been submitted to the Journal of the Air Pollution Control
 Association.

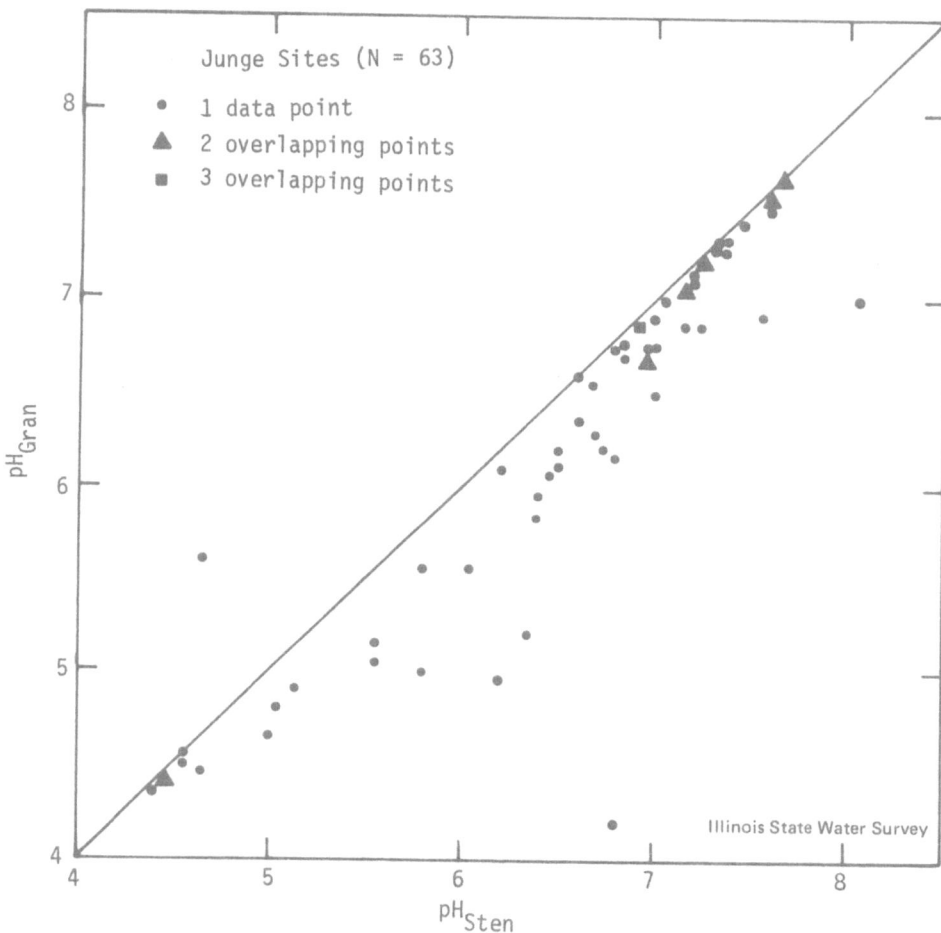

Figure 4.   The pH calculated with the assumptions made by Granat
             versus the pH calculated with the assumptions by Stensland,
             for the 1955-56 Junge data (from ref. 2).

because the manual sampling technique carefully eliminated any dry
deposition and because individual precipitation events were
analyzed.   The sampling site in both 1954 and 1977 was at the
Champaign-Urbana airport (referred to as CMI), located 7 kilometers
south of Champaign and surrounded by cultivated fields except for
a golf course to the east.   The data sets were collected from
October 26, 1953 to August 12, 1954 and from May 15, 1977 to
February 6, 1979.   For convenience the two sets will be referred
to as the 1954 data and the 1977 data.

For the 1977 study an automatic collector, which exposed the wet-side bucket only during the time of precipitation, was used. The polyethylene bucket was changed within 24 hours of the end of the precipitation event and then brought immediately to the laboratory where pH and conductivity were measured followed by sample filtration with a 0.45 micrometer membrane filter. The ions $SO_4^=$, NO_3^-, Cl^-, NH_4^+, Ca^{++}, Mg^{++}, K^+, and Na^+ were determined by standard methods on AutoAnalyzer and atomic absorption instruments. The sampling and analysis procedures for the 1954 data are described in detail in the Larson and Hettick paper.[6]

From about May 15 to September 30 the landscape around CMI is mostly green due to the corn and soybean crops. In the fall season most of the farmland is tilled, producing a brown landscape. Since the local dust conditions thus may be quite different for the two periods, the precipitation chemistry data for 1977 was divided into the *green* period events and the *brown* period events. The data for pH show that this potential dust effect was not large for the 1977 data set. Due to the relatively small number of data points for 1954 these events were not divided into *brown* and *green* periods. Three other types of data separation were carried out. First, in the 1954 data there were six precipitation events wherein consecutive samples were taken, so a volume-weighted average was computed for each ion. These average values were used in the subsequent analyses. Second, there were six events, also in the 1954 data, where the sample collection began after the precipitation had started. Because the chemical concentrations are highest at the beginning of precipitation events, these six samples were given special consideration. Third, for the 1977 samples the procedures allowed very small samples to be analyzed and these data were also noted separately.

The precipitation pH was not reported for the 1954 data, so Eqs. 5, 6, and 7 were used to calculate the pH. The frequency distributions of the calculated pH for 1954 and for the measured pH for 1977 are presented in Figure 5. It can be seen that the median pH does not change significantly when the smaller samples are included, but the *green* period pH is somewhat lower than the *brown* period pH. The 1977 combined *brown* and *green* period median pH, for events ≥0.70 mm, is 4.1. This compares with the calculated median pH of 5.9 for the 1954 data. If those 1954 samples with precipitation at the beginning are excluded, the 1954 median pH is 6.05. The more basic precipitation in 1954 could have resulted from low levels of acidic ions (e.g., sulfate and nitrate) or from high levels of basic ions (e.g., calcium and magnesium). This issue is addressed with the data in Table 1. For this table the small precipitation samples (<0.70 mm) were not included.

For sulfate in 1977 the median was 80 µeq/ℓ for the *brown* period, 65 µeq/ℓ for the *green* period, and 70 µeq/ℓ for the com-

bined *brown* and *green* period.   The median for 1954 was 50 µeq/ℓ,
but when samples with precipitation at the beginning were excluded,
the 1954 median was 60 µeq/ℓ.   As with sulfate, the 1977 nitrate
values had a somewhat larger median for the *brown* period.   The
nitrate median was 38 µeq/ℓ for the *brown* period, 28 µeq/ℓ for the
green period and 30 µeq/ℓ for the combined *brown* and *green* periods.
The 1954 median nitrate value was 18 µeq/ℓ while the exclusion of
samples with precipitation at the beginning increased the median to
20 µeq/ℓ.   The hardness, defined as the sum of calcium and magnesium

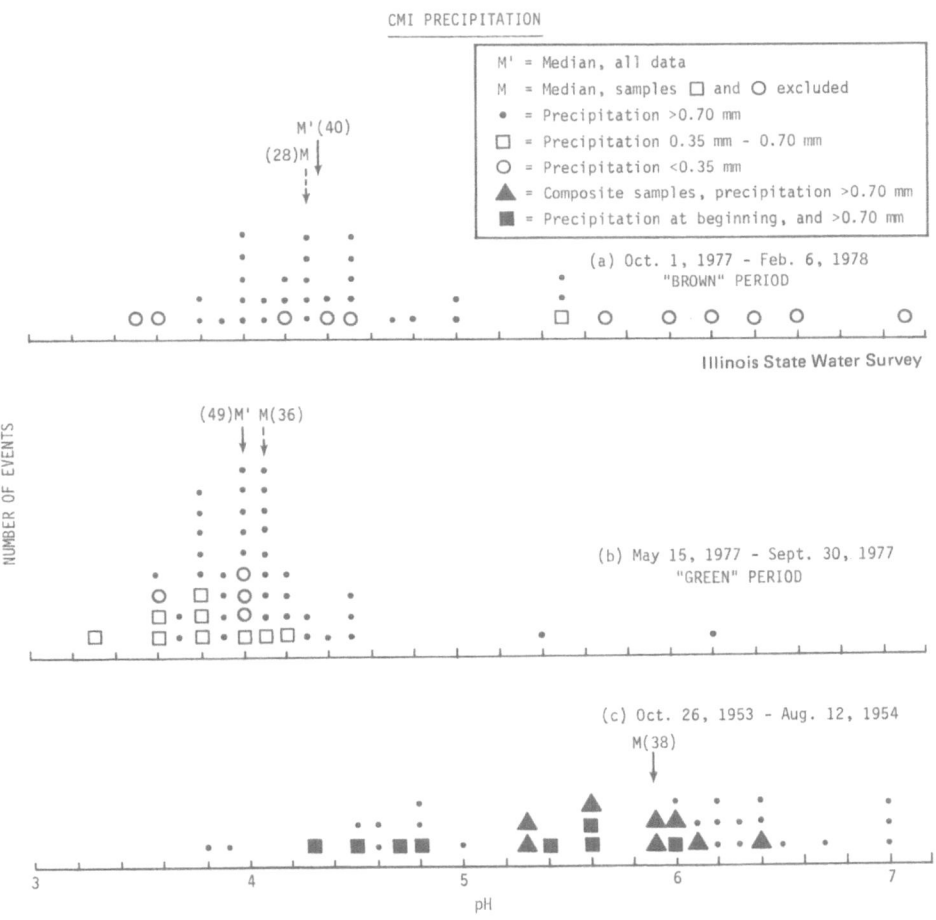

Figure 5.   Frequency distribution of measured pH for 1977
            precipitation events and calculated pH for 1954
            precipitation events.   The numbers in parentheses
            indicate the number of data points from which the
            median was obtained (from ref. 1).

concentrations (Ca^{++} + Mg^{++}), was reported for the 1954 data and therefore is the quantity presented in Table I. The median hardness value for 1977 for the combined *green* and *brown* period was 10 µeq/ℓ, with the *brown* period median being 15 µeq/ℓ and the *green* period value being 10 µeq/ℓ. For 1954, the median hardness value was 65 µeq/ℓ, but when samples with precipitation at the beginning are excluded the median was 82 µeq/ℓ. Of the nine 1954 events with hardness greater than 100 µeq/ℓ, two were in the *green* period and seven in the *brown* period.

Table I. Ion concentrations (µeq $ℓ^{-1}$) and pH for precipitation samples (<0.70 mm).

	($SO_4^=$)	(NO_3^-)	(Ca^{++} + Mg^{++})	pH
Ill., 1954[a]	60	20	82	6.05[b]
Ill., 1977[a]	70	30	10	4.1

---

[a]Median values for CMI event samples.
[b]Includes the empirical correction for calculated pH. Without this correction the value would be about 6.75.

With the data summarized in Table I there can be little question that the CMI samples in 1954 were much more basic than the 1977 samples. Although both $SO_4^=$ and NO_3^- were apparently lower in the 1954 samples, it was the high concentrations of soil related species, Ca^{++} and Mg^{++}, which produced the high pH in 1954. If the Ca^{++} + Mg^{++} concentration in 1954 had been 10 µeq/ℓ (the 1977 level), the pH would have been 4.17 instead of 6.05. The median pH = 6.05 for 1954 includes the empirical correction in Eqs. 6 and 7. Without this correction the 1954 median pH value would have been about 6.75, and 4.34 with the 1977 level of Ca^{++} + Mg^{++}.

The high 1954 Ca^{++} + Mg^{++} concentration could have resulted from problems in the chemical analysis procedures, or may be indicative of higher ambient air levels of these elements in 1954. Recent work at the Illinois State Water Survey has shown that the Ca^{++} + Mg^{++} levels in rain can rise considerably with time if the samples are not filtered.[7] The 1977 samples were filtered but, most of the 1954 samples were not. There are other preliminary data, however, which suggest that the absence of filtering will not account for the majority of the increase in the 1954 pH levels compared to the 1977 values.

Assuming now that the reported $Ca^{++} + Mg^{++}$ values were accurate
for the precipitation falling in 1954, one must then conclude that
more calcium and magnesium were present in the atmospheric aerosols
in 1954 than in 1977. A possible source could have been the dust
from rural gravel roads, if they were more common in 1954. Also
the types of crops grown and the size of the machinery used to
cultivate the fields has changed in the CMI area since 1954. This
could have modified the local dust levels to some extent, but no
quantitative assessment is available. The final suggestion offered
for the elevated $Ca^{++} + Mg^{++}$ levels in the atmospheric aerosols is
that the years 1953-1954 were very dry in many states of the Midwest
and the Plains, and thus more susceptible to soil erosion by the
wind, as compared to 1977. For instance, in 1954 the precipitation
was -33% (below normal) in Oklahoma, -25% for Kansas, -14% for
Nebraska, -13% for Missouri, +9% for Iowa, and -6% for Illinois.
Research is now in progress to more fully investigate the possible
relation between droughts and elevated basic ions in precipitation.

SULFATE AND NITRATE TRENDS IN THE NORTHEASTERN UNITED STATES

The atmospheric cycles of sulfur and nitrogen include the
following components: (a) emission of the compounds from the earth's
surface into the atmosphere; (b) transport and transformation of
these compounds in the atmosphere; (c) and wet and dry removal of
the compounds from the atmosphere. The relationship between emissions levels and air and precipitation quality is a physical fact
if the time and space boundaries are defined appropriately. For
example, if the global emissions of sulfur increase, the average
air quality for sulfur for the globe will decrease. This section
will examine annual average values for source emissions estimates
and for precipitation quality, for the northeastern United States
for the mid-1950's versus the late 1970's.

The analysis presented in this section was undertaken because
precipitation chemistry data have recently become available for the
1977-1978 period and therefore can be compared with existing data
from 1954-55/56.[8] The locations of the sampling sites used in
this analysis are shown in Figure 6. The MAP3S network[9] is still
in operation and now consists of eight sites in the Northeast.
Beginning in September 1977, the samples from this project were
analyzed for all the major cations and anions.[10] There was about
one year of data for each of the three MAP3S sites used for this
analysis (IT, SC, and CH). The Illinois State Water Survey (ISWS)
began an event precipitation chemistry collection program near
Champaign, Illinois in May 1977, and this site (CMI) provided data
for this analysis.[1] In addition, the ISWS study at the same site
from October 1953, to August 1954, provided historical data for
this analysis.[6] The Junge network was in operation from July 1955
to June 1956, and provided the remainder of the historical data.[5,11]

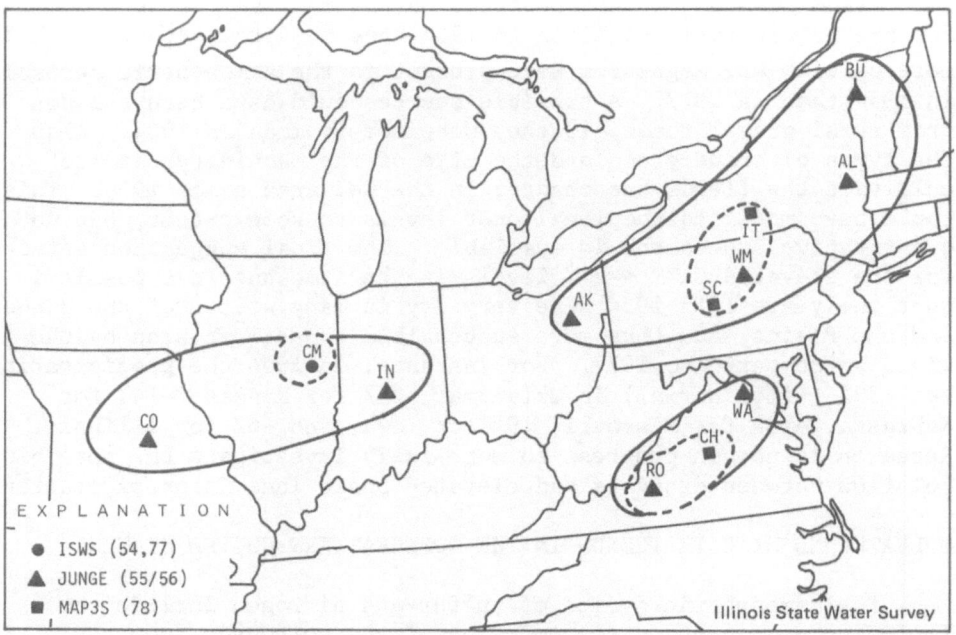

Figure 6.   Location of precipitation chemistry sampling sites in
            the northeastern United States. The dashed lines
            enclose the primary sites being compared while the
            solid lines enclose additional sites for which data
            are presented (from ref. 8).

The Junge network consisted of 62 sites distributed across the
continental United States. Data from eight of the sites will be
used in this paper. The dashed lines in Figure 6 enclose those
sites considered most important for the present analysis while the
solid lines indicate the additional sites which provide comparative
data.

     The type of precipitation chemistry data needed for this study
were ion concentrations for wet-only samples. That is, evaporation
losses and dry deposition effects had to be absent. There are data
available from some networks where automatic wet-only collectors
were used but where the protective cover did not make a tight seal
with the sample container. Evaporation and contamination during
non-rain periods probably occurred and therefore such data were not
used.

     For the MAP3S sites, an automatic wet-only sampler is used to
collect event samples, where event refers basically to a daily
sample. The collecting funnel is thoroughly rinsed after each

sample is removed. The MAP3S sites are all located at least several kilometers away from the nearest urban areas. The 1977 ISWS samples were collected in a manner very similar to the MAP3S samples, with a HASL collector being used. The 1954 ISWS samples consisted of manual collections of precipitation events, a 1.2 meter diameter stainless steel funnel being uncovered and well rinsed at the beginning of the events. Finally, the 1955/56 Junge samples were collected with a plexiglas funnel and polyethylene bottle which were exposed only during precipitation events. The event samples were sent to the analytical laboratory and composited to provide one sample per site per month. The samplers were located at Weather Bureau facilities so that personnel would be available to expose the collecting funnels during the precipitation events. Some of the Junge sites were closer to cities and industrial areas than desirable, but the consistent data patterns produced by the study indicated that most of the sites produced data representative of their region.

It is well known that the ion concentrations in rain and snow events are strongly related to the precipitation amounts; when the amount is less than about 5 mm, the concentrations are usually very high. To compensate for this effect many researchers use the sample volume weighted averages for data interpretation. Another approach to remove the effects of extreme values is to report median values. Table II displays both the median and the sample volume weighted averages for the MAP3S sites. The median values are consistently higher for each of the ions.

Table II. Median concentrations and sample volume weighted
average concentrations for the MAP3S samples
(mg/ℓ) (from ref. 8).

	Sulfate			Nitrate		
	Number of Samples	Median	Weighted Average	Number of Samples	Median	Weighted Average
IT	54	2.88	2.66	55	2.17	1.71
SC	86	2.88	2.54	86	2.26	1.67
VA	51	2.79	2.46	51	1.74	1.54

The sulfate and nitrate concentration data for the sites shown in Figure 6 are presented in Tables III and IV. Accurate sample volumes were not available for the 1954 ISWS data. Therefore, for all the ISWS data in Tables III and IV (site CMI), the median concentra-

Table III. Sulfate concentrations in precipitation (mg/ℓ as
            $SO_4^=$) and ratios of recent values to historical
            values (from ref. 8).

MIDWEST

1977/78	Ratio	1954	1955/56
CMI: 3.79----(1.2)---------------CMI: 3.10			CO: 3.10
			IN: 2.67

EAST

1977/78	Ratio	1955/56	1955/56
IT: 2.66----(0.8)--------------WM: 3.48			BU: 2.89
			AL: 2.29
SC: 2.54----(0.7)--------------WM: 3.48			AK: 3.26
CH: 2.45----(0.9)--------------RO: 2.67			WA: 2.67

Table IV. Nitrate concentrations in precipitation (mg/ℓ as
           NO_3^-) and ratios of recent values to historical
           values (from ref. 8).

MIDWEST

1977/78	Ratio	1954	1955/56
CMI: 1.85----(1.5)--------------CMI: 1.25			CO: .58
			IN: .53

EAST

1977/78	Ratio	1955/56	1955/56
IT: 1.71----(1.3)-------------WM: 1.31			BU: 1.12
			AL: .77
SC: 1.67----(1.3)-------------WM: 1.31			AK: 1.05
CH: 1.54----(1.8)-------------RO: .84			WA: .54

tions are given. For the Junge data the monthly compositing procedure produced sample volume weighted concentrations. Since the
Junge data and the MAP3S data are to be compared, the MAP3S sample
volume weighted averages are listed in Tables III and IV instead
of the median values.

The upper portion of Table III gives the sulfate data for the midwestern site and the lower portion presents the data for the eastern sites. The dotted lines connect those sites for which it is considered most appropriate to calculate ratios between the recent and the older concentrations. The right hand column provides additional data for comparison to the middle column. The same format is used in Table IV. The basic feature to note in the tables is that the sulfate ratios are near one, while those for nitrate are greater than one. These trends will now be compared to the source emissions trends.

The emissions data which are presented in this section are values for the entire United States. It is assumed that the trends for these emissions data are representative of the sources contributing to the precipitation quality sites in the northeastern United States. A study of the geographical distribution of the 1972 emissions data has shown that the EPA air quality regions I-IV (basically the eastern United States) contributed 80% of the total nitrogen oxide emissions.[12] The same study showed that six states, New York, Pennsylvania, Ohio, Michigan, Indiana, and Illinois, contributed 45% of the total. For sulfur oxides the study indicated that the northeast sector of the United States accounted for about half of the total sulfur oxide emissions.

It is generally accepted that the anthropogenic sources dominate over natural sources for industrialized areas such as the Northeast. Of the various sources categories, fossil fuel combustion at stationary sources dominates for the sulfur oxides.[13] In 1970 about 78% of the sulfur oxides were emitted from these sources. Steam electric power plants were the major stationary sources and they contributed 57% of the total sulfur oxide emissions. In total, coal combustion accounted for about 65% of the total sulfur oxide emissions. For nitrogen oxides, gasoline combustion in motor vehicles accounted for 34% of the total emissions in 1970, while coal, fuel oil, and natural gas combustion in stationary sources accounted for 44% of the total.[13]

The time trend of the estimated emissions is summarized in Table V. The EPA report providing the data for the lower half of Table V warned that the earlier estimates (upper half of Table V) had not been calculated in exactly the same way and thus would not be exactly comparable. This explains why, for instance, the two 1970 estimates for sulfur oxides do not agree exactly. The data in these tables are expressed as SO_2 and NO_2. In Table VI the data from Table V have been normalized by making the two 1970 values agree. This normalization procedure is a reasonable and practical approach that allows the 1940 to 1976 time period to be examined. Table VI shows that nitrogen oxide emissions have increased much more than sulfur oxide emissions. From mid-1950 to 1976, nitrogen oxides increased from about 11 to 23 million metric tons per year,

Table V.   Estimated nitrogen oxide and sulfur oxide emissions
           for the continental United States (millions of
           metric tons per year).

	1940	1950	1960	1970
Nitrogen Oxides[a]	7.2	9.4	12.7	20.6
Sulfur Oxides[a]	19.5	21.6	21.1	30.8

	1970	1971	1972	1973	1974	1975	1976
Nitrogen Oxides[b]	20.4	21.3	22.2	22.9	22.6	22.2	23.0
Sulfur Oxides[b]	29.1	27.9	28.8	29.7	28.2	25.7	26.9

[a]Source: Reference #13 (p. 4, Table 1)
[b]Source: Reference #14 (p. 5-1, Table 5-1)

Table VI.   Normalized emissions estimates as calculated from
            Table V (millions of metric tons per year).

	1940	1950	1960	1970
Nitrogen Oxides	7.1	9.3	12.6	20.4
Sulfur Oxides	18.4	20.4	19.9	29.1

	1970	1971	1972	1973	1974	1975	1976
Nitrogen Oxides	20.4	21.3	22.2	22.9	22.6	22.2	23.0
Sulfur Oxides	29.1	27.9	28.8	29.7	28.2	25.7	26.9

a factor of 2.1. At the same time, sulfur oxides increased from
about 20 to 27 million metric tons per year, a factor of 1.3. The
relatively small increase in the sulfur oxide emissions is explained by the fact that natural gas and low sulfur oil and coal
have become more heavily used since the mid-1950's.[15] The larger
growth rate in the nitrogen oxide emissions is due to the large
increases in energy consumption for electric power and transportation. From 1955 to 1970, the former increased by a factor of 2.6
and the latter by a factor of 1.6.[15]

CONCLUSIONS

    The results in this paper suggest that at present the best
approach to calculating the pH of a precipitation sample is to
assume that a charge balance exists and then to apply an empirical
correction, as expressed in Eqs. 6 and 7. With this approach the

1977 central Illinois samples are shown to be much more acid than those from 1954. The primary reason for this pH decrease is the decreased concentration of calcium and magnesium in the samples.

For the mid-1970's compared to the mid-1950's, the source emissions estimates for the United States were up by a factor of 1.3 for sulfur and 2.1 for nitrogen. During this same time interval the sulfate concentrations for the precipitation chemistry sites in the northeastern United States remained about the same, and the nitrate concentrations increased by factors varying from 1.3 to 1.8. Since the emissions estimates increased more than the precipitation concentrations, it would seem that a smaller fraction of the pollutants were being removed over the northeast by wet processes in the mid-1970's than in the mid-1950's. This could be a result of the trend towards the use of taller stacks. However, the data presented in this paper are very limited both spatially and temporally, so the conclusion should be considered as rather speculative.

ACKNOWLEDGMENTS

The author thanks S. A. Changnon, Jr. and R. G. Semonin for general supervision of this work. The analytical chemistry support of the team of F. F. McGurk and L. M. Skowron under the leadership of M. E. Peden is also gratefully acknowledged. This work was sponsored by the U. S. Department of Energy under contract EY-1199.

REFERENCES

1. G. J. Stensland, A comparison of precipitation chemistry data at Champaign, Illinois in 1954 and in 1977, in Chapter 3 of the 16th Prog. Rept. on Cont. EY-76-S-02-1199, Ill. State Water Survey, Urbana, IL (1978).
2. G. J. Stensland, Calculating precipitation pH, in preparation for the 17th Prog. Rept. on Cont. EY-76-S-02-1199, Ill. State Water Survey, Urbana, IL (1979).
3. L. Granat, On the relationship between pH and the chemical composition in atmospheric precipitation, Tellus, 24:550 (1972).
4. C. V. Cogbill and G. E. Likens, Acid precipitation in the northeastern United States, Water Resour. Res., 10:1133 (1974).
5. C. E. Junge, "Air Chemistry and Radioactivity," Academic Press, New York, (1963).
6. T. E. Larson and I. Hettick, Mineral composition of rainwater, Tellus, 8:191 (1956).
7. M. E. Peden and L. M. Skowron, Ionic stability of precipitation samples, Atmos. Environ., 12:2343 (1978).

8. G. J. Stensland, A comparison of sulfate and nitrate in
    precipitation in 1955 and 1977 to emissions and air quality
    trends, in preparation for the 17th Prog. Rept. on Cont.
    EY-76-S-02-1199, Ill. State Water Survey, Urbana, IL (1979).
9. Battelle, The MAP3S Precipitation Chemistry Network: First
    Periodic Summary Report (September 1976-June 1977), Battelle
    Northwest Laboratories, Richland, Washington, Report No.
    PNL-2402, October 1977.
10. T. Dana, D. Drewes, S. Harris, J. Rothert, Monthly data
    summaries of MAP3S precipitation chemistry data, Battelle
    Northwest Laboratories, Richland, Washington.
11. C. E. Junge and P. E. Gustafson, Precipitation sampling for
    chemical analysis, Bull. Am. Meteoro. Soc., 37:244 (1956).
12. National Research Council, Air Quality and Stationary Source
    Emission Control, prepared for the United States Senate
    Committee on Public Works, August (1973).
13. J. H. Cavender, D. S. Kircher, and A. J. Hoffman, Nationwide
    Air Pollutant Emission Trends 1940-1970, Office of Air
    Quality Planning and Standards, USEPA, Research Triangle
    Park, NC, Publication No. AP-115, January (1973).
14. U. S. Environmental Protection Agency, National Air Quality
    and Emissions Trends Report, 1976, Office of Air Quality
    Planning and Standards, Research Triangle Park, NC,
    Publication No. EPA-450/1-77-002, December (1977).
15. E. Cook, The flow of energy in an industrial society, Sci.
    Amer., 225, 137 (1971).

DISCUSSION

KRUPA: Dr. Stensland, I don't think you're really talking about sulfate. You're talking about sulfur equivalents. Some work done in the Lake Michigan area indicates that perhaps up to 30% in precipitation samples may be stable sulfite complexes of transition metals.

STENSLAND: The MAP3S sites have also reported instances of high sulfite levels, up to 25% of the total sulfur for the month of January. However, for a yearly average, most of the sulfur was found to be in the form of sulfate. Were the data you are referring to obtained from under plumes?

KRUPA: No, this was lower Lake Michigan precip samples. Dr. Jeremy Hales tells me the reason that many people do not see this is because, when they collect their samples, they do not prevent the complete oxidation from the initial state of sulfur emission. What I'm saying is that we should look at total sulfur as sulfur equivalents.

STENSLAND: For the purpose of my paper, where sulfur emissions are compared to sulfur in precipitation, it is important that all the sulfur compounds in the precipitation be included. For all practical purposes the time lag from collection to chemical analysis allows any sulfite to be oxidized to sulfate and, thus, the measurement of sulfate in the samples is in essence a total sulfur determination.

KRUPA: The other thing I want to point out to you is that the Argonne data I have seen, at least one paper that appeared recently, does not parallel the data you have for Champaign.

STENSLAND: I do not know what data was included in the Argonne paper. I would like to comment that Argonne, being closer to Chicago, should have rather different precipitation chemistry as compared to the downstate site at Champaign.

EVANS: Researchers at Argonne National Laboratory have argued the other way, that they're upwind from Chicago. And if they talk about SO_2 problems, they say they have none, there's no SO_2 or ozone in Argonne. Another question I have is why didn't you express everything as quantities of sulfate and nitrate rather than concentrations. If you have a very rainy year, you may obscure some interesting data. However, if you look at total input rather than concentrations, it might be interesting. Any comment?

STENSLAND: You are suggesting that I use deposition values
instead of concentrations. Well, for what I am doing, I think
I should be looking at concentrations. That is, I do not want
to be confused by the fact that it rained twice as much one
year, compared to another. The rain quality (concentration)
depends on the air quality which in turn depends on the
emissions. I do not think that the yearly rainfall amount
significantly affects the relationships for sulfur or nitrogen.

EVANS: If you want to look at total input into an area and
compare that with emissions, I think you have to take into
account differences in volumes.

STENSLAND: I am not looking at total imputs into areas.

HICKS: I just want to clarify that as far as the Argonne site
is concerned. You're quite right, it is influenced by local
sources, at times. These local sources are easily
identifiable. The problem is not Chicago as much as the
Romeoville-Lockport-Joliet area.

        A second point concerning the subject of how much local
sources interfere with precipitation concerns the effect of the
emission height. This came up earlier when we were talking
about Sudbury. I had to do a few sums first to make sure of my
ground, but in the case of a tall stack injecting into the free
atmosphere above the mixed layer, pollutants might remain there
for a considerable period. At this latitude radioactive
fallout studies indicate residence times of as much as ten
days. But a cloud gets much of its air from the mixed layer,
if it's smart, because it wants water. If it doesn't have
water it isn't much of a cloud.

        So, if we want to look at where the impact of that tall
stack is going to be on precipitation chemistry, we must look
downwind sufficiently far that we can get the material down to
the mixed layer where it is available to our cloud. Now, doing
the sums we obtain that with a residence time of about 10 days,
we might have a travel distance about 7,000 km to 10,000 km.
On top of this, we might also expect to find local effects if
material is quickly entrained in a cloud circulation pattern.
So we ought to find the effects of emissions at all ranges out
to global scales.

STENSLAND: Your point ties in with the last point I made in my
presentation.

LILJESTRAND: Concerning your indirect calculations in the pH, and the disagreement in the alkaline regions or a pH of above about 7, Lerman in "Geochemical Processes" goes through the analysis and concludes a weak acid on the order of 200 micromoles/liter, could account for the difference. It would bring down the pH. Another factor would be mis-estimating the partial pressure of carbon dioxide, e.g. if P_{CO_2} was much higher at the laboratory where pH was measured versus what P_{CO_2} was at the site.

STENSLAND: With respect to the partial pressure of CO_2, you can see with the aid of Fig. 1, that we need to have about 1600 ppm instead 320 ppm to achieve an ion balance. This high value would not be present in the laboratory. With respect to the weak acid hypothesis, such an anion could produce the required ion balance. Based on conductivity data, Granat dismissed the idea that an important ion was missing. At this time, I do not have enough conductivity data to rule out this possibility.

LILJESTRAND: Was the conductivity balance on acid samples or basic samples?

STENSLAND: The data used to produce Fig. 2 included pH values from less than 4.0 to greater than 7.0, but conductivity values were not available.

ANDERSON: Gary, you took a look at the emissions from fossil fuel burning versus concentration values. Did you do any kind of a correlation with the use of fertilizer in the Fifties versus in the Seventies?

STENSLAND: I think some recent nitrogen emission estimates that I have seen indicate that fertilizer sources are still quite a bit smaller than what is coming from other sources. I am thinking of a recent paper in Ambio and I should look at that once again.

KADLECEK: Gary, what was the difference in the total precipitation during your 1950's data and your more recent data?

STENSLAND: It, of course, varied from site to site, but a severe drought was occurring during the 1950 data period.

McLEAN: One of your slides showed a decrease in the calcium amount. Could that be caused by a decrease in dust precipitation.

<u>STENSLAND</u>: That is a possibility. I think I have seen a
comment in the literature where the magnitude of that source
was estimated to be relatively unimportant, but I should do
some additional calculations concerning this source. With all
the farmland being brown so many months of the year, this is
the first source that I consider. Also I did not mention
gravel roads. If we had a lot more gravel roads in the fifties
compared to now, then this is possibly another important source
to be thinking about.

MODELING THE CHEMICAL COMPOSITION OF

ACID RAIN IN SOUTHERN CALIFORNIA

Howard M. Liljestrand and James J. Morgan

Environmental Engineering Science
California Institute of Technology
Pasadena, CA 91125

ABSTRACT

A variety of techniques to identify and quantify acid and base
components in rainwater are applied to data for southern California.
Charge balance calculations using major cation and anion concentration data indicate southern California probably had alkaline rain
in the 1950's and the 1960's with the exception of the Los Angeles
area which probably had acidic precipitation. Measurements of the
chemical composition of precipitation collected in Pasadena, California, from February 1976 to September 1977 are compared with the
charge balance and conductivity balance constraints. A chemical
balance is used to determine the relative importances of different
sources. The pH is found to be controlled by the interaction of
bases and strong acids with nitric acid being 32% more important
on an equivalent basis than sulfuric acid. The uncertainties in
the various calculations are discussed.

Introduction

The spread and intensification of precipitation acidity through
northern Europe is a well described phenomenon(1). The temporal increase of acidity is attributed to increased anthropogenic emissions
of NO_x and SO_2(2). While sulfuric acid is considered to be the
major acid(2), the net acidity is controlled by the interaction of
sulfuric and nitric acids with basic ammonia and soil dust(3).

A similar trend in precipitation acidity has been determined
for the northeastern United States although the historical record
of direct measurements of acidity is sporadic. The more complete
records of major ion concentrations have been used with charge bal-

ance and source strength models to calculate pH values by assuming equilibrium with the atmospheric partial pressure of carbon dioxide. The acidity of precipitation in the Northeast in the 1950's and 1960's was calculated to be due to sulfuric acid (60%), nitric acid (30%) and hydrochloric acid (5%), with the sources being anthropogenic emissions downwind toward the Midwest(4).

To describe the acid precipitation phenomenon, a number of techniques have been used to quantify the acidity, identify the acidic and basic components and to identify the sources of acidity. The methodology used is often determined by the available data. Although all valid models should reach consistent conclusions, the sensitivity of each technique to random and systematic errors should be determined to provide confidence limits for the results. This paper examines the acidity of precipitation in southern California with the charge and conductivity balances, direct measurements, source strength calculations and scavenging models. Some of the limitations of each approach are also discussed.

<u>Charge Balance</u>

One necessary but not sufficient condition that all of the major anion and cation concentrations in a sample have been determined is the charge balance. The charge balance is given by equation 1 where $[Me_i^{z_i+}]$ represents the molar concentration of the ith trace cationic species of charge $+z_i$ and $[A_j^{z_j-}]$ represents the molar concentration of the jth trace anionic species of charge $-z_j$.

$$1) \quad [H^+] + [NH_4^+] + [K^+] + 2[Ca^{2+}] + 2[Mg^{2+}] + \sum_i z_i [Me_i^{z_i+}]$$

$$= [OH^-] + [HCO_3^-] + 2[CO_3^{2-}] + [Cl^-] + [NO_3^-] + 2[SO_4^{2-}] +$$

$$\sum_j |z_j| [A_j^{z_j-}]$$

If the pH has not been directly measured, the charge balance can be used to solve for the pH by assuming equilibrium with a known partial pressure of carbon dioxide. Then the following equilibrium relationships can be used:

$$2) \quad K_w = [H^+][OH^-] \qquad\qquad 3) \quad K_H = [H_2CO_3^*] / P_{CO_2}$$

$$4) \quad K_1 = \frac{[H^+][HCO_3^-]}{[H_2CO_3^*]} \qquad\qquad 5) \quad K_2 = \frac{[H^+][CO_3^{2-}]}{[HCO_3^-]}$$

If the difference in the trace anion and cation equivalent concentrations is small, then equations 1-5 can be combined to relate the the unknown $[H^+]$ with the known major ion concentrations.

6) $[H^+] - (K_w + K_1 K_H P_{CO_2}) / [H^+] - 2 K_1 K_2 K_H P_{CO_2} / [H^+]^2 =$

$[Cl^-] + [NO_3^-] + 2 [SO_4^{2-}] - [NH_4^+] - [Na^+] - [K^+] - 2 [Ca^{2+}]$

$- 2 [Mg^{2+}]$

The error in $[H^+]$ calculated by equation 6 is determined by the uncertainty in the known major ion concentrations and the validity of the assumptions. The assumptions about trace ion concentrations and equilibrium with atmospheric carbon dioxide probably give errors that are small. The uncertainty in the major ion concentrations can have a large effect on calculated pH values around 5.65 due to small differences in the relatively large charge concentrations when the alkalinity and acidity are near zero(5).

Applying historical mean ion concentration data to equation 6 requires estimates of the uncertainty in the mean concentrations. Mean values are typically given as the precipitation weighted mean ($\overline{[X]}$) defined by equation 7 where P_i is the precipitation amount during the time the ith sample was collected, and n is the number of samples analyzed.

7)
$$\overline{[X]} = \sum_i^n P_i [X]_i \ / \ \sum_i^n P_i$$

Equation 7 is valid only for conservative species. The corresponding precipitation weighted standard deviation of the mean as proposed by Miller(6) is given in equation 8.

8)
$$(s_{pwm})^2 = \frac{1}{n} \ \frac{1}{\frac{1}{n}\sum_i P_i} \ \sum_i^n P_i ([X]_i - \overline{[X]})^2$$

This standard deviation includes analytical errors as well as the natural variation in the rainwater composition of the samples collected, but it does not include variation between samplers at the same site, systematic errors due to improper sample collection and preservation(7), or spatial variability(8). The coefficient of variation ($s_{pwm} / \overline{[X]}$) is seldom less than 10% or more than 30% for major ions in precipitation collected by month, by storm and by increments within storms in the United States(5). Other averaging procedures can give variabilities of slightly less than 10% to more than 30% (9). In the absence of data on the standard deviation of the mean, 10% and 30% will be assumed to be lower and upper bounds, respectively, for the uncertainty in mean values.

The charge balance (equation 6) can be used with historical major ion concentration data to calculate mean pH values when direct

measurements of pH or acidity were not reported. The data of Junge
(10) and Junge and Werby(11) are used for 1955-1956,and the data of
Lodge et al.(12) are used for 1960-1966. Locally appropriate ratios
of magnesium to calcium and ammonium to nitrate must be estimated
for the 1955-1956 and 1960-1966 data, respectively, to use the charge
balance. By assuming 10% and 30% uncertainties in the mean ion concentrations, confidence limits in the calculated pH can be determined. Figure 1 summarizes the results for southern California,
where for each site the middle range represents the 68.3% confidence
limits for 10% uncertainty in mean values, and the bottom range represents the 95% confidence limits for 30% uncertainty in mean values.
Although the uncertainties in the calculated pH values are high,
southern California probably had alkaline rain in the 1950's and
1960's with the exception of the Los Angeles area which probably had
acidic precipitation(5).

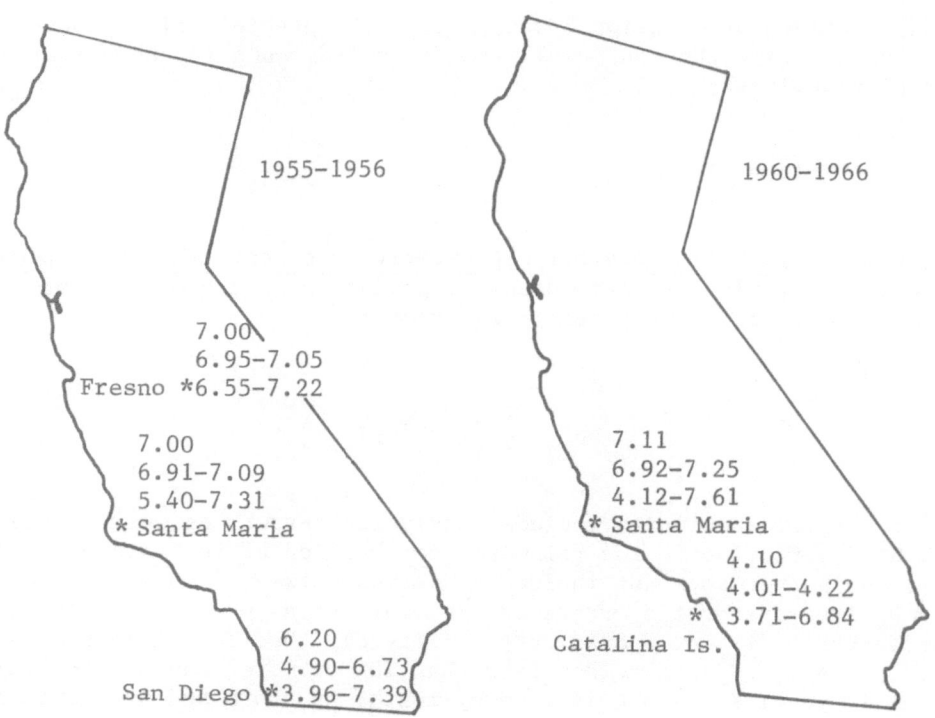

Figure 1. Calculated mean pH values for California. For each site
           the top value is calculated by the charge balance. The middle
           row is the 68.3% confidence limits for the mean pH assuming a
           10% coefficient of variation in the known mean major ion con-
           centrations; the bottom row, the 95% confidence limits assuming
           a 30% coefficient of variation(5).

Conductivity Balance

The conductivity balance is a necessary and sufficient condition that all the major ionic species have been determined. The conductivity balance is given in equation 9 where Λ is the measured specific conductance in $\mu mho/cm$, λ_i is the ionic conductance in $\mu mho/cm$-N for the ith species, n is the number of species of known concentration and m is the number of species of unknown concentration.

9)
$$\Lambda = \sum_{i=1}^{n+m} \lambda_i \; |z_i| \; [X_i^{z_i}]$$

By weighting ion normalities with their ionic conductances, ions of high mobility, such as hydrogen and hydroxide ions, have relatively more influence on the conductivity than equivalent concentrations of ions of low mobility, such as organic ions.

The conductivity balance can also be used to estimate the pH from known major ion concentrations and measured specific conductance. By assuming equilibrium with atmospheric carbon dioxide and the trace ion contribution to the conductivity to be insignificant, $[H^+]$ can be related to known major ion concentrations and measured specific conductance by the following equation:

10)
$$\Lambda - \sum_{i=1}^{n} \lambda_i \; |z_i| \; [X_i^{z_i}] = \lambda_{H^+} [H^+] + (\lambda_{OH^-}K_w)/[H^+] +$$

$$(\lambda_{HCO_3^-} K_1 K_H P_{CO_2})/[H^+] + (2\lambda_{CO_3^{2-}}-K_1 K_2 K_H P_{CO_2})/[H^+]^2$$

The usefulness of equation 10 is limited for several reasons. The assumption of insignificant trace ion contribution to the conductivity is a poor one. There is no cancellation of trace cation and anion contributions as in the charge balance. Since the measured specific conductance is not a conservative quantity, mean values cannot be used. Major ion concentrations and measured specific conductance must be available for each sample. Random uncertainty in the measured values and the problem of small differences in relatively large numbers leads to large uncertainties in calculated pH values around 6.1 where $\lambda_{H^+}[H^+]\approx\lambda_{HCO_3^-}[HCO_3^-] + \lambda_{OH^-}[OH^-] + 2\lambda_{CO_3^{2-}}$ $[CO_3^{2-}]$ for aqueous systems in equilibrium with atmospheric carbon dioxide. When the difference in measured conductivity and calculated conductivity due to the major ions is less than the minimum possible for the right hand side of equation 10, there is no valid solution. If the measured conductance is sufficiently greater than the calculated conductance due to major ions, equation 10 may still have more than one positive real root. A charge balance must still be used to choose the desired root for $[H^+]$(5).

The conductivity balance has also been used as a check that all the major ionic species have been determined(3). A comparison between measured specific conductance and specific conductance calculated from known ion concentrations, including $[H^+]$ for precipitation samples collected in Pasadena, California, from February 1976 to September 1977, shows good agreement. Most measured values agree with the calculated values within the limits of analytical errors. These errors add up to swamp the small differences due to trace ions between the relatively large calculated and measured specific conductances.

A plot of calculated vs measured conductivity values has an intercept of -0.2 and a slope of 0.99. The intercept is not significantly different from zero, but the slope may be significantly different from one. Calculated conductances that are less than measured values indicate that trace anionic/cationic species have yet to be measured. If the difference is significant, trace ion concentrations would be on the order of eight microequivalents charge per liter. Total organic acid concentrations could be even higher, depending on the percent in ionic form.

Direct Measurements of pH and Acidity

While several studies of precipitation chemistry including pH have been made in northern California, reported measurements in southern California are few. A few pH measurements were made in Los Angeles in early 1959(13). The average pH of six analyses was 4.9, with the range being from 4.15 to 5.80(14).

The pH of rainwater in Pasadena from February 1976 to September 1977 varied from 2.78 to 5.33. $[H^+]$ tends to be log-normally rather than normally distributed(15,16). The geometric mean pH was 45 μM (pH of 4.35) with σ_g equal 3.0.

The precipitation weighted mean $[H^+]$ was 87 μM or a pH of 4.06. While hydrogen ion is not a conservative species, it is approximately equal to the acidity for pH less than 5.65, assuming weak acids other than ammonium and carbonic species are present only in trace concentrations(18). Acidity is the conservative species being weighted to compute an average pH.

pH measurements determine $\{H^+\}$ which, after activity coefficient corrections, leads to $[H^+]$ or the free acidity, i.e. $\{H^+\} = \gamma_{H+}[H^+]$. Another method to determine acidity is titration of a sample with base. Two types of titration data are commonly used: 1) titrations can be to an operational endpoint to give a total acidity; 2) titration curves can give information about the strong and weak acid concentrations. Gran type functions have been used to distinguish strong and weak acids and determine acidity constants for the weak acids in precipitation samples(19-23).

The Pasadena rainwater samples were titrated and Gran function plots were used to determine total strong and weak acid concentrations. Samples of volume v_0 were placed in a jacketed plastic beaker. The initial pH and pH after degassing with high purity nitrogen gas were measured and agreed within random drift of the electrodes. Degassing removed carbonic acid species and probably a significant fraction of any sulfide species present. Removal of the carbonic acid species gave a sharper titration curve and a more accurate determination of the strong acid concentration. Of the acids given in Table I, only sulfide species were significantly removed during the stripping of carbonic acid species. pH values were measured for each volume of base (v) added. A plot of the Gran function $F_1 = (v_0+v)10^{-(pH+\log\gamma_{H^+})}$ vs volume of base (v) gave a line of negative slope equal to the normality of the base (N). Thus each equivalent of base neutralizes an equivalent of free acidity from strong acids. The intercept volume of F_1 vs v was multiplied by the normality of the base to give the equivalents of strong acid in the sample volume (v_0). Further additions of base titrated the weak acids to about pH 10.3, beyond which pH changes represented dilution of the the base; each equivalent of base added causes an equivalent increase in $[OH^-]$. A plot of the Gran function $F_2 = (v_0+v)10^{+(pH+\log\gamma_{H^+})}$ gave a straight line of slope N/K_w. The intercept volume of F_2 vs v was multiplied by the normality of the base (N) to give total equivalents of strong plus weak acid. Total weak acid concentration was determined by difference. The uncertainty in the intercept volumes was at least 3%(25).

Strong and weak acids have been implicitly defined by the Gran plotting procedure given above. In comparison with Table I, an acid with pK_a value at least two units less than the initial pH is a strong acid; less than 1% of the anionic ligand would be in the protonated form. Weak acids with pK_a values greater than 11 are not included in the total acid concentration. Their concentrations are too small to cause a noticeable change in the slope of F_2 vs v from N/K_w. Thus strong acids are those of pK_a < 2.2, and weak acids are those of 2.2 < pK_a < 12 with the exception of acids removed by degassing.

The strong acid concentrations agreed with the free acid concentrations within the error limits of the measurements. Thus weak acids, with the exception of carbonic acid, did not influence the pH. The weak acid concentration agreed with the ammonium ion concentrations within error limits. Other weak acids may have been present in concentrations below the uncertainty in the NH_4^+ and weak acid concentrations. The sum of unidentified acids could have been present at an average concentration of less than 5 μM.

Table I. Representative Values of Acidity Constants at Infinite
          Dilution in Aqueous Solutions (25°C) (26,27)

acid	$pK_a = -\log_{10}\dfrac{[H^+][A^-]}{[HA]}$	acid	$pK_a = -\log_{10}\dfrac{[H^+][A^-]}{[HA]}$
$HClO_4$	-7	H_2S	7.1
HCl	-3	HSO_3^-	7.2
H_2SO_4	-3	$H_2PO_4^-$	7.2
HNO_3	-1	Pb^{2+}	7.5
H_3O^+	0	Fe^{2+}	9.2
$HOCCOH$ (OO)	1.1	HCN	9.2
$SO_2 \cdot H_2O$	1.8	H_3BO_3	9.3
HSO_4^-	2.0	NH_4^+	9.3
H_3PO_4	2.1	$Si(OH)_4$	9.5
Fe^{3+}	2.2	⬡-OH	9.6
HF	3.2	HCO_3^-	10.3
HNO_2	3.3	Mg^{2+}	12
$HCOH$ (O)	3.8	HPO_4^-	12.4
CH_3COH (O)	4.7	$Si(OH)_3^-$	12.6
Al^{3+}	4.9	Ca^{2+}	12.6
$H_2CO_3^*$	6.3	HS^-	14
		H_2O	14

Source Strength Methods

Chemical mass balances have been used to calculate the source
contributions to rainwater chemistry(3,4,17). The source strength
model as described by Miller et al.(28) was developed to determine
the contribution of different aerosol sources to the ambient particulate aerosol from the chemical composition of the source emissions
and the ambient aerosol. The percentage of the ith element in the
ambient sample, p_i, was given by

$$11) \quad p_i = \sum_j \alpha_{ij} \, p_{ij} \, c_j$$

where α_{ij} is the fractionation factor for the ith element from the
jth source, p_{ij} is the percentage of the ith element from the jth
source and c_j is the mass fraction from the jth source to the ambi-

ent aerosol. By using characteristic tracer elements, the sea salt, soil dust, automobile particulate emissions and fuel oil fly ash contributions to Pasadena aerosol were calculated(28).

By excluding water from the mass balance, the same model has been applied to rainwater composition where the sources are typically sea salt, soil dust, ammonia, nitric acid and sulfuric acid. In addition to mass, acidity/alkalinity from each source is conserved. The net acidity or alkalinity from the sources fixes the pH for the rainwater sample for a known partial pressure of carbon dioxide(18). Fractionation factors are usually assumed to be unity for lack of specific data.

Granat's chemical relationship for European rainwater assumes the following sources and acidities/alkalinities: sulfuric acid from air pollutants with two equivalents of acidity per mole, nitric acid from air pollutants with one equivalent of acidity per mole, ammonia from air pollutants with one equivalent of alkalinity per mole, sea salt with negligible alkalinity, calcium soil dust with two equivalents of alkalinity per mole, magnesium soil dust with two equivalents of alkalinity per mole, and potassium soil dust with one equivalent of alkalinity per mole(3). Cogbill and Likens added hydrochloric acid air pollutants with one equivalent of acidity per mole for the northeastern United States calculations(4).

Granat's model was applied to the Pasadena rainwater samples to give calculated pH values. The agreement between calculated and measured pH values was good. The linear correlation coefficient(r) between calculated and measured values was 0.988. This result implies that the rainwater pH is controlled by the interaction of bases (ammonia, metal cabonates/metal oxides) and strong acids (sulfuric and nitric)(17). By including the alkalinity of sea salt and adding hydrochloric acid as an acid source, the model reduces to the charge balance equation 6(5). Since the measured cation equivalent concentration agreed with the measured anion equivalent concentration within analytical errors(17), the agreement between calculated and measured pH is expected.

The success of the model in predicting pH is, unfortunately, not a verification of the assumptions behind the calculation. Setting the fractionation factors equal to one and assuming theoretical values of source acidity/alkalinity could lead to systematic errors that balance out in the pH calculation. For example, if sea salt spray fractionated to give higher divalent ion to sodium ion ratios than in seawater, sulfuric acid and calcium and magnesium soil dust contributions would be overestimated, but the calculated pH would agree with the measured pH. Such fractionation of seawater is not expected(29), but it illustrates the ambiguity of using sources that have only one chemical species and its acidity/alkalinity in the system of simultaneous equations (equation 11).

To model Pasadena rainwater, more chemical species were used than primary sources; i>j in equation 11. Since a unique solution may not exist for the over-determined system of equations, a least squares solution was found to give the best estimate of the source strengths. The sources were assumed to be sea salt, soil dust, fuel oil fly ash, automobile aerosol, cement dust and gaseous air pollutants (sulfur oxides, nitrogen oxides, and ammonia). Since the source strength model is a mass balance, the chemical composition of total residue of the rainwater, rather than the soluble or particulate fractions separately, must be used. The results (on a mass basis) showed 35% of the total residue due to nitrate from NO_x air pollutants, 20% due to sulfate derived from SO_2/SO_3 air pollutants, 4.4% due to ammonium from ammonia, 17.2% due to soil dust, 13.6% due to sea salt aerosol, <7% due to fuel oil fly ash, 1.5% due to automobile aerosol and <2% due to cement dust(17). The measured chemical composition and calculated composition are compared in Table II. The agreement between calculated and measured nitrate, sulfate and ammonium mass ratios is exact since their mass sources have only one chemical species. For gaseous sources, p_{ij} is the delta function ($p_{ij}=0$ for $i \neq j$, and $p_{ij}=1$ for $i=j$). Moreover, there is no distinction between natural, industrial and mobile sources of gaseous species.

One possibility for separating the contributions of primary sources with the same mass composition is to include stable isotope measurements. Isotope balances in addition to mass balances have been used with some success in areas where local sources dominate long-range transport effects(30). For isotope balances, fractionation factors cannot be considered unity. There is a fractionation effect between light and heavy isotopes for each physical-chemical change. For equilibrium between gas and liquid species and species of different oxidation states, the fractionation factors are fixed.

Table II. Mass Ratio, Excluding Water, of Chemical Species in
           Pasadena Precipitation(19)

species	measured	calculated	species	measured	calculated
Al	1.5×10^{-2}	2.0×10^{-2}	Mn	3.3×10^{-4}	2.1×10^{-4}
Br^-	9.0×10^{-4}	1.4×10^{-3}	Na^+	4.8×10^{-3}	4.8×10^{-3}
C	4.9×10^{-2}	1.8×10^{-2}	NH_4^+	4.4×10^{-2}	4.4×10^{-2}
Ca^{2+}	1.7×10^{-2}	1.6×10^{-2}	NO_3^-	3.5×10^{-1}	3.5×10^{-1}
Cl^-	7.7×10^{-2}	7.5×10^{-2}	Pb	5.7×10^{-3}	5.7×10^{-3}
Fe	7.0×10^{-3}	6.9×10^{-3}	Si	3.6×10^{-2}	3.7×10^{-2}
K^+	8.7×10^{-3}	4.2×10^{-3}	SO_4^{2-}	2.2×10^{-1}	2.2×10^{-1}
Mg^{2+}	8.4×10^{-3}	8.0×10^{-3}			

During precipitation scavenging, equilibrium conditions may not
exist. Even for systems that should reach equilibrium such as the
dissolution of sulfur dioxide and ammonia, there should be a Ray-
leigh distillation type effect similar to that for hydrogen and
oxygen isotopes in rainwaters(31), so that the heavier isotopes are
gas scavenged preferentially and nonlinearly as the percent scavenged approaches one. Thus to apply isotope balances, the extent of
gas scavenging, aerosol scavenging and chemical state of the source
emissions must be estimated to distinguish natural and anthropogenic
sources.

While organic compounds were not individually identified in the
Pasadena source strength calculation, they are potentially good
source tracers. Volatile organics are not as useful primary source
tracers since there would be a delta function p_{ij} for the gaseous
sources to give perfect mass balances. Nonvolatile organics can be
traced by stable carbon isotope ratios, mass balances and stereoiso-
mer mass balances to distinguish natural and anthropogenic sources.
Few studies have been made of the organic compounds in rainwater
due to their low concentrations.

One of the advantages of the source strength model is its pre-
dictive ability. A number of statistical analyses have been used
to identify correlations between elements and identify their probable sources from ion ratios(12,32,33). The source strength model
can estimate the concentration of species that have not been measured in the precipitation from the source strengths or mass contributions to the rainwater and the sources' chemical compositions.
Unfortunately, the source strength does not distinguish between
long- and short-range sources or identify how the chemical composition will change with changes in weather conditions. These questions require models of the mechanism of transport and scavenging.

Scavenging Models

Fundamental gas scavenging models have been reviewed by Hales
(34), and aerosol scavenging models have been reviewed by Slinn(35).
These scavenging models require field measurements of air pollutant
concentrations and atmospheric conditions to be applicable. Very
little data is available for comparison with the chemical composition of Pasadena rainwater collected from February 1976 to September
1977.

Since nitric acid concentrations are high in Pasadena samples,
the mechanism of nitric acid formation in rainwater is of interest.
The equilibrium models for the gas scavenging of NO_x species are
given in equations 12-16. Other equations may be written for intermediate reactions in both the gaseous and liquid phases, but equations 12-16 determine the overall equilibrium.

12) $2 \ NO_{2(g)} + H_2O = HNO_{2(aq)} + H^+_{(aq)} + NO^-_{3(aq)}$

13) $NO_{(g)} + NO_{2(g)} + H_2O = 2 \ HNO_{2(aq)}$

14) $HNO_{2(aq)} = H^+_{(aq)} + NO^-_{2(aq)}$

15) $HNO_{3(g)} = H^+_{(aq)} + NO^-_{3(aq)}$

16) $HNO_{2(g)} = HNO_{2(aq)}$

Data for P_{HNO_3} and P_{HNO_2} during precipitation are unknown. At high relative humidity, these species or their ammonium salts will become aerosols(36). Aerosol nitrate data during the storms is too meager for statistical analysis. NO and NO_2 concentrations are monitored, and equations 12 and 13 are far from equilibrium for Pasadena rainwater. Equations 12 and 13 may be applicable to gas-aerosol equilibrium(37). Of measured gaseous nitric oxide, nitrogen dioxide and oxidant concentrations, oxidant has the highest, but not significant, correlation with measured nitrate values in rainwater. Since equations 12 and 13 show a change in the nitrogen redox state from II(NO) and IV(NO_2) to III(HNO_2) and V(NO_3^-), the processes may be limited by chemical kinetics in the gas phase (formation of N_2O_3 and N_2O_4) or liquid phase (formation of NO_2^- and NO_3^-). The high nitric acid concentrations in Pasadena rainwater are probably a consequence of photochemical smog reactions. It is still unclear if the oxidant effect is a liquid phase catalysis or, more probably, concurrent with the formation of gas/aerosol nitrogen III/V species in the air due to photochemical smog reactions. A pH of 3.6 was measured in California in 1921 during an electrical storm caused by formation of nitric acid by the discharge(17). Lightning was not the cause of the nitric acid acidity measured in Pasadena precipitation.

The mechanism of sulfuric acid formation and scavenging in the Pasadena precipitation samples is also poorly understood. Aerosol sulfate measurements were too few for comparison with the excess sulfate concentrations, the sulfate in excess of that expected from sea salt. Sulfur dioxide concentrations during the storm were consistently 10 or 20 ppb with an uncertainty of 5 ppb. Gas scavenging of sulfur dioxide to form sulfite species and oxidation to sulfate before analysis of the sample could account for a large fraction of the excess sulfate. Equilibrium total sulfite was calculated from P_{SO_2} measured at ground level and measured pH, by equations 17-20.

17) $K_{H_{SO_2}} = [SO_{2(aq)}] \ / \ P_{SO_2}$

18) $K_{1_{SO_2}} = [H^+][HSO_3^-]/[SO_{2(aq)}]$

19) $K_{2_{SO_2}} = [H^+][SO_3^{2-}]/[HSO_3^-]$

20) Total Sulfite $= [SO_{2(aq)}] + [HSO_3^-] + [SO_3^{2-}]$

The calculated total sulfite could account for 20±10% of the excess sulfate where the uncertainty is due to uncertainty in the partial pressure of sulfur dioxide. Oxidation of sulfite species to sulfate in or below the cloud would increase the percent of excess sulfate due to sulfur dioxide scavenging.

Conclusions

Precipitation in southern California was probably alkaline in the 1950's and 1960's with the possible exception of the Los Angeles area. Direct measurements of pH showed rainfall acidity in Los Angeles as early as 1959, and the weighted mean pH of rainwater in Pasadena was 4.06 from February 1976 to September 1977. Titrations showed the acidity to be predominantly due to strong acids. A source strength model calculated the acidity due to nitric acid to be 32% greater than that due to sulfuric acid. Nitric acid formation appears to be kinetically controlled. Scavenging of NO_2^-/NO_3^- species formed by photochemical reactions may be the mechanism of nitric acid formation in rainwater. Sulfuric acid formation may be a consequence of both gas and aerosol scavenging.

References

1. S. Odén, _Statens Naturvetenskapliga Forskningsråd_, Ekologikomittéen, Stockholm, Bulletin 1 (1968).
2. S. Oden, _Water Air Soil Pollut._, 6:137-166 (1976).
3. L. Granat, _Tellus_, 24:550-560 (1972).
4. C. V. Cogbill and G. E. Likens, _Water Resour. Res._, 10:1133-1137 (1974).
5. H. M. Liljestrand and J. J. Morgan, _Tellus_, in press.
6. J. M. Miller, _in_:"Precipitation Scavenging (1974)," R. G. Semonin and R. W. Beadle, ed., ERDA CONF-741003, 639-659 (1977).
7. J. N. Galloway and G. E. Likens, _Tellus_, 30:71-82 (1978).
8. L. Granat, _in_:"Precipitation Scavenging (1974)," R. G. Semonin and R. W. Beadle, ed., ERDA CONF-741003, 531-548 (1977).
9. D. F. Gatz, _in_:"Study of Atmospheric Pollution Scavenging," R. C. Semonin, prin. invest., Illinois State Water Survey, Urbana, Ill., COO-1199-59, 1-14 (1978).
10. C. E. Junge, _Eos Trans. Am. Geophys. Union_, 39:241-248 (1958).
11. C. E. Junge and R. T. Werby, _J. Meteorol._, 15:417-425 (1958).
12. J. P. Lodge, Jr., et al., "Chemistry of United States Precipitation," National Center for Atmospheric Research, Boulder, Colorado (1968).
13. L. G. M. Bass Becking. I. R. Kaplan and D. Moore, _J. Geol._, 68:

243-284 (1960).

14. D. Carroll, U. S. Geol. Surv. Water-Supply Pap. 1535-g (1962).
15. C. V. Cogbill, Water Air Soil Pollut., 6:407-413 (1976).
16. G. T. Wolff, et al., Environ. Sci. Tech., 13:209-212 (1979).
17. H. M. Liljestrand and J. J. Morgan, Environ. Sci. Tech., 12: 1271-1273 (1978).
18. W. Stumm and J. J. Morgan, "Aquatic Chemistry," Wiley-Interscience, New York (1970).
19. C. Askne and C. Brosset, Atmos. Environ., 6:695-696 (1972).
20. A. Liberti, M. Possanzini and M. Vicedomini, Analyst, 97:352-356 (1972).
21. S. Krupa, M. R. Coscio and F. A. Wood, J. Air Pollut. Control Assoc., 26:221-223 (1976).
22. M. D. Seymour, J. W. Clayton, Jr. and Q. Fernando, Anal. Chem., 49:1429-1432 (1977).
23. M. D. Seymour, et al., Water Air Soil Pollut., 10:147-16J (1978).
24. C. Brosset and M. Ferm, Atmos. Environ., 12:909-916 (1978).
25. S. L. Burden and D. E. Euler, Anal. Chem., 47:793-797 (1975).
26. L. G. Silén and A. E. Martell, "Stability Constants of Metal-Ion Complexes," Spec. Pub. 17, Chemical Society, London (1964).
27 L. G. Silén and A. E. Martell, "Stability Constants of Metal-Ion Complexes, Supplement No. 1," Spec. Pub. 25, Chemical Society, London (1964).
28. M. S. Miller, S. K Friedlander and G. M. Hidy, J. Colloid Interface Sci., 39:165-176 (1972).
29. E. J. Hoffman and R. A. Duce, Atmos. Environ., 11:367-372 (1977).
30. J. O. Nriagu and R. D. Coker, Nature, 274:883-885 (1964).
31. W. Dansgaard, Tellus, 16:436-438 (1964).
32. K. Sekiguchi, et al., J. Japan Soc. Air Pollut., 12:466-469 (1977).
33. E. J. Knudson, et al., in:"Chemometrics: Theory and Application," B. R. Kowalski, ed., ACS Symp. Ser., 52:80-116 (1977).
34. J. M. Hales, Atmos. Environ., 6:635-659 (1972).
35. W. G. N. Slinn, Water Air Soil Pollut., 7:513-543 (1977).
36. A. W. Stelson, S. K. Friedlander and J. H. Seinfeld, Atmos. Environ., 13:369-371 (1979).
37. A. E. Orel and J. H. Seinfeld, Environ. Sci. Tech., 11:1000-1007 (1977)

DISCUSSION

STENSLAND: Why have you resisted back calculating to get ammonia? Don't you think there is equilibrium?

LILJESTRAND: I believe there could be equilibrium for ammonia. I did not discuss the redox kinetics involved in the nitrogen system. The ammonium concentrations may be decreased by a number of oxidation-reduction reactions. Measured ammonium concentrations may be lower than that due P_{NH_3}.

The second problem is the calculation would have to be only for the samples that were collected at the beginning of the storm. During the storm, ammonia would be washed out of the atmosphere, and the partial pressure of ammonia would go down. The mean of all samples would tend to give spuriously low partial pressures of ammonia in the atmosphere. Even during the first increment, usually a quarter inch of rain, there might be an exponential decrease that would give spuriously low values for the partial pressure of ammonia.

JURIS: You showed in one of your slides a source grid arrangement where you have various types of sources, sea salt, road dust and other things, and then you showed the receptor. You said you calculated what the concentration would be at the receptor. How did you get from the source emissions to the concentration within the solution on the other side? That was rainwater?

LILJESTRAND: It is just linear algebra. In a rainwater solution, water is excluded from the mass balance because you obviously have a big dilution effect. It is dry residue that you are working with for mass balances between sources and the receptor - rainwater. The calcualtion does not predict concentrations from the sources but rather back calculates the source impacts from the rainwater concentrations.

MUEHLBAIER: Can you go over the source contributions again?

LILJESTRAND: On a mass basis, 35% due to NO_x, 20% sulfate, 4.4% ammonium, 17% soil dust, 14% sea salt, 7% fuel oil fly ash and 1.5% automobile aerosol.

DETERIORATION OF ARCHITECTURAL STRUCTURES AND MONUMENTS

K. Lal Gauri

Department of Geology
University of Louisville
Louisville, Ky. 40208

ABSTRACT

Natural stone, concrete and mortar are the common materials exposed at the facade of architectural structures. Carbonate and silicate minerals are the essential constituents of these materials. These minerals are susceptible to attack by atmospheric CO_2. The weathering of these minerals has increased at an alarming rate in the industrial countries due to NO_2 and SO_2 emanations.

The SO_2 attack has produced sulfate crusts on ancient buildings. The continuing reactivity behind these crusts has resulted in the removal of stone in layers obliterating the original sculptural details and causing serious damage to the structures.

Most ancient buildings and monuments contain florescences. Evaporation of water at the surface tends to accumulate the florescences in subsurface regions of the stone; their migration is facilitated by increased ionic concentration resulting from atmospheric pollution. Repeated dissolution and crystallization of the florescences in subsurface regions and the accelerated oxidation of reinforcing metals generate stresses which disintegrate the stone.

Weathering also changes the physical properties such that the stone becomes more susceptible to atmospheric attack. These properties relate to mechanical strength, water absorption, and permeability of the stone. Design of conservation treatment should include regeneration of these properties so that the stone performs as though it had been established anew in the existing environment.

Introduction

A study of the chemical deterioration of marble in urban industrial environments of western Europe and North America and in the urban environments of less industrial countries reveals distinctly that sulfur dioxide (SO_2) and carbon dioxide (CO_2) are the major chemically active gases responsible for stone decay. The study further reveals that while SO_2 has been the main cause of accelerated decay in the industrial regions, the weathering in less industrial regions has been due to carbon dioxide. This is illustrated by comparison of the type and degree of deterioration of marble at the Acropolis in Greece and at the Field Museum of Natural History, Chicago, with the marble that has been exposed at the Taj Mahal in Agra, India. These marbles are similar in their mineralogical composition as well as in their physical properties. Yet, the corrosion of the marble in Athens and in the much younger Field Museum in Chicago is far more advanced than that of the marble which forms the facade of the Taj Mahal. The sulphates that form due to SO_2 attack on marble are altogether absent at the Taj, while at the Acropolis and at the Field Museum they form black crust and have penetrated into the intergranular space of marble.

The accelerated decay is due to the increased acidity of precipitation in industrial environments. This precipitation, in equilibrium with atmospheric CO_2, acquires a pH value of 5.6. But SO_2 has caused the precipitation to locally achieve values as low as 2.1 in the northeastern (NE) United States (13, p. 1176). Figure 1 shows that the 1972-1973 pH isopach for the value 4.5 covered a much larger area than the same isopach in the year 1955-56; in addition the new 4.2 isopach also made its appearance. The presence of these acids is primarily related to SO_2 emissions generated by the burning of fossil fuels.

Nitrogen dioxide produced primarily during combustion processes by the oxidation of atmospheric nitrogen is the main cause of the acidity of precipitation in the Los Angeles Basin; here, NO_3^- is more than twice as concentrated as SO_4^{-2} (14, Table 1). The NO_3^- is also present to a lesser extent in the NE United States. But due to higher solubility its lodgement time in the atmosphere is much shorter (4). It is perhaps due to this that the nitrates have not yet been identified in the stone structures. In addition, highly alkaline ammonium products in the combustion process must also have contributed to the accelerated decay of the masonry materials.

The deterioration of masonry materials involves more than the reactions of stone materials with atmospheric gases. Mechanical effects of weathering due to crystallization and hydration of florescences, freezing of water, and expansion of reinforcing iron

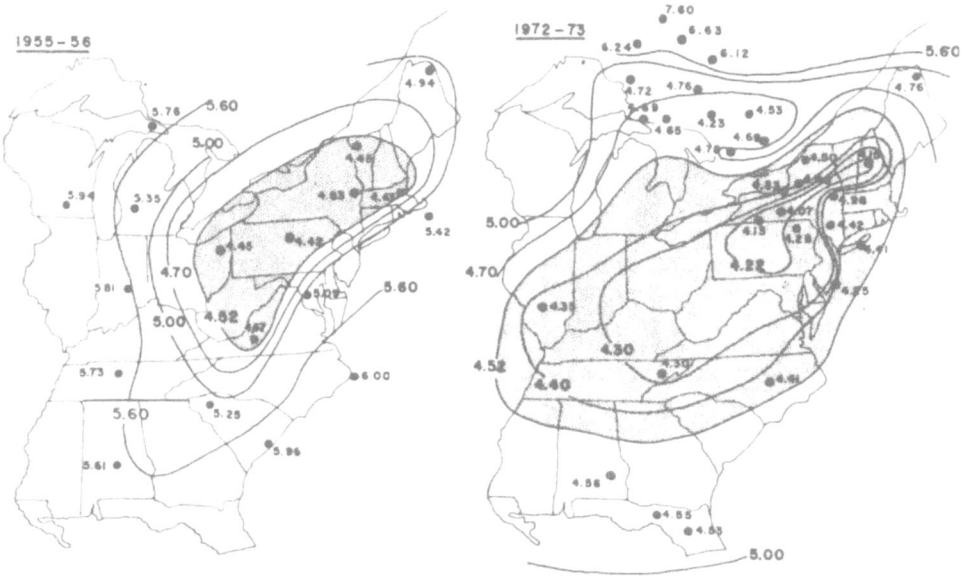

Figure 1.   The weighted annual average of precipitation pH in
            1955-56 and 1972-73 (13).

bars on oxidation are equally important in the process of stone
decay. The major purpose of this article, however, is to present
the effects of increased atmospheric toxicity so that attempts will
be made to reduce it, and so that means will be sought to save the
monuments of human patrimony from further deterioration.

Masonry materials and their decomposition

The common masonry materials include such natural building
stones as limestones, marbles, sandstones, basalt and granites, as
well as such artificial materials as brick, concrete, mortar and
terra-cotta. All of these materials are composed primarily of
carbonate and silicate minerals. The grains of quartz (SiO_2) in
sandstone may be bonded together with calcite ($CaCO_3$), while the last
is an essential constituent of limestone and marble and, to a lesser
extent, of the lime-mortar. Natural silicate minerals compose basalt,
granite, and porphyries, and complex silicates are formed in the
process of fabrication of concrete and terra-cotta. The decomposition of the masonry materials may thus be considered in terms of
attack of atmospheric CO_2 and SO_2 on calcite and the silicate
minerals.

Decomposition of stone containing silicate minerals

The silicate minerals crystallize from magma following the fundamental chemical principle that they possess an electrical neutrality.
Two basic complex ions, namely $(SiO_4)^{-4}$ and $(AlSi_3O_8)^-$, form during
the early cooling of the magma. They combine selectively with the
common cations viz. Mg^{2+}, Ca^{2+}, $Fe^{2+, 3+}$ in magma to yield two major
groups of silicate minerals.

Fe and Mg react with $(SiO_4)^{-4}$ to yield an isomorphic series
of Fe, $Mg (SiO_4)$. These are the first formed ferro-magnesian (mafic)
silicates. Later, the $(SiO_4)^{-4}$ polymerize and combine selectively
with above cations and other ions in magma such as K^+, Na^+, $(OH)^{-1}$,
as well as with early formed minerals to form a complex array of
mafic silicates.

At about the same temperature where first mafic minerals form,
at first Ca^{2+} reacts with $(AlSi_3O_8)^-$; later Na^+ and K^+ also react
with it, forming the group of feldspar minerals.

In the last stages of the cooling of magma, residual $(SiO_4)^{-4}$
link with each other in a spatial relationship to form quartz
(SiO_2) which, due to the absence of soluble cations, is highly
resistant to atmospheric attack.

Although the positive and negative charges are equated in these
minerals, there exist free valences at the crystal surfaces. These,
as well as the ionic potential, i.e. the ratio of charge to ionic
radius, determine the stability of these minerals when exposed to
atmospheric agents.

Jenny (11) postulated that the hydration of the above silicate
minerals occurs through the polarization and the ensuing dissociation of the water dipole into H^+ and OH^- due to the attractive
forces of the free valencies. He explained this, for K-feldspar,
by the model given on the next page.

In this interaction the oxygens are converted to hydroxyl
groups and at least part of the potassium is removed in solution.
A partial or total cationic (K^+, Na^+ Ca^{2+}) depletion decomposes the
feldspars. Since the neutral water now achieves an increased pH,
introduction of acids, such as carbonic and sulfurous acid formed
by CO_2 and SO_2 dissolution in water, neutralize these alkalies and
thus facilitate a further decay of these minerals. Two typical
equations, one for a feldspar and another for a mafic mineral, may
be written as follows:

$$2KAlSi_3O_8 + 2CO_2 + 11H_2O \longrightarrow Al_2Si_2O_5(OH)_4 + 2K^+ + 4H_4SiO_4 + 2HCO_3^-$$

Potash   Kaolinite
Feldspar

$$2CaFeSi_2O_6 + 1/2O_2 + 10H_2O + 4CO_2 \longrightarrow Fe_2O_3 + 4H_4SiO_4 + 2Ca^{2+} + 4HCO_3^-$$

Hedenbergite

The clay minerals, such as kaolinite in the above equation, that form by the dissociation of feldspars and certain aluminous mafic minerals, further enhance the deterioration of the igneous building stones as given in a following section.

Jenny's model for the decay of feldspar

Feldspar Surface	Water		Feldspar Surface	Potash
O	H⟍O H⟋		O-H	
Si			Si-OH	
O	O⟍H ⟋H		O-H	
Al	+ H⟍O H⟋	=	Al-OH	+ KOH
O			O-H	
K	O⟍H ⟋H		H	
O			O-H	

Decomposition of stone containing carbonate minerals

Calcite ($CaCO_3$) and dolomite $Ca\,Mg(CO_3)_2$ are the common carbonate minerals of sedimentary (limestone, sandstone) and metamorphic (marble) rocks used as building stones. Not only are these minerals highly susceptible to attack by acid precipitation, but the rocks which are made of these minerals also contain variable quantities of clay minerals. The chemical decomposition of the carbonate minerals as well as the contribution of clay minerals to the disintegration of these rocks is the subject of discussion in this and in the next section.

$$CaCO_3 \xrightarrow{H_2O,\ CO_2} Ca(HCO_3)_2 \xrightarrow{\hspace{2cm}} Ca^{2+} . + HCO_3^-$$

$$CaCO_3 \xrightarrow{H_2O,\ SO_2} \begin{array}{c} CaSO_3 . 1/2\ H_2O + CO_2 \\[1em] CaSO_3 . 2H_2O + CO_2 \end{array} \xrightarrow[H_2O]{O_2} CaSO_4 . 2H_2O$$

In the case of CO_2, the reaction occurs only when CO_2 is dissolved in water. The resultant calcium bicarbonate occurs in the
ionic state as Ca^{2+} and HCO_3^-. Immediately on drying, $CaCO_3$ is
precipitated from the solution. The effectiveness of the CO_2
reaction thus is confined to the period of wetness of the structure
which essentially, at the exposed surface, is only slightly longer
than the actual period of the rainfall. Depending thus on the
duration and intensity of the rainfall, the $Ca(HCO_3)_2$ may be completely drained from the stone's surfaces during sustained heavy
rains, or it may be precipitated as $CaCO_3$ on the stone surface during
short and mild rainfalls. Figure 2 shows such recrystallized calcite
forming a whitish encrustation at the facade of St. Paul's Cathedral,
London. Figure 3 shows that the penetration of the solution into
the intergranular space, and its crystallization have produced a
pore-plugged region in the vicinity of the stone surface.

The SO_2 reaction, however, continues to occur in the presence
of liquid water as well as water vapor. The calcium sulfate that
forms is less preserved in the regions of structures exposed to
direct rainfall. In protected regions, however, SO_2 continues to
attack almost perpetually. The $CaSO_4.2H_2O$ that forms from this
reaction is able to form crusts on protected surfaces (6).
Figure 4 shows the occurrence of gypsum crust at the Field Museum
of Natural History, Chicago; Figure 5 shows that the gypsum has
penetrated into the intergranular space; this space must have first
formed due to the dissolution of calcite.

The occurrence of recrystallized $CaCO_3$-crusts is a rather rare
phenomenon. However, the black appearance of most ancient buildings
in the industrialized countries is due to gypsum crusts which have
incorporated soot in the process of crystallization. It seems para-
doxical at first that the facades directly showered with the acid
precipitation are largely clean, whereas the covered areas of the
building are coated with black gypsum crusts. The explanation of
this, however, is quite simple in that the condensed water vapors
in a shaded area keep the building from drying, and once the crust
has formed it continues to grow inwardly for not having been washed
away by driving rain. These crusts of weathering are much less

Figure 2
St. Paul's Cathedral, London.
Notice the white crust along
the cornices formed by the
recrystallization of calcite.

Figure 3
St. Paul's Cathedral, London.
Scanning electron micrograph
of Portland Limestone with
white crust. Upper portion
of the micrograph is the
surface view, and the lower
portion is the transverse
section showing that the pores
are plugged with recrystallized calcite.

Figure 4
Field Museum of Natural
History, Chicago. The
Caryatid on the left has
a black gypsum crust; the
Cayratid on the right
lacks such a crust because
it is continually washed
by the rain.

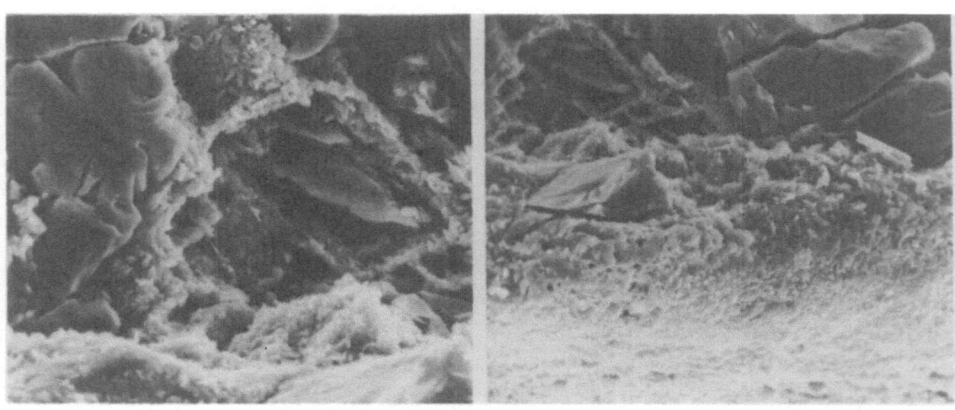

A ——————— 20 μ   B ——————— 10 μ

Figure 5. Field Museum of Natural History, Chicago. Scanning
           electron micrographs of Georgia Marble with black
           crust. Showing gypsum at surface and in the inter-
           granular space. Figure 5B is an enlargement of a
           portion of Figure 5A.

permeable to water than is the marble itself. The longer periods of
wetness in these areas, in the long run, allow the moisture to
penetrate through these crusts. This moisture then stays much longer
in the marble due to reduced evaporation at the surface and to the
rather impervious nature of the crust. The transport of SO_2 along
the water films corrodes the marble behind the crust, rendering these
regions highly friable. An erroneous inference is often drawn that
the crusts form a protective coating on the marble. In actuality,
such crusts have accelerated the decay phenomenon in that, here,
layers of surface fall off as opposed to grain-by-grain dissociation
of marble in unprotected areas of the facade. In the latter process,
the acidic water corrodes around the grain bounderies, leaving the
grains unsupported so that they eventually dissociate from the main
structure.

Figure 6 illustrates this phenomenon. Note that the angel under
the dome has been rendered unrecognizable by the exfoliation of
gypsum crusts formed by weathering. The angel in the open, on the
other hand, has been much less severely deteriorated, although it is
made of the same marble.

The reaction equilibria for the CO_2 - calcite system have been
determined by several scientists. Hougen et. al. (10) obtained
a value of 0.6×10^{-6} for the equilibrium constant from thermodynamic
calculations. Garrels and McKenzie (9, p. 146) reported a value
of 1.58×10^{-6}. Gauri et. al. (8, Table 1), reacting Green Vermont
Marble in dynamic atmospheres with variable partial pressures of
CO_2, obtained an average value of 1.48×10^{-6}.

The necessary reaction equilibria for SO_2 calcite, however,
have not been determined. It is required that CO_2, NO_2, and SO_2
reactions be studied individually and in combinations of these
gases for a judicious simulation of reactions in the ambient. It
is especially necessary to do so because the reaction equilibrium
in any combination of gases is likely to be quite different from
the simple sum of reactions with single gas species.

In addition to the chemical composition, several other properties
of the stone determine its actual rate of decay. These
properties relate to the porosity and pore structure which govern
the transport of water and chemically active gases into the stone.
Also, the aspect of exposure influences the decay rate. Therefore,
determination of the carbonate-acid reaction in reaction-kinetic
studies yields only partial information on the actual weathering
rates.

Geologists have computed surface reduction rates for certain
carbonate rocks on the basis of the occurrence of resistant minerals
in these rocks. Figure 7 shows aggregates of a mica, Phlogopite,
projecting above the present stone surface. This reveals that the

Figure 6
Cave Hill Cemetery, Louisville.
Angels showing exfolliation of
gypsum crust in areas under the
dome and grain by grain dissociation of marble exposed to
direct impact of rain.

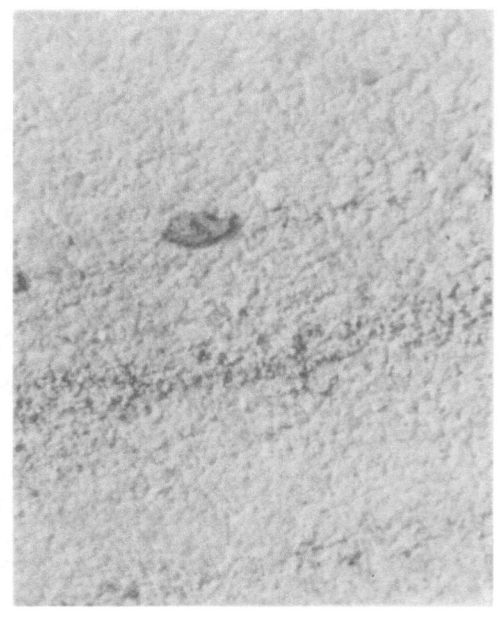

Figure 7
Field Museum of Natural
History, Chicago. Crystals of
Phlogopite projecting nearly
2mm above marble surface.

marble in these areas has experienced nearly 2mm of surface reduction since the erection of the Field Museum in 1912. Winkler (18, p. 349) deduced the surface reduction of a fine-grained marble in a suburban South Bend graveyard to be 1.5mm in 43 years as measured against a quartz vein. Due to the variety of controls involved, the surface reduction rates thus determined permit only qualitative comparison and lack universal applicability for even one variety of rock.

Yet, controlled studies of weathering promise the generation of data usable,not only for the determination of weathering rates, but also for the reconstruction of the history of the acidity of precipitation. As shown earlier in the introduction, the absence of nitrates and sulfates in the weathered rock indicates that CO_2 is the cause of decay; this reveals that the pH of the precipitation has been nearly 5.6. Since the presence of NO_2 and SO_2 considerably reduce the pH of the precipitation, the occurrence, quantities, and the depth to which the weathered products have penetrated in dated stones (e.g. monuments in graveyards) may form bases for determinations ofactual acidity when the pH of the precipitation is below 5.6. It will, however, be necessary to determine the reaction rates in laboratory conditions at known acidity levels and correlate them with reaction rates in ambient conditions,both for stones which are directly exposed to precipitation and those which weather due to acid aerosols while protected under the dome.

Decay of stone due to clay minerals

Clay minerals are hydrated silicates of aluminum formed by a partial or complete leaching, in situ, of alkali metal and metal cations in feldspars and aluminous mafic minerals of most igneous and metamorphic rocks. Because sedimentary rocks form by the weathering of igneous and metamorphic rocks, they contain mostly clay minerals. The grey veins in limestone, sandstone and occasionally in marble are due to the aggregation of clay minerals.

The clay minerals are sub-microscopic (grain size: <0.004mm) particulates having thus a high surface/volume ratio. Therefore, they have large unsatisfied charges at their surfaces. This provides them with a capacity for adsorptions, and exchange of ions from circulating waters. The resulting expansion of these minerals makes them a potential source of stone decay. Further, the adsorbed sulfate ions in clay minerals make them sites for nucleation of gypsum. This is why most bricks show large-scale gypsum florescences which are highly detrimental to the structures.

Decay of stone due to florescenses

The florescences are water soluble salts present within the stone and occur as encrustations at the surface; in the latter case they are termed efflorescences. They produce discolorations at the facades of buildings, and, upon repeated crystallization and hydration, they mechanically disintegrate the stone.

The florescences may be inherently present in the stone or may be deposited on the stone surface from sea-sprays, or, in areas of high water table, they may migrate into the stone from sub-surface waters. They may also form as a result of the attack of chemically active gases on masonry materials. Also, more than one type of florescence may interact to produce salts which are even more complex.

A majority of the sedimentary rocks have formed in a marine environment. These rocks therefore have incorporated some evaporite minerals in their structures. Some such common minerals are the chlorides, sulfates, and nitrates of alkali and alkaline-earth metals. Gypsum ($CaSO_4.2H_2O$), for instance, is ubiquitous in sandstones and shales. Also such metal sulfides as pyrite (FeS_2) are common constituents of shales. Therefore such synthetic masonry units as brick, concrete and mortar, which are manufactured using sandstone and shale, inherit these deleterious florescences.

Gypsum is added to portland cement to expedite the setting of concrete; many concrete structures therefore exfoliate due to the crystallization of gypsum in subsurface regions. NaCl used as a deicing agent migrates into the stone structure: most granite failures at street levels of buildings are associated with the repeated crystallization of this salt.

Gypsum and magnesium sulfate form due to atmospheric SO_2 reaction with the parent calcite and dolomite $CaMg(CO_3)_2$ of limestone, marbles, and the lime-mortars. The black crusts on surfaces of most ancient buildings consist of gypsum which has incorporated soot in the process of transformation of the parent calcite.

Complex salts such as $Na_2(CaSO_4)_2$ in the cast stone (5) may have formed by the reaction of NaCl, deposited from sea sprays, with existing $CaSO_4$ which may have been added to the mortar, and may have additionally formed due to the SO_2-$CaCO_3$ reaction.

Lastly, such backing materials as gypsum dry-walls appear to be a major source of florescences in all kinds of masonry materials in the humid climates.

The florescences by themselves are not highly injurious to the stone. However, the water movement through the pore structures of

the masonry materials and its evaporation at the facades tend to
concentrate these salts at and near the stone surfaces. When
concentrated in sub-surface regions, they become very detrimental to
the stone. Also, most of these florescences form such hydrated
salts as Mirabilite ($Na_2SO_4.10H_2O$) and Epsomite ($MgSO_4.7H_2O$). In
these cases humidity alone may be adequate to accomplish the hydration. Repeated crystallization of these salts and their hydration
in the wake of wet and dry climatic episodes generate stresses in
the pore space, resulting in the crumbling of the stone.

The crystal pressure (Eq. 1) and hydration pressure (Eq. 2) are
calculated from the following equations:

$$P = \frac{RT}{V} \cdot \ln \frac{C}{C_s} \quad (1): \quad \text{Correns and Steinborn (1)}$$

Where

   P = pressure exerted by growing crystal, in atm,
   R = gas constant. 0.082 liter-atm/mole ^{o}K,
   T = absolute temperature, ^{o}K,
   V = molar volume of solid salt, liter/mole,
   C/C_s = degree of supersaturation, where C is the existing solute
       concentration and C_s is the saturation concentration

and $$P = \frac{RT}{V} \cdot \ln \frac{P_o}{p} \quad (2): \quad \text{Mortensen (15)}$$

Where

   P, R, T, V are same as above
   P_o/p = ratio of vapor pressure of ambient air to vapor pressure
       salt at t ^{o}C.

Applying equation 1, Winkler and Singer (19) calculated the crystallization pressures for several salts at various C/C_s and T values.
Table 1 is condensed from their Table I (p. 3512).

Table II is generated (3, p. 223) on the basis of equation 2.
It is evident from this table that the salts with fewer molecules
of hydration exert larger hydration pressures.

The American Society of Testing Material's test, ASTM C-88,
for determining soundness of aggregates is based on the principle
of generating stresses by artifically depositing sodium or magnesium
sulfate in the pore space. In this procedure, the specimen is
impregnated with a concentrated solution of the salt at 20°C. The
specimen is then dried at 110°C and cooled to 20°C, after which the

Table I. Crystallization pressures (Atm) of common florescenes

Salt	Molar Volume cm^3/mole	Crystallization pressure.			
		$C/C_S = 2$		$C/C_S = 10$	
		$0^{\circ}C$	$50^{\circ}C$	$0^{\circ}C$	$50^{\circ}C$
Anhydrite $CaSO_4$	46.00	335	398	1120	1325
Gypsum $CaSO_4 \cdot 2H_2O$	54.00	282	334	938	1110
Halite NaCl	27.85	554	654	1845	2190
Therandite Na_2SO_4	53.00	292	345	970	1150
Mirabilite $Na_2SO_4 \cdot 10H_2O$	220.00	72	83	234	277
Epsomite $MgSO_4 \cdot 7H_2O$	147.00	105	125	350	415

Table II. Hydration pressures of common florescences

Anhydrous Salt	Hydrated Form	Volume Increment (%)	Hydration Pressure (Atm.)
$CaSO_4$	$CaSO_4 \cdot 2H_2O$	32	1100
Na_2SO_4	$Na_2SO_4 \cdot 10H_2O$	208	240

specimen is again impregnated with the concentrated solution. This procedure is repeated until enough salt is deposited in the pore space so that the specimen begins to rupture in the following cycle.

We subjected specimens of Indiana Limestone to this test, but by forcing the solution into pore space by the use of a vacuum pump. We determined that the grains began to dissociate in the second cycle immediately after the specimen had been placed in the solution; 0.68% salt -- by wt. of dry specimen -- was adequate to cause the failure of the stone. A repeated cycling of much lesser quantity of an efflorescence such as calcium sulfate as given in Table III shall be adequate to cause disruption of stone structures.

The problem of florescences is germane to atmospheric pollution. First of all, the florescences may originate in the attack of atmospheric gases on masonry materials. Once they have formed, they increase the concentration of ions in solution and thereby increase the solubility of potential efflorescences dispersed in the matrix of the stone, and thus concentrate them in the near-surface regions where they are most effective in causing the disintegration of the stone.

Table III. Ionic composition (wt. % of dry stone) specimens from Field Museum of Natural History, Chicago: expressed as percent weight of dry specimen

Composition	1*	2*	3*
Ca^{2+}	5.1	3.9	0.07
Mg^{2+}	0.02	0.02	0.0009
Na^{2+}	0.11	0.13	0.01
K^{+}	0.013	0.006	0.0006
SO_4^{-2}	10.6	7.8	0.05

* 1, 2. Scrappings from black-crusted surfaces; 3. Grains from naturally cleaned surface.

Disintegration of stone due to corrosion of metallic reinforcements

More damage to stone has occurred due to the corrosion of metallic bars than by any other mechanisms of stone decay. Fortunately, such bars are commonly present in structural units of the building and not in the sculptured regions where their effect would be even more serious.

Iron bars have been widely used for anchoring blocks of stone with each other and with the structural framework. Also, monuments in cemeteries have frequently been attached to pedestals by iron bars. Unfortunately, an ancient practice of coating iron bars with lead to inhibit the corrosion has been ignored in recent times. The results have been rather alarming. Figure 8 shows a lead-coated iron bar projecting from the limestone surface. It is obvious that this bar has not caused any damage to the stone.* Conversely, the use of uncoated iron bars in the early 20th century restoration of the Erechtheum at the Acropolis has seriously damaged the stone, as seen in Figure 9.

Oxidation of iron produces the mineral limonite (FeO.OH) which has greater volume than the parent iron. This volume increment generates stresses sufficient to disintegrate the stone.

The presence of SO_2 in the atmosphere has decidedly accelerated the decay phenomenon. The mechanism is described by the following equations (12):

$$4Fe + 4H_2SO_4 + 2O_2 \longrightarrow 4FeSO_4 + 4H_2O$$

$$4FeSO_4 + O_2 + 6H_2O \longrightarrow 4FeOOH + 4H_2SO_4$$

Further, the concentration of ions increases the electrical conductivity of the water film, thereby increasing the oxidation rate of the iron bars in the masonry.

---

*Schaffer (17, p. 21) points out that serious damage to the stone had occurred due to oxidation of iron clamps and dowels at St. Paul's Cathedral. However, in all instances the author of this paper has seen, lead coated iron bars were not rusted.

Conservation of architectural structures and monuments

The net effect of the entire deterioration process is to alter chemical and physical properties such that the weathering stone becomes even more susceptible to atmospheric attack. The function of conservation treatments is to regenerate, at least partially, the original properties so that the stone acts as though it had been established anew in its existing environment.

The chemical properties worthy of consideration relate to the crusts, florescences, and the reactivity of the stone to ambient gases. The stone surfaces thus must be cleaned of discoloring crusts at surfaces as well as of florescences lodged in the subsurface of regions. Since sulfates are the predominant constituents of florescences they must be diminished so that the cleaned stone has less than 0.03 weight percent of these ions. In case of stone containing chlorides and nitrates, specifications for their minimum content must be developed so that the stone does not succumb to pressures generated by crystallization in the pore space. The conservation treatment should also enable the stone to resist attack by NO_2 and SO_2.

The physical properties relate to the mechanical strength, water absorption, and permeability of the stone. The mechanical strength of the stone becomes highly reduced in the zone of weathering, while the water absorption greatly increases; the permeability is reduced in crusted stone, but it increases in naturally-cleaned surfaces. A conservation treatment should bring these values back almost equal to the corresponding values of unaltered regions of the stone (7).

Professor John B. Patton (16) exemplifies this philosophy. Commenting upon the conservation treatment of the Christ Church Cathedral in Indianapolis he states, "Now that the work is completed and the scaffolding down, Christ Church Cathedral looks much as it did before the renovation -- somewhat cleaner and the ivy missing for the present, but much the same." He continues, "I proposed a goal: that 100 years hence, Christ Church Cathedral look more like the building of 100 years ago than it did when the renovation started."

Figure 8.   St. Paul's Cathedral, London. Lead coated
            iron pins projecting above the weathered
            surface. Note that the irons pins have not
            oxidized and therefore the stone has not
            suffered disintegration.

Figure 9.   Erechtheum, Acropolis, Athens. Fragmenta-
            tion of marble due to oxidation of iron
            bars clearly visible at joints where the
            bars were used to fasten the marble blocks.

Bibliographical references

1. Correns, C. W. and Steinborn, W. Experimente zur Messung und Erklaerung der sogennanten Kristallisationskraft. Z. F. Kristallografie, 101, p. 117-133, Frankfurt/Main, 1939.

2. Double, D. D. and Hellawall, A. The Solidification of Cement. Scientific American, p. 82-90, July 1977.

3. Fitzner, B. Die Pruefung der Frostbestaendigkeit von Naturbausteinen. Geologische Mitteilungen, 10, p. 205-296, Aachen, 1970.

4. Galvin, P. T., Samson, P. J., Coffey, P. E., and Romano D. Transport of Sulfate to New York State. Env. Sci. Tch., 12, 5, p. 580-584, 1978.

5. Gauri, K. L. Conservation of the California Building, San Diego, U.S.A. A Case History. International Symposium: Deterioration and Protection of Stone Monument, June 5-9, 1978, 19 p. UNESCO/RILEM, Paris, 1978.

6. — Effect of Acid Rain on Structures. Am. Soc. Civil. Eng. Preprint, 3598, 1979.

7. — Gwinn, J. A., and Popli, R. K. Performance Criteria for Stone Treatment. 2nd International Symp. on the Deterioration of Building Stones, Sept. 27-Oct. 1, 1976, p. 143-152, National Technical Univ., Athens, Greece, 1976.

8. — Tanjaruphan, P., Appa-Rao, M., and Lipscomb, T. Reactivity of Treated and Untreated Marble in Carbon Dioxide Atmospheres. Trans. Ky. Acad. Sci. 38, p. 38-44, 1977.

9. Garrels, R. M. and Mackenzie, F. T. Evolution of Sedimentary Rocks. 397 p. W. W. Norton & Co., New York, 1971.

10. Hougen, O. A., Watson, K. M., and Ragatz, A. Chemical Process and Principles (Part 2), 1072 p., John Wiley & Sons, New York, 1959.

11. Jenny. H. Origin of Soils. Applied Sedimentation (P. O. Trask ed.), p. 41-61, Wiley, New York, 1950.

12. Kucera, V. Effects of Sulfur Dioxide and Acid Precipitation on Metals and Anti-Rust Painted Steel. AMBIO, 5, p. 243-248, 1976.

13. Likens, G. E. Acid Rain: A Serious Regional Environmental Problem. Science, 184, p. 1176-1179. 1974.

14. Liljestrand, H. M. and Morgan, J. J. Chemical Composition of Acid Precipitation in Pasadena, Calif. Environ. Sci. & Tech., 12, 12, p. 1271-1273, 1978.

15. Mortensen. H. Die Salzsprengung und ihre Bedeutung fuer die regional klimatische Gliederung der Wuesten. Dr. A. Petermanns Mitt. J. Perthes Geogr. Anstalt, 79. Jg., p. 130-135. Gotha, Leipzig, 1933.

16. Patton, J. B. New Skills for Architects, Landscape Archi-
     tects, Planners and Preservationists. Proceedings,
     Hist. Preserv. Conference Dec. 2-3, 1977, Eds. Johnson
     and Hermansen, p. 57-76. Indiana Geological Survey,
     Bloomington, 1977.
17. Schaffer, R. J. The Weathering of Natural Building Stones.
     Building Research Establishment, Special Report No. 18,
     149 p. Garston, England, reprinted 1972.
18. Winkler, E. M. Important Agents of Weathering for Building
     and Monumental Stone. Eng. Geol., 15, p. 381-400, 1966.
19. ― and Singer, P. C. Crystallization Pressures of Salts
     in Stone and Concrete. Geol. Soc. Am. Bull. 83, 11,
     p. 3509-3513, 1972.

DISCUSSION

SCRUDATO: I was curious about the carbonate mineralogy with regards to the materials that were made out of, let's say, a higher magnesium calcite or dolomite.

Gauri: We have purposely selected pure calcite materials in order to obtain a fairly accurate idea of reaction rates. The dolomite certainly has a lower reactivity.

TOMLINSON: You said you're getting calcite deposits on St. Paul's Cathedral in London. They had tremendous sulfur dioxide problems there. Is is because they have managed to get their stacks higher and why is it that we are not finding calcium sulfate there now?

GAURI: The X-ray diffraction of the weathered crusts reveals only calcite. If very small quantities of sulfates were present, they were not resolved by this technique.

TOMLINSON: There must be a fair amount of sulfate in the rain in London. Is there not?

GAURI: I believe this is true, but even though this building has been up for about 400 years, I have found only recrystallized calcite in the weathered regions of the stone. This is unusual because most limestone and marble buildings in the industrial countries have gypsum crusts. I am not quite sure but the explanation lies perhaps in a combination of factors such as relative solubilities of calcite and gypsum and the pore characteristics of this limestone.

THE MERCURY PROBLEM: INTRODUCTORY REMARKS

George H. Tomlinson

Domtar Inc., Centre de Recherches
Senneville, Quebec, Canada

This morning we heard some interesting papers about acid rain, of its nature, and of its serious effect on limestone buildings. This afternoon we're going to talk about another aspect of acid rain and its effect on a certain type of geographical area where the pH of water is becoming low. Our speakers are from different areas - Ronald McLean is from the Province of Quebec, Jay Bloomfield is from New York State, Arne Jernelöv is from Sweden and John Wood is from Minnesota. Certain sections of the diverse regions form part of the Precambrian Shield where there is very little calcium present in the soil which can buffer acid now contained in the rain.

Now acid rain as such, if you consider acid in its literal sense, either sulfuric or carbonic acid, has been going on for billions of years. The concentration of sulfuric acid in the rain was probably quite high at one time in the ancient past, judging from the sulfur we find in shales and in coal. The Precambrian Shield, in contrast with more recently formed areas covered with sedimentary deposits has been subjected to rainfall for over three billion years, minus a few millenia when it was covered with ice during the various ice ages that occurred. The Shield area covers much of Northern Canada, extends down into the Adirondacks of New York State and swings around Lake Superior into Minnesota. Similar ancient areas are found in Scandinavia. Calcium has been leached from the soil by sulfuric and carbonic acids over vast periods of time, and now, with increasing concentrations of sulfuric and nitric acids in the rain, we find little calcium left in the soil and in the lakes and rivers. This situation is quite different from that in geologically younger areas where plenty of calcium remains, neutralizing and buffering the current inputs of acid.

I made a quick calculation recently. The Ottawa River, which
runs down into the St. Lawrence through the Province of Quebec,
currently carries about two million tons of calcium, calculated as
calcium carbonate. And this extraction has been going on for an
awfully long time, no doubt with a continuously decreasing load as
the calcium is depleted. So it means that water in the lakes up in
the North contain only 1.5 to 2 parts per million of calcium, which
is actually lower than the calcium present in the rainfall that you
get here in Rochester. And, of course, these lakes are not subject
to the same inputs of dust as occur in Illinois where it's coming
off the Great Plains areas, as discussed this morning.

Carbonic acid, resulting from solution and reaction of the
carbon dioxide from the air, is a relatively weak acid. Although
it dissolves and extracts calcium from rock and soil, it cannot, by
itself, reduce the pH of water below about 5.6. At this pH, fish
can survive. However, once the calcium has been largely removed,
the stronger acids, sulfuric and nitric, now contained in rain, can
drop the pH to 4.5 to 4.0 or even lower. Large amounts of acid are
released during snow melt and in sensitive calcium-depleted waters
which have not yet become permanently acid, the pH can drop rapidly
for a relatively short period in the spring. Under these conditions,
fish reproduction is impaired, fish stocks decline, and surviving
fish have been found to contain elevated levels of mercury. The
speakers this afternoon will deal with the serious problem triggered
by this chain of events.

DETERMINATION OF MERCURY IN NATURAL WATERS

SAMPLING AND ANALYSIS PROBLEMS

R.A.N. McLean, M.O. Farkas and D.M. Findlay

Domtar Inc., Research Center
Exit 40 Trans Canada Highway West
Senneville, Quebec H9X 3L7

ABSTRACT

Many studies have been carried out on the factors involved in the collection of water and precipitation samples for the later determination of trace heavy metal concentrations. Problems which have been identified in the use of sampling vessels for mercury determinations include (a) contamination from the vessels and from the atmosphere, (b) loss of mercury to the vessel walls, and (c) loss from solution to the air due to chemical or microbiological reduction. An assessment of some earlier data on mercury in natural waters and precipitation will be given. In our initial studies on mercury transport in natural waters, we encountered a number of problems in sampling, principally contamination. Fortunately, the automatic method used for the determination of total mercury in water could be adapted to carry out determinations without the intermediate use of sampling vessels. The equipment can be used on site and the water to be analyzed is pumped directly into the analytical system. It was originally used to evaluate different sampling techniques, sampling vessels and storage methods. Some data is shown to illustrate resultant improvements in the sampling methods. The method has also been used as a continuous monitor for determining total mercury concentrations. A number of measurements of mercury in precipitation have also been carried out. The method is sufficiently versatile to be adapted for speciation of mercury in natural waters, and some results are shown to illustrate this use.

Introduction

In recent years, there has been a growing interest in the determination of trace elements and their species in the natural environment and a number of excellent reviews have appeared on this subject (31). The many precautions required to obtain represent- ative, reliable data and to interpret this data have been stressed a number of times (39). Much unreliable data has been produced because of inadequate procedures to avoid contamination.

An automatic analysis system has been used regularly over the past three years for the determination of mercury in natural waters. Some results will be presented to show the possible applications of this system to determine directly, without the use of sample vessels, the concentration of mercury and its species in natural waters. In this way, contamination - mainly due to sampling procedure - can be considerably reduced.

A number of fatalities occurred at Minamata and Niigata, Japan between 1955 and 1962, caused by the consumption of fish and shell- fish, which had been contaminated by the direct release of methyl mercury species to the waterways (33). It was discovered in Sweden in 1966 that even in remote areas, methyl mercury species form the major part of the mercury in the fish muscle tissue (38), and that the less toxic inorganic mercury species can be converted in sediments to the more toxic methyl mercury species (22). Since then, many studies have been carried out on mercury in the aquatic environment (29).

Until recently, most of these studies were carried out in areas which received direct inputs of mercury from effluents. Of the media of interest in the transport and transformations of mercury in the aquatic environment, fish have been of the greatest concern because of their importance in the human diet and also because of their ability to bio-magnify methyl mercury, thus facilitating analytical determination for both total and methyl mercury. In addition, a considerable amount of attention has been given to the determination of total mercury in sediments, because they act as the sink for the inorganic forms of mercury, and thus are good indicators of elevated inputs. Less attention has been paid to determinations of mercury and its forms in other aquatic media, particularly water, partly because of the lower concentrations observed, and thus the analytical difficulties involved. Because of this, there are still large gaps in our knowledge of the transport of mercury in the aquatic environment, even in clearly contaminated areas. Furthermore, it has now been discovered that there are many areas remote from point sources where elevated levels of methyl mercury are found in fish flesh (30). The range of supposedly background levels of mercury in fish of the same size and age can

TABLE I   Mercury concentration in pike muscle tissue in
           some Ontario and Quebec lakes

	Lake	No.	Mean Weight (kg)	Mean Length (cm)	Mean Hg Conc. (ppm)	Ref.
	Remote Areas:					
A1	L. Chibougamau	29	2.6	69.5	0.73	30
A2	L. Waconichi (15 km North of A1)	6	2.6	67.9	0.21	30
A3	L. Chensagi (150 km West of A1)	12	2.4	63.2	1.51	30
A4	L. Maicasagi (Adjoining A3)	17	3.2	74.7	2.58	30
	Urban/Industrial Areas:					
B1	Lake St. Clair (1972)	38	2.5	71.3	4.3	3
B2	Lake St. Clair (1976)	50	2.4	68.4	1.6	3
B3	Lake Simcoe (1971)	12	2.2	-	0.19	14
	(1975)	3	2.5	-	0.16	14
	(50 km North of Toronto)					

be quite high (Table I). The mercury concentrations in fish in
remote areas (Table I, A3 and A4) can be as high as that in fish
in waterways close to urban areas and even in those fish downstream
from point sources (Table I, B1 and B2).

There is now a considerable effort going into studying the
factors involved in the variation of mercury concentration in
freshwater fish in remote areas (6,24). Such studies will require
an examination of often quite small differences in the many factors
which contribute to the bio-accumulation of methyl mercury in fish.
This will involve the determination of mercury and its forms and
other species involved in the transport of mercury through to its
uptake as methyl mercury in fish. Clearly, the sampling and
analytical techniques, used in studying differences in background
areas require to be much more refined than for a comparison of
mercury levels in sediments, upstream and downstream from a point
source. For instance, the concentration of mercury in water is
measured in units of ng/liter (ppt), units one millionth the size
of ppm, the units normally used for expressing concentration of
mercury in fish and water.

Analytical Methods for the Determination of Mercury in Environmental Samples

There are a considerable number of reviews on the subject of
the determination of mercury in enviornmental samples (34). A

number of analytical techniques with varying degrees of sensitivity
have been used, but of these methods, neutron activation analysis
(NAA), flameless atomic absorption (FAA), and atomic fluorescence
(AF) have been the techniques of choice. Activation analysis has
the advantage of requiring no sample preservation, of having great
sensitivity, and of allowing multi-element determinations. However,
because of the cost of the technique, its lack of flexibility and
the length of time required for each sample, FAA and AF based
methods have received much wider acceptance, especially for determinations carried out in routine monitoring and pollution control
programs. In addition, FAA and AF methods have the capability with
suitable chemical pretreatments of being used for determining
different species of mercury in the environment. A critical
examination of some of the flameless atomic absorption methods in
general use in Canada for the determination of mercury in environmental samples has been produced recently (10).

Problems with Historical Data
on Total Mercury Concentrations in Water

Because of the sudden awareness of the toxic effects of mercury
pollution problems in several parts of the world, many
environmental studies were carried out in the early 1970's.
However, a number of the precautions in trace element determinations
were not adequately recognized. As a result, much of the data on
mercury concentrations in the environment requires reassessment and
many earlier conclusions on the accumulation of methyl mercury in
the environment have required recent reevaluation (6, 24). The
lack of reliability of some of the data is particularly evident in
determinations of the "total" mercury concentrations in natural
waters.

In recent years, there have been a number of reports of an
apparent decrease in the total mercury concentration in some
natural waters. We illustrate a few of these in Table II. In one
instance, this decrease was considered to be due to a decrease in
the circulation of mercury in the environment, though the lack of
point sources on the river in question makes this seem highly
unlikely. In light of more recent experience, it is much more
likely that improvements in the sampling and analytical techniques
have caused the apparent decrease in mercury in water. From 1970
to 1975, the reported mercury concentrations in river and lake
waters in the area of interest in Northwestern Quebec were quite
variable, but with a mean of around 500 ng/liter. However, on
improving the analytical technique, as shown below, in 1975-1976,
it was discovered that the mercury concentration in this river was
always less than 100 ng/liter and was rarely higher than 20 ng/liter.
It was recognized that the earlier higher concentrations were caused
by a lack of sensitivity in the analytical technique, contamination
from the sample vessels, and contamination during the analytical
procedure. Little of the data on mercury concentration in natural

TABLE II   Total mercury concentration in water -
before and after 1975

Location	Pre-1975 Mean Hg. Conc.(ng/l)	Post-1975 Mean Hg. Conc.(ng/l)	Ref.
B.C. Canada Columbia River	2000	< 50	19
Manitoba, Canada Assiniboine River	930 - 2000	< 50	19
N.W. Quebec, Canada Bell River	500	< 10 - 20	11
Washington, U.S.A. Columbia River	50 - 500	10 - 20	4,20

freshwaters collected in Canada and the United States before 1975,
can be considered reliable.

In much of the earlier published data on the mercury concentrations in water, few details were given on the methodology used
to obtain the data so that later evaluation is rather difficult.
However, fortunately in one of the more widely quoted papers on
mercury in the environment (35), adequate details on the methodology were given. In this paper, it was argued that an increase in
the mercury concentration close to the surface of cores from the
Greenland Ice Sheet indicated a recent increase in mercury deposition due to man's activities (Table III). The data has been
criticized since the deeper core (deposited from 800 AD to 1944 AD)
was obtained from a different site than that of the surface core
(1952 to 1965). Therefore, one of the important requirements of
trace analysis - that all samples be treated in an identical manner
has not been satisfied (9). It now seems that there were other
shortcomings in the techniques used for handling samples such as
follows: (1) The samples were stored as water in polyethylene
bottles, (2) They were stored for some time without an adequate
preservative, (3) No details were given of adequate pretesting of
the sampling vessels, and (4) One of the most fundamental requirements for trace element analysis, the procedural blank, did not
undergo the same treatment as the samples. The distilled water
blank was commenced at the subsampling stage prior to irradiation
and chemical separation in the activation analysis method and did
not include storage in the same vessels as the samples for the
same period of time.

Because of the possibilities of both contamination and loss of
mercury from solution, the use of polyethylene vessels is now

TABLE III   Total mercury concentration in the Greenland Ice
Sheet and Arctic snows

Publication Date	Year of Deposition of Ice (A.D.)	Mercury Concentration (ng/l)			Ref.
		Range	Mean	(S.D.)	
1971	800-1944	30-75	60	(17)	35
	1952-1965	87-230	125	(52)	
1975	1802-1844	48-78	52	(16)	36
	1960-1971	31-73	48	(14)	
1977	1600-1860	263-823	513	(246)	18
	1971-1973	290-880	494	(212)	
1978	1974	<5	<5		37

not recommended for storage of natural water samples for mercury
determination (10). In addition, it has been shown that some type
of chemical preservation is necessary to prevent losses of mercury
from sub-ppb solutions (see below). More recent studies by the
same group, which have received much less attention than the
original study, produced somewhat different results (Table III).
The authors recognized (18) that there may have been some methodological problems with the two previous studies, and suggested that
the true concentrations may have been considerably higher and more
in agreement with their later data. However, the study on the
Arctic snows produced a much lower concentration of mercury, casting
some doubt on the validity of any of the data on the Greenland Ice
Sheet. Therefore, in contrast to the fairly clear picture which we
have of the accumulation of lead in recent times in the Greenland
Ice Sheet, methodological uncertainties have left us with a rather
hazy picture of the mercury accumulation.

The determination of all trace metals in natural waters present
a challenge to the analytical chemist, but the unique properties of
mercury (volatility, strong complexes with natural organic compounds,
etc.) require extra precautions in addition to those taken for other
heavy metals. This makes the role of the analytical chemist even
more crucial in studies of mercury in the environment.

Forms of Mercury in Water

Before discussing methods for the determination of mercury in
natural waters, we first give some consideration to the mercury
species which should be determined in evaluating mercury transport
and transformations with particular reference to availability for
bioaccumulation. Various compounds of mercury can be present in
water in dissolved form or associated with particulate matter. The
dissolved mercury can be present as inorganic complexes (more

likely in seawater) or as complexes with organic species such as humates, fulvates (more likely in freshwater). Little is known at present on the availability of the various forms of mercury for transformation into readily bioaccumulable forms.

Different proportions of the total mercury in water will have widely varying availabilities depending on the concentration of the individual mercury species and on the concentration of other ions, etc. For example, in an area where mercury is mainly in the form of mercury (II) sulfide, and the water has a low concentration of humates and a relatively high pH, the mercury will be much less available for methylation and metallation than in waters with little sulfide, high humate and low pH. This situation leaves the environmental analytical chemist with a presently difficult problem. It is impossible to make determinations of the fractions of the total mercury present which are available to different degrees. Certain working definitions can be used, e.g. "total" (extractable by oxidizing acid), "dissolved" ($<0.45\ \mu$), "particulate" ($>0.45\ \mu$), "free-ionic", etc. Various combinations of these alternatives could be used to give an indication of the availability of mercury for further transformation (8). However, as yet, there is little indication of which of these definitions is most generally useful. For the immediate future, it will be desirable to have some measure of the total mercury concentration in the water, of some of the individual species, and finally, of their association with molecular or particulate fractions of various sizes. Hence, we shall commence with the measures of the "total" mercury in water and then deal with possible techniques to determine the various species of interest.

The Determination of Mercury in Natural Waters using Flameless Atomic Absorption (FAA)

In the last decade, the reduction aeration technique has become a part of many methods for the determination of mercury in water using FAA. The modification by Hatch and Ott (17), involving initial oxidative acid digestion, followed by a two-stage reduction, has been widely used and an automated version of the technique by Goulden and Afghan (15) has become standard in North America (10). This method has been the subject of a number of interlaboratory quality control tests and has been shown to be adequate in most laboratories down to a concentration of 100 ng/liter. A number of laboratories are able to obtain satisfactory results in the mercury concentration range between 10 and 100 ng/liter. With appropriate purification of the reagents, as given in Table IV, and maximizing other parameters in the system, it is possible to maintain a detection limit close to 5 ng/liter. However, this limit of detection is still at or higher than the concentration of mercury in many river and lake waters, and at least a five fold increase in the sensitivity of the method is required. Later, we shall discuss

TABLE IV   Total mercury determination in water -
                       purification of reagents

	Reagent	Purification Method	Typical Post Purification Hg Conc. (ng/1)	Frequency of Necessity for Purification**
1)	Conc. H_2SO_4	add 0.5 mg $SnCl_2$ to 2 l reagent, aerate at 2 l/m for 2 hrs. with air scrubbed by 5% $KMnO_4$	20	3
2)	0.5% $KMnO_4$	closed vessel, stand in sunshine	10	1
3)	0.5% $K_2S_2O_8$	extracted with 0.1% dithizone in $CHCl_3$	20	3
4)	3% $NH_2OH.HCl$	as (1)	10	2
5)	3% NaCl	as (1)	10	3
6)	10% $SnCl_2$ in 2 N H_2SO_4	as (1), but aerate 10 mins.	10	3
7)	5% $K_2Cr_2O_7$	extract with 0.1% dithizone in $CHCl_3$	100	3
8)	Air	Scrub with 5% $KMnO_4$	1	1

* as sample preservative
** Purification required (1) always, (2) sometimes, (3) rarely

further modifications to the method and other methods, which can be
used to carry out routine monitoring of mercury at the levels found
in natural waters.

Sampling and Storage of Waters for Total Mercury Determination

In our laboratory, prior to the use of the automatic analysis
system for the determination of total mercury in water, acid washed
polyethylene vessels were used as storage vessels and 1% v/v conc.
sulfuric acid was used as a preservative. At that time, the manual
method used for the determination of total mercury in water had a
detection limit of 100-200 ng/liter. In the rivers of the region
of interest, Northwestern Quebec, the reported concentrations of
mercury in the water were highly variable with a mean value of
$\approx$ 500 ng/liter.

When the improved automatic analysis system was set up and a
much more rapid turnover of samples was possible, problems were

recognized with the existing sampling, storage and analysis techniques (11).

It was obvious that contamination had been occurring during the analytical procedure. In addition, adequate pretesting of the polyethylene vessels revealed that contamination of the sample preserved with 1% v/v conc. H_2SO_4 could occur both by desorption of mercury from the walls of the vessel (despite previous concentrated acid washes) and by absorption through the walls of the vessel (5). In addition, it was recognized that volatile organic compounds were being desorbed from the walls of the vessel and were interfering in the analytical procedure. Because of these problems and the similar observations appearing in the literature (2,21) it was decided to use borosilicate glass for further sampling.

A number of methods have been used to prevent the loss of mercury from natural water and low concentration standard solutions (2,21). Until recently, the recommended method in Environment Canada involved the use of a 1% v/v conc. H_2SO_4 solution. In addition, freezing the sample and various complexing agents had been used to preserve mercury in solution. It has been recognized that bacterial activity (21) and the ready reduction of mercury species to Hg^0 can take place even in acidic solution. The use of oxidizing agents has now become accepted to destroy the bacterial activity and to prevent mercury reduction in natural water samples. The preservation methods presently recommended by Environment Canada and E.P.A. are respectively solutions of 1% sulfuric acid + 0.05% potassium dichromate and 5% HNO_3 + 0.05% potassium dichromate. These methods have been adequately tested in individual laboratories but have yet to be the subject of extended interlaboratory quality control tests. In our experience, standard solutions containing from 0.1 to 10 ppb preserved by 1% sulfuric acid + 0.05% potassium dichromate are stable for a period of several weeks providing no positive contamination is introduced. In addition, in natural water samples with mercury concentrations >0.01 ppb, no losses have been observed over a period of five weeks.

A more rigorous sampling procedure was developed and is detailed in Table V. To test this procedure, a comparison of the total mercury concentration in the water in the sampling vessels and that obtained by direct determination was carried out.

Direct Determination of Total Mercury in Water

The automatic analysis system discussed above is used for the direct determination of mercury in water without using sampling vessels. In remote areas, this requires the use of an electrical generator with a voltage stabilizer and, if the air temperature is low, some insulation of the absorption cell of the mercury monitor. The system can be run on a normal 30 sec sampling-150 sec distilled

TABLE V   Sampling of natural waters for total mercury

1. Sample Bottle: All Borosilicate Glass

2. Sample Bottle Cleaning: 5% $KMnO_4$, 1:1 conc. H_2SO_4; HNO_3, Distilled Water Rinse, Bake at 400° for 16 hours

3. Storage Before Use: Filled with 1% H_2SO_4 : 0.05% $K_2Cr_2O_7$

4. Preservative: Conc. H_2SO_4 + 5% $K_2Cr_2O_7$ diluted 1:100 in sample: added immediately prior to sampling

5. Number of Samples/Site:

Expected Conc. (ppb Hg)	Number of Samples
< 0.1	10
0.1-1	6
> 1	3

6. Recovery and Blank Tests: One for every ten samples

water wash cycle with the regular periodic testing of standard mercury solutions (concentration of 50 ng/l - 1000 ng/l). This method is suitable for determining water close to the surface since the transfer from the tested water to the wash must be relatively fast. The water to be tested is pumped into the system through the required length of prewashed teflon tubing.

The system can also be used as a continuous monitor by pumping from the test water for longer periods of time, and comparing the absorption intensity with that of standards after a longer wash period (generally five minutes for concentrations less than 1 ng/ml). This method is used if depth profiles of mercury in water are required and if testing is to be carried out without the attention of the operator.

Comparison of the Total Mercury in Water obtained by Direct Determination in Samples Collected in Sample Vessels

At a number of times during the testing of various sampling and storage procedures, comparisons were made of the total mercury concentration in water obtained by direct determinations with those obtained by determination using the sample vessels. This facilitated the observation of improvements in the sampling procedure. A comparison of mercury concentrations in water obtained by direct determination and using the sampling method of Table V are shown in Table VI. The water bodies studied were: (a) a shallow pond having a direct input of mercury from an incoming stream, (b) two sites on a river with a high flow (sites 1 and 2), and (c) a third site on another river of low flow with a direct input of mercury (site 3).

TABLE VI   Total mercury in water - comparison of direct
determination with determinations using recommended sample vessels

	Direct Determination[o]			With Sample Vessels		
Water Body	Number of Determin- ations	Hg Conc. (ppb) Mean (±S.D.)	No. of Vessels	Number of Determin- ations	Hg Conc. (ppb) Mean (±S.D.)	
Shallow Pond	3	0.45 (±0.02)	4	8	0.48 (±0.06)	
River Site 1	14	0.015(±0.010)	18	57	0.015(±0.014)	
River Site 2	13	0.019(±0.008)	19	58	0.018(±0.016)	
River Site 3	4	0.078(±0.010)	10	33	0.082(±0.017)	

[o] Direct determination - pumping the water directly into
the Automatic Sampling System.

The mean values obtained by both methods show excellent agreement,
though for river sites 1 and 2, the values were close to the
detection limit of the method. The relative standard deviations
are somewhat higher for the method using sample vessels, presumably
because losses and contamination have not been completely eliminated.
The direct determination gave more precise results, but in this study,
satisfactory accuracy was obtained for the total mercury concentration
using sample vessels.

It was considered that the recommended sampling technique could
satisfy the purpose of the study, i.e. to find a sampling method
which would give a close approximation to the mercury concentration
in natural waters when it is significantly different from the
detection limit of the analytical method.

Improving the Sensitivity for the Determination of Mercury in Water

Since the optimum detection limit of the automated method for
the determination of mercury in natural waters used in this work is
at or close to the level of mercury in natural waters, some improvements in the method are necessary before it can be used on a routine
basis for monitoring the mercury concentrations in natural waters.
Fortunately, the FAA technique is flexible enough to allow greater
sensitivity in a number of ways. The two principal methods of
improving the sensitivity have involved preconcentration (27) and
the maximization of the physical parameters in the aeration-atomic
absorption system (25).

Preconcentration methods have generally been used in studies
involving the determination of mercury in seawater but there has
been little standardization of the techniques and many different
preconcentrating methods have been used. Most recent studies (27)

have used methods based on preconcentration on noble metals. In this way, detection limits down to 0.5 ng/liter and less can be obtained. However, preconcentration methods are prone to contamination and require considerably longer analysis time per sample than the direct determination method referred to above.

It has been shown that optimization of the parameters in the existing aeration-atomic absorption system can produce considerable improvements in the sensitivity over those obtained in the standard method discussed above. The aim in optimizing the parameters is to aerate as much mercury as possible, in the shortest time possible, with the lowest volume of air possible, in the longest pathlength, lowest volume cell possible. A number of detailed studies have been carried out on the examination of the method parameters including the absorption cell dimensions, the liquid sample volume, the aeration rate, drying agents, etc. (25). By maximizing these parameters, detection limits of 0.5 to 1 ng/liter have been claimed.

Some recent advances in methodology have been made by Goulden, one of the originators of the automated analytical technique for the determination of total mercury in water (16), using an automatic analysis manifold similar to that already used in the standard technique. The method, which enables speciation, is a more sophisticated automated modification of the selective reduction method devised by Magos (10,26). A detection limit of 1 ng/liter can be obtained readily, using this method.

Speciation of Mercury in Water

Although many determinations of the total mercury concentration in water have been carried out, few attempts have been made to determine the mercury species which are present in natural waters. Filtration through 0.45 μ filters has been carried out regularly, but, in many cases, it is suspected that not only the particulate-associated, but also some of the non-particulate-associated mercury is trapped by the filter material. In many studies carried out on seawater samples, the so-called reactive Hg is measured by reducing with stannous chloride in acid solution, aeration and concentration on Au, Ag, acid permanganate or in a liquid N_2 trap (13, 27).

A number of studies have been carried out on the separation and determination of mercury species at concentrations considerably higher (usually two orders of magnitude) than those observed in natural water samples. The methods have included thin layer chromatography, the selective reduction of inorganic and organic mercury species and other chemical treatments, and various ion exchange processes. However, few, if any, of these techniques have been able to satisfactorily determine the species of mercury in waters

TABLE VII   Survey of reported methyl mercury concentration in natural waters

Location	CH₃Hg Detection Limit ng/l	CH₃Hg Conc (ppt) Mean/Range ng/l	Total Hg Conc (ppt) Mean/Range ng/l	CH₃Hg/Total Hg %	Ref.
Canada:					
Most Lakes and Rivers	<0.25	<0.25	-	-	7
Lake St. Clair) Clay Lake) Pinchi Lake)		0.5-1.7	-	-	23
Sweden:					
Uncontaminated Lakes	0.1	0.1	< 10	< 1	23
U.S.A.:					
Mississipi River	1.0	1	30-40	1-3	1
N.W. Quebec:					
Rivers	<0.5	<0.5	5-50	<10	28

at natural levels. Dithizone extraction has been used to pre-concentrate various forms of mercury in solution and sensitivities down to 0.01 ppb have been claimed (32). It would be difficult to fit these methods to an automatic analysis system. We shall now illustrate some modifications of the automated analysis which allows some speciation. In the very near future, it is likely that the method developed and carried out by Goulden (16) will be used routinely for determination of various species of mercury in water samples.

Determination of Methyl Mercury in Natural Waters

Methyl mercury species are the major species of interest in the transport of mercury in the natural environment since methyl mercury is bio-accumulated in the fish tissue and is also extremely toxic to humans. Very few attempts have been made to determine the level of methyl mercury in natural waters and Table VII illustrates the few results which are available in the literature.

It is very difficult to evaluate these results in terms of normal quality control criteria since the methods have not been generally accepted and have not been used in many laboratories.

TABLE VIII   Methyl mercury concentrations in water and snow

| | | Concentrations ng/l | | | |
| | Conc of CH$_3$HgCl Added | Total Hg By FAA | Methyl Mercury By FAA | Methyl Mercury By GC | CH$_3$Hg/ Total Hg % |
Sample					
Distilled Water	12.5	10($\pm$5)	0($\pm$10)	10($\pm$1)	-
	125	115($\pm$12)	85($\pm$30)	110($\pm$8)	-
Pond Water	-	2000($\pm$18)	100($\pm$60)	55($\pm$3)	2.8%
River Waters- N.W. Quebec	-	<5 - 50	<5	<0.5	<10%
Snow Near Sewage Treatment Plant	-	200($\pm$30)	100($\pm$40)	125($\pm$40)	63%
Snow-Remote Area	-	<10 - 80	<5	<0.5-2.5	<3%

However, since these techniques have produced, for the most part, concurring data, the overall picture is probably correct. There has been some confusion in the literature because a supposed determination of organic mercury in water was actually a measure of both organic mercury and inorganic mercury, bound strongly to organic material from which it must be photo-labilized (12). As a result, very high organic to total mercury ratios in natural waters have been reported.

The method of preference for methyl mercury determination has involved determination of methyl mercury halides by gas chromatography after certain preconcentration extractions (1,5,23,28). However, we have discovered that by the removal of the potassium persulfate reagent and the reduction of the digestion temperature to 80° in the standard method (10), it can be used for obtaining an estimate of the methyl mercury species in water. An evaluation of this method compared to results using the GC method is illustrated in Table VIII. Clearly the GC method is more selective but the FAA method appears to give a reasonable, though lower concentration for methyl mercury in water.

The GC method is extremely slow requiring at least one whole day for each determination. Therefore, any improvements in the much faster automated FAA technique to include determination of methyl mercury species would be most welcome.

Using the modifications to the automated FAA technique discussed above, Goulden (16) has been able to determine directly the methyl mercury concentration in natural waters down to 2 ng/liter. This should encourage further attempts at speciation of mercury in natural waters using the FAA method and from our experience, it is

TABLE IX   Determination of different fractions of mercury in water samples by selectively eliminating reagents from the automatic flameless atomic absorption method

Sample // Reagent Eliminated		Hg Concentrations in ng/l					
		None	$K_2S_2O_8$	All Oxidants	All Oxidants & Reductants	None^X	None^O
1) Effluent from a Chemical Plant	a)	2,200	2,300	2,300	1,206	–	2,000
	b)	1,000	980	950	420	700	960
2) Effluent After Mixing with a Source of Organic Matter		500	510	150	30	460	250
3) Natural River Water		30	30	–	–	40	10

^{X} Sample aerated at 2 1/min prior to Determination.

^{O} Passed through 0.45 μ filter.

possible to carry out these determinations directly without using sample vessels. This may initially be a prerequisite for the speciation of mercury in natural waters using FAA, since any use of preservatives in sample vessels will almost certainly change the form of mercury in the water. Therefore, it would appear that direct determination by pumping straight into the automated analysis system will be, in the near future, the only method for satisfactorily speciating mercury in natural waters.

Determination of other Mercury Species in Natural Waters

In addition to our attempts to determine methyl mercury species in natural waters by eliminating the potassium persulfate reagent, some further studies have been carried out by selectively eliminating other reagents from the automated FAA procedure. The effect, of omitting groups of reagents from the procedure , on the mercury concentrations determined in effluents from a chemical plant and in natural waters is illustrated in Table IX. The mercury concentration in solution after aerating the sample for some time, and the total dissolved mercury concentration obtained by filtering the sample through a 0.45 μ filter are also given. As expected, the mercury from the chemical plant is almost all in solution and much of it is in the elemental state. On mixing with the source of organic matter, much of the mercury becomes particulate associated and requires an oxidant to determine the true mercury concentration. Oxidants are also required to recover all the mercury from natural river water samples, and much of the mercury is associated with particulate matter.

It seems from these fairly preliminary results that this method can be used in the speciation of mercury in water samples. With the more sensitive technique becoming available, speciation should soon be possible at the levels of mercury found in natural water samples.

Conclusions

In this paper, we have attempted to illustrate some of the flexibility which exists in the automated FAA method for the determination of mercury in natural waters. The system, at present, can be used for the direct determination of mercury in some natural waters though greater sensitivity will be required to extend the use to all natural waters. In addition, it has been shown that modifications can be made to the chemical pretreatment to allow for speciation of mercury in natural waters. Some recent developments have improved the detection limit of the method by a factor of 10 and speciation down to levels of 1 ng/liter will be possible. Because of the problems of maintaining species in collected samples, it seems likely that direct determinations of the individual species, as described here, will be necessary. Therefore, this method has the potential for the on-ship determination of mercury species.

References

1. A. W. Andren and R. C. Harriss, Observations on the association between mercury and organic matter dissolved in natural waters, Geochim. Cosmochim. Acta. 39, 1253 (1975).
2. G. E. Batley and D. Gardner, Sampling and Storage of Natural Waters for Trace Metal Analysis, Water Research 11, 745 (1977).
3. J.N. Bishop and B. P. Neary, The Decline in Mercury Concentration in Fish from Lake St. Clair, 1970-1976, Ministry of the Environment, Ontario, (1977).
4. M. Bothner, PhD Thesis, University of Washington, Seattle, (1973).
5. M. H. Bothner and D. E. Robertson, Mercury Contamination of Seawater Samples Stored in Polyethylene Containers, Anal. Chem. 47, 592 (1975).
6. R. J. P. Brouzes, R. A. N. McLean and G. H. Tomlinson The Link Between pH of Natural Waters and the Mercury Content of Fish, Domtar, Research Center, Senneville, Quebec, (1977).
7. Y. K. Chau and H. Saitoh, Determination of Methyl Mercury in Lake Water, Intern. J. Environ. Anal. Chem. 3, 133 (1973).
8. A. Demayo, A. R. Davis and M. A. Forbes, Forms of Metals in Water, Scientific Series No. 87, Inland Waters Directorate Ottawa, (1978).
9. E. M. Dickson, Mercury and Lead in the Greenland Ice Sheet: A Re-examination of the Data, Science 177, 536 (1972).

10. Environment Canada, Mercury: Methods for Sampling,
     Preservation and Analysis, (1977).
11. D. M. Findlay and R. A. N. McLean, Mercury in Water in
     the Quevillon Area, Domtar Research Report, Senneville,
     Quebec, (1976).
12. W. F. Fitzgerald and W. B. Lyons, Organic Mercury Compounds in
     Coastal Waters, Nature 242, 452 (1973).
13. W. F. Fitzgerald, W. B. Lyons and C. D. Hunt, Cold-Trap
     Preconcentration Method for the Determination of Mercury
     in Seawater and in other Natural Materials, Anal. Chem.
     46, 1882 (1974).
14. R. Frank, M. W. H. Holdrinet, R. L. Desjardins and D. P. Doge,
     Organochlorine and Mercury Residues in Fish from Lake
     Simcoe, Ontario 1970-1976, Env. Biol. Fish. 3. 275 (1978).
15. P. D. Goulden and B. K. Afghan, An Automated Method of
     Determining Mercury in Water, Tech. Bulletin No. 27, Inland
     Waters Branch, Dept. of Energy, Mines & Resources, (1970).
16. P. D. Goulden, Private Communication (1979).
17. W. R. Hatch and W. L. Ott, Determination of Sub-Microgram
     Quantities of Mercury in Atomic Absorption Spectrophotometry,
     Anal. Chem. 40, 2085 (1968).
18. M. M. Herron et al, Vanadium and Other Elements in Greenland
     Ice Cores, Unpublished Manuscript, (1977).
19. Inland Waters Directorate, Mercury Levels in the Rivers
     of Western Canada 1970-1976, Social Science Series No. 16,
     Ottawa, Canada, (1976).
20. E. A. Jenne, Mercury in Waters of the United States,
     Open-file Report, Geological Survey, U.S. Dept. of the
     Interior, Menlo Park, Calif., (1972)
21. E. A. Jenne and P. Avotins, The Time Stability of Dissolved
     Mercury in Water Samples - 1. Literature Review, J. Envir.
     Qual. 4, 427 (1975).
22. S. Jensen and A. Jernelov, Biological Methylation of Mercury
     in Aquatic Organisms, Nature 223, 753 (1969).
23. A. Jernelov, L. Landner and T. Larsson, Swedish Perspectives
     on Mercury Pollution, J. Water Poll. Control Fed. 47, 810
     (1975).
24. A. Jernelov, The Effects of Acidity on the Uptake of
     Mercury in Fish, In: Polluted Rain, T.Y. Toribara, ed.,
     Plenum Press, New York, In Press, (1979).
25. S. R. Koirtyohann and M. Khalil, Variables in the Determination
     of Mercury by Cold Vapor Atomic Absorption, Anal. Chem. 48,
     136 (1976).
26. L. Magos, Atomic Absorption Determination of Total Inorganic
     and Organic Mercury in Blood, J. Assoc. Off. Anal. Chem.
     55, 966 (1972).
27. K. Matsunaga et al, Possible Errors Caused Prior to Measurement
     of Mercury in Natural Waters with Special Reference to Sea-
     water, Environ. Sci. Tech. 13, 63 (1979).

28. R. A. N. McLean, Unpublished Work (1978).
29. J. O. Nriagu, ed., Biogeochemistry of Mercury in the
    Environment, Elsevier, New York, Vol. 1,2, in press, (1979).
    A. F. Penn, The Distribution of Mercury, Selenium, and
    Certain Heavy Metals in Major Fish Species from Northern
    Quebec, Report prepared for Fisheries and Environment
    Canada by the Grand Council of the Crees of Quebec, (1978).
31. J. P. Riley and G. Skirrow, ed., Chemical Oceanography,
    Academic Press, London, Vol. 3, Ch. 19, (1975).
32. J. Stary et al, Determination of Phenylmercury, Methylmercury
    and Inorganic Mercury in Potable and Surface Waters,
    Intern. J. Environ. Anal. Chem. 5, 89 (1978).
33. T. Tsubaki and K. Irukayama, ed., Minamata Disease,
    Kodansha, Tokyo, Elsevier Scientific Amsterdam, (1977).
34. J. Uthe and F. A. Armstrong, Microdetermination of Mercury
    and Organomercury Compounds in Environmental Materials,
    Toxic and Environ. Chem. Reviews 2, 45 (1974).
35. H. V. Weiss, M. Koide and E. D. Goldberg, Mercury in a Green-
    land Ice Sheet: Evidence of Recent Input by Man, Science
    174, 692 (1971).
36. H. V. Weiss, K. Bertine, M. Koide and E. D. Coldberg, The
    Chemical Composition of a Greenland Glacier, Goechim.
    Cosmochim. Acta. 39, 1 (1975).
37. H. V. Weiss, M. M. Herron, and C. C. Langway, National
    Enrichment of Element, in Snow, Nature 274, 352 (1978).
38. G. Westoo, Determination of Methyl Mercury Compounds in Food-
    Stuffs: 1. Methyl Mercury Compounds in Fish, Identification
    and Determination, Acta. Chem. Scand. 20, 2131 (1966).
39. M. Zief and J. W. Mitchell, Contamination Control in
    Trace Element Analysis, Wiley-Interscience, New York, (1976).
    D. N. Hume, Pitfalls in the Determination of Environmental
    Trace Metals, In: Chemical Analysis of the Environment and
    Other Modern Techniques, S. Ahuja et al, Plenum Press, New
    York, (1973).

DISCUSSION

CLARKSON: Mercury in the ocean is present on the order of 100 or 150 million tons, and the last year this estimate seems to be coming down remarkably in the calculations. Is this the result of the problem you have been discussing that the latest analytical numbers are lower?

McLEAN: That's certainly the case. I think that much of the data on the mercury concentrations in atmospheric precipitation are too high. As far as the oceans are concerned, I think oceanographers are the leaders in analytical methodology and contamination control for the determination of mercury in water. The levels of mercury which they are finding in the oceans are now much lower (∿1-10 ng/liter) than what were being quoted five years ago.

CLARKSON: Does this change our view about fish in the ocean presenting mercury, as to whether in fact it's always been there, that fish levels of mercury have always been high because of this vast reservoir of mercury or are the latest results perhaps changing our ideas?

McLEAN: The concentration in water and precipitation used in the models for global mercury transport and for mercury methylation and uptake into biota are too high. Some reassessment of these models is certainly necessary. In addition, the concentration of methylmercury in water for the fish to accumulate is also lower than was first considered.

TOMLINSON: This means they bioaccumulate from a lower level than we thought possible before. Analytical values for mercury in water initially gave values of about 0.5 parts/billion. However, after analytical techniques were improved to minimize various sources of error, it was found that the value was nearer 0.03 parts/billion. The theory still stands but differences in the concentrations of mercury in water and in fish are greater than originally thought.

JERNELOV: I think it's important to remember that the part of mercury that the fish actually does accumulate is methylmercury which is only a very tiny fraction of the total mercury present. It's the total mercury that we've talked about here.

McLEAN: On the next slide Table A1, I show some typical concentrations of methylmercury in various media which we have studied. We've been using a technique for determination of methylmercury in natural waters, which is not very satisfactory since it's detection limit is higher than the natural

TABLE A-1

METHYLMERCURY IN ENVIRONMENTAL SAMPLES OBSERVED IN THIS STUDY

Sample Type	Observed Concn. Range (ppb)	% of Total Mercury
Water, Remote Rivers and Lakes	< 0.0005	< 5*
Water, Stagnant Pond	< 0.0005-0.06	≤ 3
Snow, Remote Areas	< 0.0005-0.0025	≤ 3
Snow, Near Sewage Treatment Plant	125	60
Sediment and Soil	< 0.1-100	< 0.1
Fish - Muscle	10-4500	75-100
Fish - Liver	20-5000	30-100
Fish - Stomach Contents	100-1500	50-100 (Pisciverous)
		0-70 (Non-Pisciverous)
Macroinvertebrates	< 10-500	~ 1
Mosses, Aquatic & Other Vegetation	< 2-5	~ 1

*Where total mercury is above detection limit

background level. However, it can be seen that the
methylmercury concentration level in natural waters is always
less than 0.5 nanograms/liter, except in areas which are highly
contaminated.

KRUPA: Can you comment how you quantify mercury in
precipitation and in snow?

McLEAN: We use the same analytical method as was used for
determining total mercury in water.

KRUPA: So you're essentially quantifying the methyl mercury by
quantifying total mercury.

McLEAN: No, until now I was speaking mainly about total
mercury. One cannot use total mercury to quantify
methylmercury in any medium other than perhaps fish.

TOMLINSON: About 3% of it is methyl mercury in water and snow.

McLEAN: In almost all the samples we have examined, it is even
less than 3% and only in the snow samples collected close to a
sewage treatment plant did we observe a higher proportion of
methylmercury (60%). The normal level of methylmercury in a
remote area would be fairly close to the detection limit of
0.5-2 ng/l.

KRUPA: The point I was getting at is all the procedures
available now are for measuring total mercury. They may have a
recovery rate of 60-70% of perhaps what's there because you
lose a lot more by volatilization. I'm talking about water,
snow or rain.

McLEAN: We routinely check the recovery of total mercury in
spiked natural water samples and, as for all other media, it is
always greater than 90%. The recovery of methylmercury is also
pretty good, except for sediment and soils, where we often get
poor recoveries (50-60%).

BERG: If you took a mercury mineral appearing somewhere in the
precambrian shield, reputed to be highly insoluble, let it
stand with a clean water sample, what kind of concentrations
would you expect?

McLEAN: I do not know but if one used distilled water for
dissolved mercury, I would expect to find no more dissolved
mercury than I was talking about just now - 1 to 100 ng/liter.

TOMLINSON: Except you'd find most of it in natural water is coming in from the rain.

McLEAN: Are you interested in the concentration of mercury dissolved in water in mineralized areas?

BERG: I would like to compare the experimental situation with the natural. Is there agreement or not?

McLEAN: I think that there would be reasonable agreement, but there is very little data on dissolved mercury in water in undisturbed mineralized areas.

BERG: Well, is it correct to conclude then from this that the leaching is a minor factor?

McLEAN: I think leaching of mercury from ores is a minor factor in affecting the bioaccumulation of methylmercury in fish in undisturbed mineralized areas.

TOMLINSON: One of the curious things is that in this area, these very remote areas, the mercury in these sediments is not particularly high, and yet you run $2\frac{1}{2}$ parts/million in your fish. That's why the aerial transport via precipitation became of such interest.

McLEAN: I give can you one example from my Table 2. The first two lakes (A1 and A2) are in a mineralized area, where the mercury concentrations in the bedrock are quite high (for Lake Waconichi, 0.5-1.0 ppm). The mercury concentration in the fish in these two lakes are much less than the concentrations in the next two lakes A3 and A4 which are not mineralized and have low sediment mercury concentration (0.05-0.1 ppm). The minerals are sulfide-containing, so presumably, this prevents the release of mercury to the water.

ANDERSON: Back about a decade ago, there was a great hue and cry throughout North America about the mercury in fish as being detrimental to public health. Concentration limits were issued by various health departments advising how many pounds of fish to eat; bans were made on eating fish from certain waters. Would you, based on your kind of findings and your own analytical technique developed over the years, tend to classify those findings of a decade ago as rather inconclusive. Were the techniques were really wrong?

McLEAN: I don't think there was much wrong with the findings of high mercury concentrations in the fish. In some places,

they had trouble with analytical techniques. Some of the
earlier Japanese data from Minamata in the 1960's is not good.
However, most of the fish data in the recent literature is
fairly reliable.

ANDERSON: But what about analyzing low concentrations in the
waters themselves?

McLEAN: I would be suspicious of almost all of the data on
mercury concentrations in natural waters in North America
before 1975, because it has since been found that contamination
of and loss of mercury from samples were serious problems with
the techniques used.

TOMLINSON: You see that's in parts/trillion in the water as
against parts/million in the fish, and several orders of
magnitude difference can create serious analytical problems.

ATMOSPHERIC AND WATERSHED INPUTS OF MERCURY TO

CRANBERRY LAKE, ST. LAWRENCE COUNTY, NEW YORK

Jay A. Bloomfield, Scott O. Quinn

New York State Department of Environmental Conservation
50 Wolf Road
Albany, New York 12233

Ronald J. Scrudato, Dean Long

State University Research Center
State University College
Oswego, New York 13126

Arthur Richards, Frank Ryan

New York State Department of Health
Empire State Plaza
Albany, New York 12201

ABSTRACT

Cranberry Lake is a large (28.2 km^2) reservoir located in the northwestern Adirondack Mountains in New York State. The area surrounding the lake is primarily wilderness with no major industry. In 1969, fish collected from Cranberry Lake and nearby Stillwater Reservoir as part of a Statewide sample collection program yielded anomalously elevated mercury levels when compared to similar sized and aged fish of the same species from other lakes in the State. This paper documents studies conducted by New York State over the last ten years concerning the Cranberry Lake situation.

Results to date indicate that levels of mercury in atmospheric fallout at Cranberry Lake are low; rainfall and snowpack levels are usually less than 25 ng Hg/l of liquid sample. The Cranberry Lake watershed contributes more mercury to the lake than it receives from the atmosphere, as stream inputs to the lake exceed

atmospheric inputs to the watershed by a ratio of 1.4:1. Although
the pH of Cranberry Lake is generally above 6, measurements taken
of streams and lake water during spring runoff yielded pH values
of less than 4.5 It is hypothesized that the acidity of the Cranberry Lake system is causing an increase in the availability of mercury.
to the biota, rather than an absolute increase in water mercury
concentrations due to atmospheric or other inputs.

INTRODUCTION

Mercury Distribution

Worldwide lithospheric mercury concentrations vary considerably but are usually at the part per billion (ppb) level.
Most igneous rocks contain less than 200 ppb; exceptions include
alkalic igneous and certain deep-seated, ultrafamic rocks including
eclogites and kimberlites, which may contain several hundred ppb
(Fleischer, 1970). Except for the finer-grained varieties, the
mercury content of most sedimentary rocks is less than 100 ppb.
Organic, fine-grained sedimentary rocks such as shales and sedimentary iron and manganese claystones, are commonly mercury-
enriched and may contain several hundred ppb (Fleischer, 1970).
Certain geographic anomalies exist which contain notably higher
lithospheric mercury concentrations including the Donets Basin,
Kerch-Toman area and the Crimea of the Soviet Union (Panov, 1959,
Bulkin, 1962) where igneous and sedimentary rocks contain up to
20,000 ppb. The Franciscan Formation of California contains soil
concentrations of up to 100,000 ppb and localized ore mineralization may result in soil and rock mercury concentrations far above
the part per million level.

Under natural aqueous conditions, mercury may occur in one or
more of three oxidation states including the most reduced metal
form (Hg^0), the more reduced of the two ionic forms (Hg_2^{+2}) and
Hg^{+2}, the most stable form under oxidizing conditions, particularly
at low pH (Hem, 1970). Organic complexing of inorganic mercury by
anaerobes may occur by microbial action (Jensen and Jernelov,
1969). It is well documented that methylmercury is an integral
part of the mercury cycle (Jensen and Jernelov, 1969; Bisogni and
Lawrence, 1975) and once produced it is known to accumulate
rapidly in the cells of higher organisms.

Methylmercury demethylation may occur by sediments, microorganisms, soils and by fecal organisms (Furukawa and Tonomura,
1971, 1972 ; Summers and Silver, 1972; Summers and Sugarman,
1974; Schottel et al., 1974; Tezuka and Tonomura, 1976) resulting
in hydrolysis of the mercury-carbon bond producing methane and
mercuric ion (Wood, 1974).

Mercury, therefore, occurs in a variety of chemical forms including: a) metallic mercury (Hg^0); b) mercurous ion (Hg_2^{+2}); c) mercuric ion(Hg^{+2}); d) methylmercury (CH_3Hg); and/or e) dimethyl-mercury ($(CH_3)_2Hg$). The mercury species, microbial community, size, type and concentration of suspended materials, and the Eh-pH characteristics of the system will influence the sediment-water-biological mercury concentration in any natural or perturbed aquatic environment.

The annual natural flux of mercury to the atmosphere has been estimated at 25 to 30 x 10^9g/yr (Wollast, et al., 1976) which includes both terrestrial and oceanic contributions. Although data are scarce and from widely scattered locations, it is estimated that the total mercury input deriving from all atmospheric sources, including anthropogenic and natural, amounts to about 40 to 50 x 10^9g/yr.

Remote rural area mercury concentrations are highly variable, but are usually less than 10 ng/m^3. Volcanic eruptions, mining, industrial and municipal activities can all contribute to localized higher atmospheric mercury concentrations, and therefore affect global concentrations as well as atmospheric variations.

The redistribution and availability of mercury in the aquatic environment is directly affected by the role suspended and bottom sediments play in the adsorption-desorption process. Sorption phenomena are affected by a variety of bio-geochemical factors including, but not limited to, mercury speciation, associated ionic concentrations, oxygen concentrations, H_2S availability, the pH and Eh of the bulk and interstitial waters, bacterial activity, chloride concentrations and sediment type and size.

Although it is generally agreed that organic suspended sedi-ments are strong mercury sorbers, the role inorganic particles play in mercury sorption is still not clear. Reimers and Krenkel (1974) concluded that inorganic mercury is adsorbed by sands, clays and various organic particulates. They also reported methylmercury sorption by dodecanethiol, clay minerals and fine sands. Sorption was found to be directly dependent on chloride concentrations and pH of the aqueous system.

Farrah and Pickering (1977), however, discount inorganic particulate sorption of mercury as being a major factor in the transport and redistribution of mercury in natural water systems and suggested that precipitation/dissolution processes, rather than sorption, are the dominating factors in mercury water-sediment partitioning.

Kudo and Hart (1974) reported mercury uptake by Ottawa River bottom sediments which included wood chips, clay and sand. Their

study indicated river bed sediments effectively removed soluble
mercury; uptake processes and rates were dependent on mercury
concentrations, bed-sediment surface areas, type of bed sediment
and water movement.

A study of mercury sorption by Mississippi River sediments conducted by Khalid, et al. (1975) found a strong correlation between
mercury sorption, pH, Eh and total mercury availability. They
found maximum adsorption occurred under strongly reducing conditions; desorption was found to increase with increased oxidation
and, in general, retention was favored under reduced, alkaline
conditions.

Studies conducted by Kudo, et al. (1975) on factors controlling
mercury desorption from Ottawa River bed sediments indicated total
mercury availability is dependent on the Eh of the aqueous system
and also on the total depth (volume) of mercury contaminated bed
sediments. More mercury was desorbed under anaerobic conditions
and also at higher sediment volumes.

Review of the mercury sorption literature indicates large
variation in opinion and results on the role played by organic
and inorganic sediments in the adsorption-desorption process.
Some of the differences are no doubt related to the complexity
of the mercury system including the various forms of mercury
produced under varying environmental conditions and also on the
lack of laboratory and field control, particularly in relation
to detailed and accurate characterization of sediments. Sorption
phenomena are strongly dependent on the type (mineralogy,
chemical species, etc.) and also on the total available surface
area of adsorbing materials. These variables must be accurately
determined if reproducible results are to be found in any
mercury sorption study.

Since the turn of the century the combustion of fossil fuels
has significantly increased the total atmospheric concentration of
nitrous oxides, sulfur oxides, chlorine gases, particulate matter
and trace elements and compounds. Acidic gases produced by the
release of large quantities of SO_2 and NO_x into the atmosphere
have increased the acidification of precipitation by as much as
100 times (Likens, et al., 1972; Likens and Bormann, 1974; Galloway
et al., 1976).

More than fifty percent of the U.S. fossil fuel combustion
occurs within the northeast and upper midwestern sections of the
country. Since most of the cyclonic storms move from west to east,
acid precipitation and deposition of associated trace metals and
compounds is focused in a relatively small regional area of the
country (Cogbill and Likens, 1974). The Adirondack Mountains of

New York State have been significantly affected by acid precipitation, most of which is believed to originate from the Ohio Valley.
Lake acidification and resultant effects on aquatic populations is
particularly extensive in higher altitude Adirondack Lakes (Schofield, 1973; 1976). The effects of acid precipitation in Adirondack
Mountain Lakes and streams is particularly acute because of the poor
buffering capacity of most Adirondack soils and susceptibility of
fish to increased acidity and increased and potentially lethal
dissolved aluminum levels resulting from accelerated chemical
leaching (Schofield, 1977, Cronan and Schofield,1977, Cronan and
Schofield, 1979). Shan-Ching (1975) found in a laboratory bioassay
of selected species of freshwater fish that the uptake of mercury
by fish was greatly enhanced at pHs less than 5.5. In addition to
mercury contributions deriving directly from precipitation, the
increased acidity and the presence of humic material of Adirondack
surface waters might favor increased mobility or bioavailability
of soil and bedrock metallic cations, including mercury.

It has been well-documented that fossil fuel combustion is a
source of mercury contamination. The mercury content of thirty-
six American coals ranged from about 70 ppb to as much as 33,000
ppb (Joensuu, 1971). Studies conducted by Gladney and Gordon (1978)
indicate that as much as 93 percent of the total mercury liberated
from a coal-fired power plant was concentrated in the < 2 µm
particulate size fraction. Mercury data for oil is limited although
crude oil samples from California contained from 1,900 to 21,000
ppb mercury (Bailey, et al., 1961).

Statement and Extent of N.Y. Mercury Problem

From 1969 to 1972, over 3,500 fish from New York State lakes
and streams were collected and analyzed for mercury content
(Eisenbud et al., 1978; Boulton and Hetling, 1972). Less than ten
percent of these fish had mercury levels above 1.0 mg Hg/Kg body
weight, the present United States Food and Drug Administration
"actionable level" for commercially caught fish. A sizable proportion of these fish were collected from Onondaga Lake, near
Syracuse, New York, a body of water contaminated by discharges
from a chlor-alkali plant.

Many of the fish with mercury levels over 1 mg/Kg were larger
individuals of predatory species such as Walleye (Stizostedion
vitreum), Northern Pike (Esox lucius),lake trout (Salvelinus
namaycush) and smallmouth bass (Micropterus dolomieui).
Two northwest Adirondack Mountain reservoirs, Cranberry Lake
and Stillwater Reservoir (Figure 1) produced fish with
high mercury levels despite the undeveloped nature of their watersheds. Additional sampling conducted between 1972 and 1975
(Harris, unpublished data) for predator species such as smallmouth

Figure 1.   Cranberry Lake and Stillwater Reservoir locations in relation to the Adirondack Forest Preserve, New York.

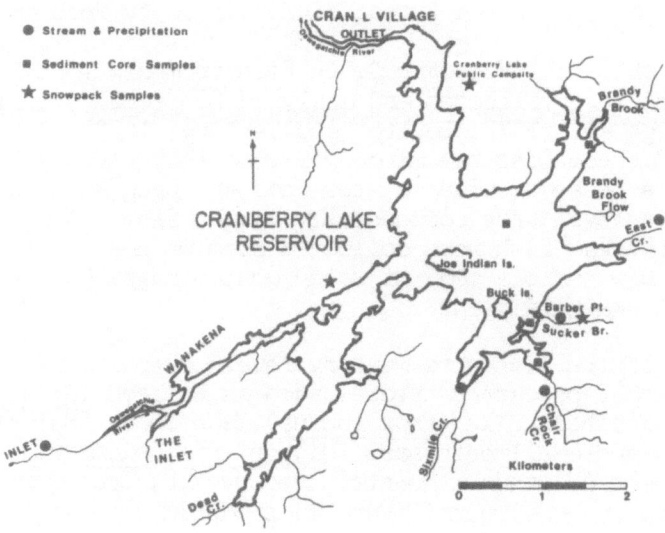

Figure 2.   Cranberry Lake stream and precipitation, sediment and
snowpack sampling locations.

bass and splake (<u>Salvelinus namaycush</u>) <u>Salvelinus fontinalis</u> gave results consistent with earlier samplings.

Between October, 1972 and July, 1974, Earl Harris of the New York State Department of Environmental Conservation (NYSDEC) conducted a study of atmospheric fallout at Cranberry Lake (NYSDEC, 1978). This study gave an atmospheric deposition rate of 44 µg, Hg/m^2-yr, with over eight percent of the mercury associated with particulate material >0.45 µm. Curiously, one sample location at a magnetite sintering operation about 20 km to the northwest of Cranberry Lake had an atmospheric deposition rate of 280 µg/m^2-yr. This sintering operation was terminated in 1975.

In 1978, a study was initiated at Cranberry Lake with several objectives, including:

1) to determine if mercury levels remain elevated in Cranberry Lake fish;
2) to conduct additional field sampling in order to better define the mercury budget for the lake;
3) to examine the mercury history of the lake by collecting and analyzing cores of lake bottom sediment; and
4) to collect fish from adjacent, previously unsampled, lakes in order to determine the spatial extent of the mercury phenomenon.

This paper documents preliminary results available from this study.

Possible Causes and Mechanisms

There are three possible causes of the high mercury values in fish which may be relevant to the Cranberry Lake situation, including:

1) there may be a natural input of mercury to the watershed;
2) there may be an input due to the present or past activities of man; or
3) the high mercury levels in fish may be an indirect effect of the acidity of the lake.

There may be several issues in the Cranberry Lake situation which remain unresolved. It is unclear whether past mining, lumbering or the previously mentioned anthracite sintering operation have played a role. Additionally, there may also be something quite unique about the aquatic community of the reservoir. Investigations to date have revealed nothing unusual about the fish community and few historical mining operations in the watershed. The area was extensively logged during the late nineteenth and early twentieth century and there are records of a sawmill about 1 km northeast of the lake. Although mercury-containing fungicides

are commonly used in pulp and paper production, no record of such
use could be found for the Cranberry Lake watershed area.

MATERIALS AND METHODS

Location and Geology

Cranberry Lake is located in the northwest portion of the
Adirondack Forest Preserve (Figure 1). The lake was enlarged to
its present 28 Km² size with the completion of the 1867 dam on
the Oswegatchie River outlet. Most of the Cranberry Lake shore-
line is State controlled; the area is sparsely populated and there
are no industrial discharges into the lake watershed.

The dominant rock types found within the Cranberry Lake watershed area consist of granitic gneiss and quartz syenitic gneisses
(Buddington and Leonard, 1962). Localized basin-like meta-sedimen-
tary and metavolcanic rocks are found within the watershed. Several
large faults and fractures occur within the watershed area which
significantly influence tributary drainage patterns and the lake
configuration. Variable thicknesses of glacial till and outwash
deposits cover bedrock in most places. In general, bedrock deposits
of the western Adirondacks consist of deeply eroded, high grade
metamorphics. Although the bedrock underlying Cranberry Lake is not
considered a suitable geological setting for mercury mineralization,
minor amounts of mercury are associated with Adirondack zinc ore
bodies located in St. Lawrence County. The Belmat-Edwards Zinc
District is located about 30 km from the Cranberry Lake watershed.
In addition, anomalously high mercury concentrations were observed
in soils overlying a known sphalerite vein exposed near Rossie,
St. Lawrence County, located about 70 Km from the Cranberry Lake
watershed (Brown, 1970).

Even though no known sulfide ore mineralization is known to
exist within the Cranberry Lake watershed, this possible source of
the abnormally high concentration of mercury identified in Cranberry Lake smallmouth bass cannot be completely discounted until
more extensive studies are conducted.

Field Study

Sampling stations were established in July, 1978 at six
Cranberry Lake watershed sites (Figure 2). Rain and stream water
samples were collected and analyzed for the following chemical
parameters: pH, alkalinity, sulfate, Ca, Na, Mg, total nitrate
and total mercury.

Figure 3 and Table 1 show the locations of the sampled watersheds and relative drainage areas. Note that sampling was conducted

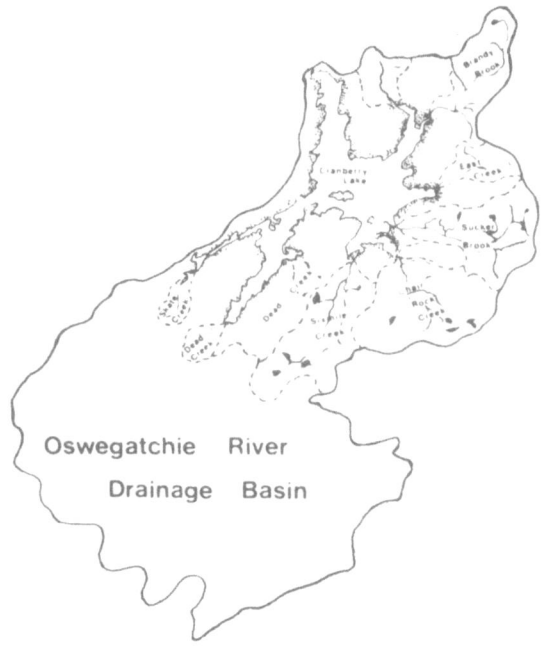

Figure 3.   Cranberry Lake watershed including locations of sampled
            tributaries.

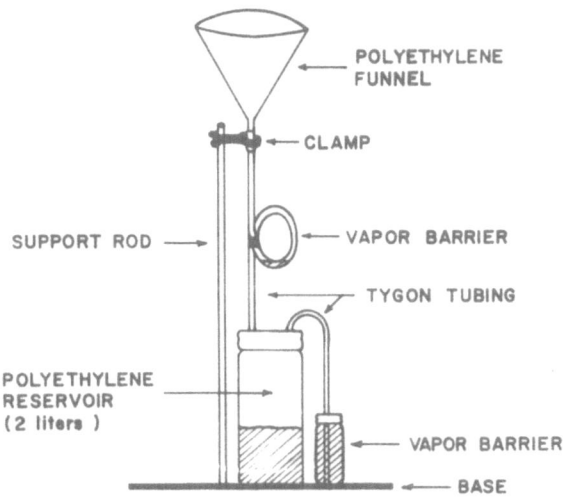

Figure 4.   Modified Hubbard-Brook sampler used to collect precipi-
            tation samples. The diameter of the collecting funnel
            is 28 cm.

Table I.   Cranberry Lake Watershed Collecting Stations Sampled
           During Period July 29, 1978 to November 16, 1978

Area	Km^2	% Watershed	Approximate Flow[1] (CFS)	(m^3/sec)
Cranberry Lake	28.23	7.6	–	
Total watershed	344.8	100.0	307	8.7
East Creek	4.9	1.3	12	0.3
Sucker Brook	16.8	4.5	30	0.8
Chair Rock	12.7	5.8	36	1.0
Six Mile Creek	17.8	4.8	31	0.9
Oswegatchie River @ Wanakena	192.3	51.6	187	5.3

[1] Gregory and Walling (1973)

within about sixty-eight percent of the Cranberry Lake watershed
system.

     Two rain samples were collected at each of five of the six
watershed sampling sites and included samples located beneath a
forest canopy (subsequently referred to as canopy) and samples
located in open fields. One open rain sample was located at
Wanakena; a similar canopy sampler was located at the Oswegatchie
sampling site, both within the Oswegatchie Watershed. A
modified Hubbard-Brook rain sampler was used to collect all rain
samples (See Figure 4). The collecting funnels on the rain samplers were mounted on steel stakes; collecting funnels were
elevated about two meters above ground level. All collected rain
samples were split; one portion was acidified for mercury analyses,
the other portion was saved for other analyses.

     All stream water samples were collected near mid-stream. A
minimum of one liter of sample was collected from each site. On-
site pH measurements were made within ten minutes of sample collection. Each 250 ml water sample concentrated to be analyzed for
mercury was acidified with 1.5 mls of nitric acid at the time of
collection. All samples were kept chilled (4°C) until analyzed.

     In addition to watershed precipitation and stream samples,
Cranberry Lake water, snowpack and sediment core samples were collected (See Figure 2 for locations). Open lake water samples were
collected with a two liter PVC Kemmerer water sampler; samples
were collected at one meter intervals from the surface to lake
bottom.

     Sediment core samples were collected with a one meter, 4.45

cm diameter gravity corer. All collected sediment samples were kept frozen until analyzed. Snow samples were collected at three sites on December 1, 1978, February 8 and 9, 1979, and March 3 and 10, 1979 (See Figure 2 for sampling locations). Each snow sample was collected with a 7.0 cm snow corer. Collected samples were melted, pH analyzed and split into two portions. One aliquot was acidified (1.5 ml to 250 ml) for mercury analyses; the other portion was kept frozen for future chemical analyses.

Fish were collected with gill and trap nets during the period 1969 to 1975 by Department of Environmental Conservation regional fish management units as part of their routine fish management program. Additional fish were collected by the authors during 1978. Each fish was tagged; its length and weight were recorded at time of capture, and they were frozen as soon as possible. The fish were thawed prior to analysis, and species and sex were then determined. Age was determined by reading scale annuli (Lagler, 1956); species and sex were determined at the NYSDEC laboratories.

Chemical Analyses

Mercury concentrations in water were determined by flameless atomic absorption spectrophotometry after pretreatment with sulfuric acid, concentrated nitric acid and potassium permanganate (EPA, 1978). Sediment samples were air-dried, sieved (100 mesh) and digested (Krishnamurty and Reddy, 1975). Whole fish were ground and digested without addition of peroxide, with concentrated sulfuric acid and potassium permanganate (Harris et al., 1970). Sediment and fish samples were analyzed for mercury by flameless atomic absorption spectrophotometry. Detection limits for water samples were 400 ng Hg/l, for sediment 80 μg Hg/Kg dry weight and for fish 0.05 μg Hg/Kg body weight. Water samples that were below the atomic absorption detection limit were analyzed by Atomic Fluorescence spectrometry (Mitchell et al., in prep.). The instrument is based on a Jarrel Ash 1/4m monochromator fitted with a 1000 μm entrance slit and a 250 μm exit slit. Fluorescence radiation was excited using a quartz lamp (Ultra Violet Products Inc., San Gabriel, Calif.). A Hamamatsu R-106 photomultiplier was used with a Hewlett Packard 0-30004, 0-6 mA DC power supply. The photomultiplier signal was amplified and the output displayed on a stripchart recorder. The samples were digested with persulfate reagent and sulfuric acid. The sample was passed through the spectrometer gas train using purified nitrogen. The detection limit for water samples was 25 ng Hg/l. This limitation is introduced by mercury levels in distilled water blanks in the range of 10 to 20 ng Hg/l. All mercury analyses in aqueous samples were conducted within 14 days of sample collection.

All pH measurements were conducted on site utilizing a Model

407A/F Orion pH meter. Sulfate and alkalinity analyses were conducted following Standard Methods (American Public Health Association, 1976); Ca, Na, Mg concentrations were determined by atomic
absorption spectrophotometry (Perkin-Elmer 290B) following EPA
(1978) recommended procedures.

RESULTS AND DISCUSSION

Water Chemistry

Table II summarized the chemical data for all open forest
precipitation samples collected during the period August 28, 1978
to November 16, 1978 for all ten sampling sites. As can be seen
from Table II, calcium concentrations ranged from a low of 0.05
mg/l to a maximum of about 1.2 mg/l. The mean magnesium concentration for all sites and all collection dates was 0.19 mg/l;
samples ranged from 0.02 mg/l to 0.4 mg/l. Mercury concentrations
ranged from <25 ng/l to 27 ng/l. Nitrate, expressed as total
nitrate, ranged from 0.03 to 1.6 mg/l and averaged about 0.94 mg/l
for all samples. The pH of all open rain samples averaged about
3.4 and ranged from about 2.4 to 4.9. These data were collected
during the period August 29, 1978 to March 10, 1979. The lowest
pH value (2.4) recorded was from a sample collected at the
Oswegatchie River - Wanakena site on March 10, 1979. This site
was the only sampling site occupied during the period November
16, 1978 to March 10, 1979 because of its accessibility during
winter months. This low pH value is possibly in error.

Sulfate concentrations ranged from a minimum of < 1 to a
maximum of 12.0 mg/l and averaged about 6.0 mg/l for the seven
month sampling period. Sodium concentrations were normally below
0.25 mg/l and ranged from 0.01 to 0.49 mg/l. The mean value for
the seven month sampling period was 0.27 mg/l.

Forest canopy precipitation chemistry data are summarized in
Table 3. Calcium concentrations in samples collected from the open
precipitation sites were similar to those collected from the canopy
sites although the canopied sites averaged overall higher concentrations (0.05 to 1.9 mg/l). There was a significant difference in
magnesium concentrations between the open and canopy precipitation
samples. Open precipitation concentrations averaged about 0.19
mg/l and ranged from 0.02 to about 0.4 mg/l, whereas canopy samples
averaged about 0.50 mg/l and ranged in concentration from 0.05 mg/l
to a maximum of 1.3 mg/l. Overall, the magnesium concentration of
samples collected from the canopy sites were higher than those collected from the open sites. Canopy mercury concentrations ranged
from <25 ng/l to a maximum of 83 ng/l. Of the seventeen canopy precipitation samples analyzed for mercury, fourteen of the samples
were at <25 ng/l concentrations. The highest concentration of 83

Table II.   Cranberry Lake Open Rain Samples Collected
8/28/78 to 3/10/79

	Range	No. of Samples	$\bar{x}$	SD
Calcium (mg/l)	0.05 - 1.2	22	0.52	0.40
Magnesium (mg/l)	0.02 - 0.4	26	0.19	0.19
Mercury (ng/l)	< 25 - 27.0	28	13(1-25)*	3(5-0)*
Nitrate (mg/l)	0.03 - 1.6	18	0.94	0.48
pH (units)	2.4 - 4.9	27	3.4	0.25
Sulfate (mg/l)	< 1 - 12	22	6.0	2.0
Sodium (mg/l)	0.01 - 0.49	28	0.27	0.15

Table III.   Cranberry Lake Canopy Rain Samples Collected
8/28/78 to 11/16/78

	Range	No. of Samples	$\bar{x}$	SD
Calcium (mg/l)	0.05 - 1.9	26	0.82	0.45
Magnesium (mg/l)	0.05 - 1.3	23	0.50	0.16
Mercury (ng/l)	<25.0 - 83.0	21	17(6-8)*	16(19-13)*
Nitrate (mg/l)	0.94 - 1.2	8	0.74	0.44
pH (units)	2.4 - 4.4	21	3.50	0.30
Sulfate (mg/l)	<7.0 - 16.0	16	8.00	4.00
Sodium (mg/l)	0.07 - 0.41	24	0.17	0.07

Table IV.   Cranberry Lake Watershed Stream Water Quality Data
for Period 7/21/78 to 11/16/78

Parameter	Range	No. of Analyses	$\bar{x}$	SD
Alkalinity (mg/l of CaCO$_3$)	<1.0 - 14.5	24	5.10	4.00
Calcium (mg/l)	0.06 - 1.8	33	1.32	0.38
Magnesium (mg/l)	0.43 - 0.83	36	0.57	0.08
Mercury (ng/l)	<25.0 - 72.00	55	33(27-39)*	28(32-24)*
Nitrate (mg/l)	0.10 - 2.00	30	0.69	0.50
pH (units)	5.30 - 7.00	24	6.5	0.36
Sulfate (mg/l)	5.00 - 12.00	35	8.0	1.40
Sodium (mg/l)	0.44 - 1.20	39	0.88	0.15

* First number is determined by using one-half the detection limit for
  values below the detection limit. Numbers in parenthesis were
  calculated using zero and the detection limit value respectively for
  values below the detection limit.

ng/l was from a sample collected from the Chair Rock site on
September 20, 1978. Nitrate concentrations ranged from 0.94 mg/l
to 1.2 mg/l with a mean value of 0.74 mg/l. There is no significant difference noted between the open and canopy precipitation
nitrate concentrations.

Canopy pH values did not significantly differ from open precipitation values and averaged about 3.50. The lowest value of 2.4
was from a sample collected on September 20, 1978 from the East
Creek watershed sampling site.

Canopy precipitation sulfate concentrations did not differ
appreciably from open samples with concentrations ranging from
<1.0 mg/l to 16 mg/l. Sodium concentrations also did not differ
appreciably from open rain samples although then precipitation
samples did average about 0.10 mg/l greater than canopy samples.

Table IV summarizes Cranberry Lake watershed stream data for
the preiod July 21 to November 16, 1978. During this period
alkalinity ($CaCO_3$) concentrations ranged from <1.0 to 14.5 mg/l with
an average value of about 5.1 mg/l. Calcium concentrations of collected stream samples were significantly above open and canopy precipitation concentrations and ranged from 0.06 to 1.8 mg/l with an
average concentration of about 1.3 mg/l. Magnesium concentrations
ranged from 0.43 to 0.83 mg/l; mean concentration of about 0.6 mg/l
which was approximately equal to the mean magnesium concentration
of canopy precipitation values.

Stream mercury levels varied from a low of < 25 ng/l to as
much as 72 ng/l. Of the forty-two mercury analyses, twenty-three
were above the analytical detection limit of 25 ng/l. In order to
calculate mean values, all samples which were reported as < 25 ng/l
were consistently assessed a 12.5 ng/l value. From available data,
it appears that samples collected on October 3, 1978 and October
10, 1978 contained above average mercury levels. In addition, the
November 16th sampling period also was consistently above mean mercury values for all collected stream samples. Nitrate concentrations ranged from a low of 0.1 to a maximum of 2.0 mg/l, with a
mean value of 0.69 mg/l which was comparable to precipitation
concentrations.

Are these mercury values for Cranberry Lake streams high or
low? They are similar to levels reported for Lake Windemere (12 -
29 ng/l) by Gardner (1978) and the unpublished work of Harris (1971).
on the Oswegatchie River (17 - 72 ng/l) but somewhat lower than
Onondaga Lake, near Syracuse, New York contaminated by chlor-alkali
plant residues. Buller's synoptic survey (1972) of mercury in surface waters of the Adirondack Mountain region yielded no samples
above 500 ng/l total mercury and 100 ng/l soluble mercury.

Stream pH values varied from a low of 5.3 to a high of 7.0 for the sampling period October 3 to November 16, 1978. The pH of all stream waters analyzed during the entire six week sampling period was slightly acidic with a mean value of approximately 6.5; Oswegatchie pH values measured on March 10, 1979 were 4.2 at Wanakena and 4.5 at the Cranberry Lake Village dam (see Figure 2).

Sulfate concentrations ranged from 5 to 12 mg/l and averaged about 8 mg/l which was not significantly below canopy and open precipitation sulfate average concentrations of 8.0 mg/l and 6.0 mg/l, respectively.

Stream sodium concentrations were significantly above precipitation concentrations and ranged from 0.44 mg/l to as much as 1.2 mg/l as compared to average canopy and open precipitation sodium concentrations of 0.17 and 0.27 mg/l.

Snowpack samples were collected from three sites within the Cranberry Lake watershed during the period December 1, 1978 to March 10, 1979 (see Figure 2 for sampling locations). Table V summarizes the snow chemical data. Calcium concentrations ranged from < 0.1 to 7.1 mg/l and averaged about 0.85 mg/l which was comparable to canopy and open precipitation concentrations. Magnesium concentrations also closely paralleled canopy and open precipitation (rain) concentrations and averaged about 0.3 mg/l. Mercury concentrations in all collected snow samples were < 25 mg/l. Nitrate ranged from 0.1 to 1.8 mg/l and averaged about 0.85 mg/l which was comparable to precipitation (rain) concentrations. Snowpack pH values ranged from 2.6 to 5.6, with an average of about 3.6; sulfate concentrations were similar to open and canopy precipitation values and averaged about 5.0 mg/l. The water content of the collected snow samples varied from 3.0 - 5.4 ml/cc and averaged about 4.0 ml/cc.

Open water samples were collected form Cranberry Lake during the period September 20 to November 16, 1978 from the lake surface to depths of nine meters. These data are summarized in Table VI. Lake alkalinity values were significantly lower relative to stream samples collected during the same period. Stream alkalinities averaged about 5.1 whereas lake alkalinities had a mean value of 2.1 mg/l. Calcium lake concentrations were similar to stream values; little to no depth variation in calcium concentrations was observed during the approximate seven and one-half week period. Magnesium lake concentrations also closely paralleled stream water concentrations; stream concentrations averaged about 0.6 mg/l; the lake mean value for all depths was 0.5 mg/l. The mean mercury lake concentration was 29.1 mg/l which compared with the 33 mg/l mean value observed for all stream samples. Lake nitrates were similar to stream concentrations whereas lake pH values were

Table V.   Cranberry Lake Snow Samples Collected During Period
12/1/78 to 3/10/79

	Range	No. of Analyses	$\bar{x}$	SD
Calcium (mg/l)	0.06 - 7.10	12	0.85	1.97
Magnesium (mg/l)	0.08 - 0.92	13	0.30	0.51
Mercury (mg/l)	< 25.00	19	<25.00	--
Nitrate (mg/l)	0.10 - 1.80	16	0.85	0.35
pH (units)	2.60 - 5.60	18	3.60	0.41
Sulfate (mg/l)	1.00 - 8.00	18	5.00	1.10
Water content (ml/cc)	3.00 - 5.40	7	4.04	0.8

Table VI.   Summary of Cranberry Lake Water Quality Data Collected
September 20, 1978 to November 16, 1978 for All Depths

	Range	No. of Analyses	$\bar{x}$	SD
Alkalinity ($CaCO_3$, mg/l)	1.00 - 3.30	5	2.10	1.10
Calcium (mg/l)	0.89 - 1.30	7	1.13	0.19
Magnesium (mg/l)	0.50 - 0.61	6	0.54	0.04
Mercury (ng/l)	<25.00 - 45.00	13	29(25-33)*	13(18-7)*
Nitrate (mg/l)	0.10 - 1.80	5	0.70	0.70
pH (units)	5.90 - 7.50	12	6.90	0.60
Sulfate	5.00 - 12.00	7	7.00	2.40
Sodium (mg/l)	0.78 - 0.86	5	0.80	0.03

Table VII.   Sediment Core Sample Mercury Concentrations

	Core Depth (cm)	Mercury Concentration (mg/kg) range
Sucker Brook - 1	44	0.08 - 0.32
Sucker Brook - 2	32	0.08 - 0.08
Cranberry Lake	32	0.10 - 0.18
Chair Rock	28	0.08 - 0.10
Chair Rock	39	<0.08 - 0.10
Brandy Brook - 1	20	<0.08 - 0.08
Brandy Brook - 2	26	<0.08 - 0.10

* See note from Table II.

slightly higher relative to sampled streams. Open lake pH values
ranged from 5.9 to 7.5 adn averaged about 6.9 compared to a mean pH
stream value of about 6.5.

Sulfate and sodium concentrations of lake and stream water
samples were comparable for all sampling depths and sampling periods.

Seven 4.45 cm diameter core samples and four bottom grab samples were collected from within the Cranberry Lake watershed. Core
lengths which varied from 28 to 44 cm were vertically sectioned at
observed lithologic changes and analyzed for total mercury concentrations utilizing flameless atomic absorption spectrophotometry
following Environmental Protection Agency (1977) procedures. Table
VII summarizes the mercury concentrations of all collected core
samples. The four bulk bottom sediments were all collected from
the same location and varied in mercury content from < 0.08 - 0.10
mg/kg.

As can be seen from Table VII, the mercury concentration in all
seven core samples ranged from < 0.08 mg/kg to a maximum concentration of 0.32 mg/kg.   The maximum concentration (0.32 mg/kg)
occured at a depth of 24-31 cm at the mouth of the Sucker Brook
watershed system. All other samples were generally at concentrations near the 0.08 mg/kg detection limit; no discernible difference
in mercury concentration was observed with depth and the variation
that does exist is more likely attributed to sediment type
rather than to a depth (time) factor.

Cranberry Lake watershed stream concentrations of calcium,
magnesium, sodium and pH values were generally higher than canopy
and open precipitation sampling during the sampling period July
21, 1978 to November 16, 1978. Figure 5 represents a plot of
stream water to canopy and open precipitation (stream: precipitation) ratios for calcium, magnesium, sodium, pH and sulfate concentrations. With the exception of sulfate, stream values were
generally greater than precipitation concentrations. It is recognized that these data are based on instantaneous stream samples
compared to composite rain samples, although both rain and stream
samples were collected within 24 hours of each other. Throughout
the sampling period, precipitation pH values were consistently
lower than stream and lake waters. Note also the considerable
range in open precipitation magnesium concentrations relative to
the more consistent values observed for the canopy samples.

Figure 6 is a plot of mean pH and alkalinity values for all
streams against time in days. As can be seen from Figure 6, pH
and alkalinity changes were closely paralleled during the study
period October 3, 1978 to November 16, 1978. A comparison of
stream and precipitation magnesium and mercury variations

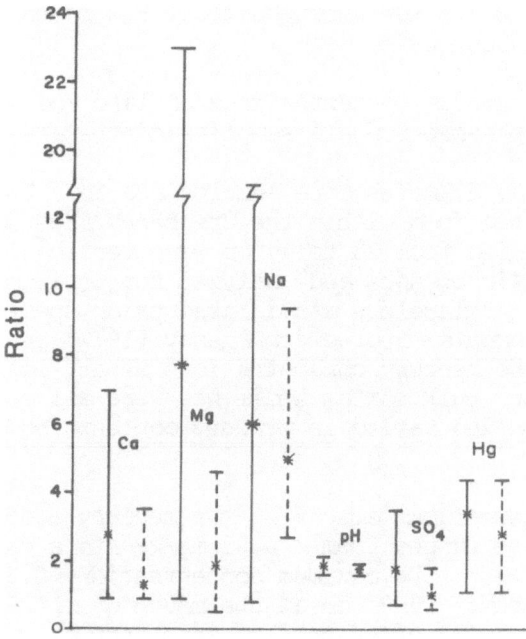

Figure 5. Ranges and averages of the ratio of stream chemistry to
           precipitation chemistry for six constituents. Dashed lines
           represent total fallout and solid lines represent forest
           throughfall.

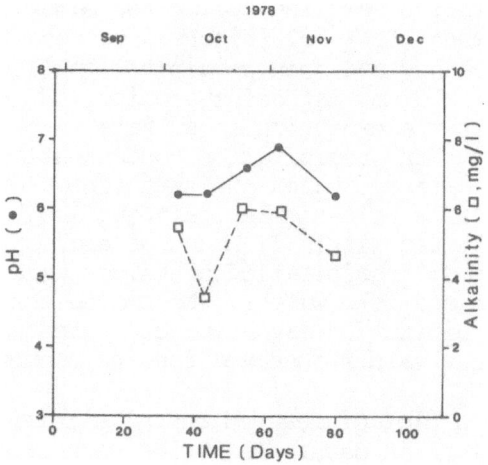

Figure 6. Plot of average pH and total alkalinity (as $CaCO_3$) for
           all stream water samples collected and analyzed from the
           Cranberry Lake watershed versus time.

through time is illustrated in Figure 7. It is interesting to note the wide fluctuation in magnesium open and canopy precipitation concentrations through time whereas stream magnesium concentrations were less varied. Mercury stream concentrations also varied widely from < 25 ng/l to more than 50 ng/l.

Fish

Mercury levels for smallmouth bass from several lakes around the State of New York were compared. Smallmouth bass are used for comparison based on the following assumptions: First, smallmouth bass represented the largest sample size of all species and displayed the greatest Statewide distribution as a result of its wide range of habitat preference. Secondly, smallmouth bass are considered a top level predator and game species, likely to contain appreciable amounts of mercury if present in the environment.

Figure 8 is a bivariate plot, depicting mercury versus length for smallmouth bass in several New York State lakes. The Adirondack Lakes category was created to provide a statistically adequate sample size for those Adirondack lakes which tended to have similar physical and chemical characteristics, but lacked adequate sample sizes of smallmouth bass on an individual lake basis. It is apparent that smallmouth bass from Onondaga Lake and the combined waters of Cranberry Lake and Stillwater Reservoir exhibit much higher average mercury values for a given length than other New York State lakes.

The higher mercury values in Onondaga Lake fish are caused, for the most part, by mercury contributions from a chlor-alkali plant during the 1960's. Other areas identified, such as Lake Ontario, Lake Champlain, and the St. Lawrence River either have or have had in the past sources of mercury contamination. The high levels of mercury found in the smallmouth bass of Cranberry Lake and Stillwater Reservoir cannot be so easily interpreted as those in the other lakes described, due in part to the sites' remoteness and absence of anthropogenic mercury sources.

Species Comparison

When comparing mercury concentrations in several different fish species, it becomes necessary to differentiate between the broad categories of species composition; fish range from being planktivores (eg., cisco; Coregonus) to omnivores (eg., bullhead; Ictalurus) to predators (eg., black bass; Micropterus).
At each level in this progression, the fish, by virtue of its food composition, is exposed to increasing amounts of mercury.

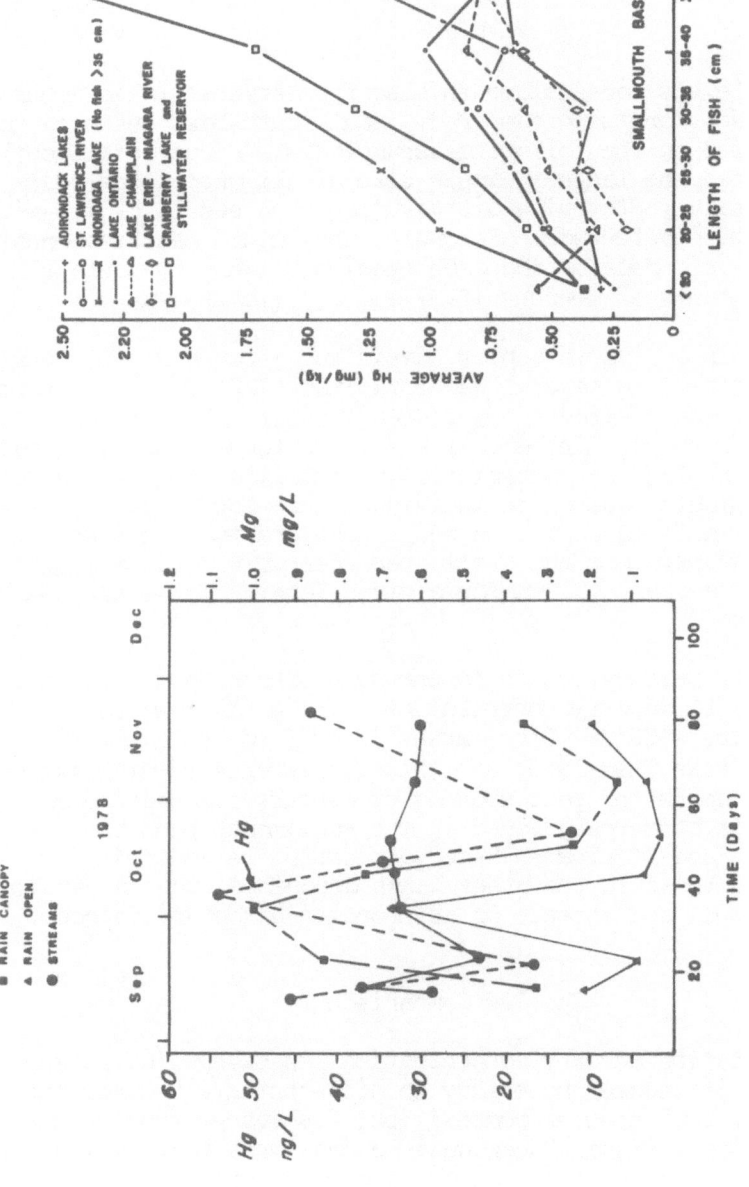

Figure 8.   Average mercury concentrations versus length
class in New York State
Small mouth bass.

Figure 7.   Comparison of stream and precipitation
magnesium and mercury concentrations with
time.

Hamilton (1971) summarizes the biological magnification of mercury in the aquatic food chain. It appears evident that at each trophic level the organism is being exposed to a larger dose of mercury contamination, resulting from its food composition and quantity (the latter of which, is governed somewhat by metabolic rate). A range from 0.01 mg/kg for planktivores to 5.82 mg/kg for predators is noted. Jernelov (1972) and Jernelov and Lann (1971) estimated that fish at the higher trophic levels obtain half of their mercury burden through their food and the remainder from the water; fish at the lower trophic levels receive as much as 75 percent of their mercury burden from the water. However, Hannerz (1968) reported that animals bioaccumulate mercury in direct relation to such factors as metabolic rate and food habits rather than trophic position.

Figure 9 is a bivariate plot which represents a composite of various fish species from Cranberry Lake and Stillwater Reservoir. As expected, higher trophic level fish (smallmouth bass and splake) have much higher mercury values than lower trophic fish (brown bullhead Ictalurus nebulosus and white sucker Catostomus commersoni). There also appears to be a sharp upward trend with mercury and length in fish of higher trophic levels, but not so pronounced a trend with fish of lower trophic levels. The inconsistency realized in mercury levels for fish of similar length, but different species, therefore, may be due to the fishes' food preference (trophic level). Adult white suckers for instance, are primarily bottom feeders, feeding on insect larvae (eg. chironomid, tricopteran, etc.) and molluscs. Adult smallmouth bass, however, are predatory fish which feed on a greatly diversified forage base including, fish (eg. white suckers, golden shiners Notemigonus crysoleucas, yellow perch Perca flavescens, etc.), crayfish and large invertebrates.

Hannerz (1968) and Jarvenpaa et al. (1970) have shown that fish take in mercury from food (via the gastrointestinal tract) and directly from the water via the gills and skin. Uptake, as described above, is a function of metabolic rate and therefore should be correlated with it (Fagerstrom, 1973). Mercury uptake and bioconcentration have been correlated with body weight and water temperature under experimental conditions (Freitas et al. 1972, MacLeod et al. 1973, and Cember et al. 1978).

In fish, one result of metabolic turnover is growth (weight and length). Since weight is an increase in growth in three dimensions and length is an increase in only one dimension, then weight should be more linearly related to metabolic turnover than length. However, Olsson (1976) in studying mercury contaminated northern pike from Swedish lakes, found the highest correlation coefficients were with mercury levels and length. This relation-

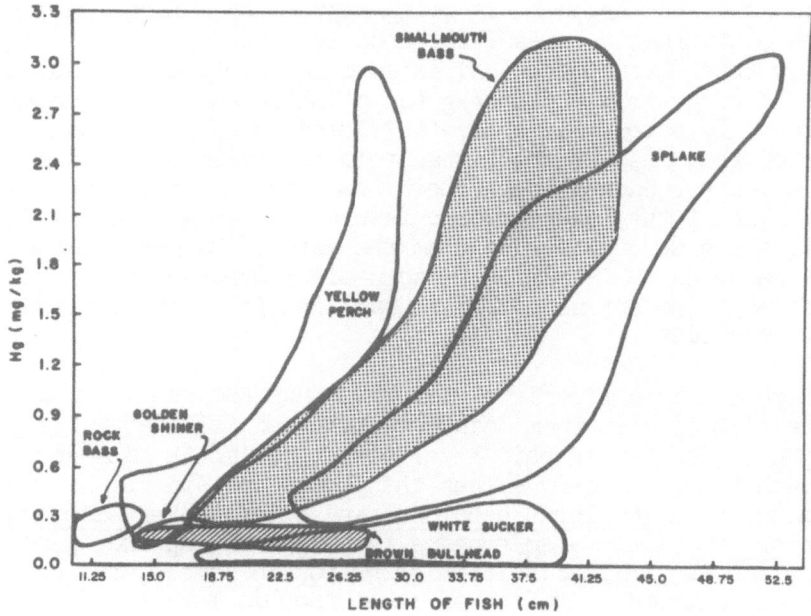

Figure 9. Scatter diagrams of mercury concentration versus length
for seven species of fish from Cranberry Lake and Stillwater Reservoir.

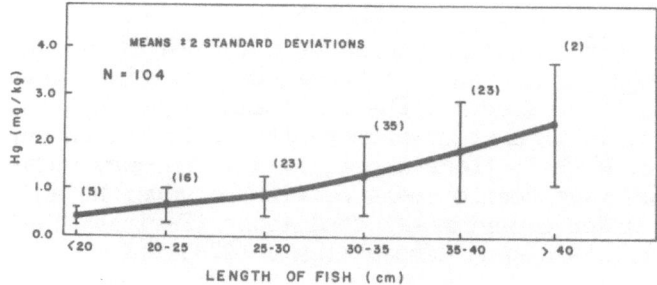

Figure 10. Plot of average mercury concentrations versus length
class for smallmouth bass from Cranberry Lake and Stillwater Reservoir. Vertical bar represents ± two standard
deviations. Number in parentheses represents sample size
for that length class.

ship was attributed to the fact that when fish starve (catabolism $\geq$ anabolism) they rarely decrease in length and to some extent continue to ventilate and consume food, resulting in a continual uptake of mercury.

Figure 10 depicts the mercury - length relationship in smallmouth bass for Cranberry Lake and Stillwater Reservoir. There is an obvious upward trend with mercury throughout the range of lengths depicted. It is also important to note that the increase of mercury with length is curvilinear, increasing in larger increments as the fish grows.

A high correlation (r = .92 and .96) was observed between age and growth (length) of smallmouth bass from Cranberry Lake and Stillwater Reservoir, respectively. It may be expected that a longer exposure time (greater age) would lead to a higher mercury concentration. Figure 11 depicts the relationship between length, age, and mercury concentration for smallmouth bass in Cranberry Lake. Confidence limits for age groups could not be fitted, due to a small number of fish in some age groups. An increase in mercury concentration with age and length is, however, evident. In addition, there seems to be little overlap in length between fish of different age groups, indicating a relatively constant growth (productivity) for Cranberry Lake smallmouth bass.

The fact that certain fish species (eg. smallmouth bass) in Cranberry Lake exhibit disproportionately high levels of mercury in their flesh when compared to other species (eg. white sucker), is probably related to the fish's status in the food chain. Table VIII summarizes the cumulative uptake rates for mercury in fish from Cranberry Lake and Stillwater Reservoir. Uptake rates were derived by dividing the mercury concentration by the age of the fish, resulting in mg/kg body weight-year. Whereas this method of calculating uptake rates is useful in determining trends of mercury accumulation in fish for a number of successive year classes, it does not reflect true instantaneous accumulation rates from year to year due to the nature of the calculation. It is apparent that average uptake rates for fish in both lakes increases as one proceeds from lower to higher trophic levels. A lack of fish from a diverse number of age groups precludes the use of uptake rates in defining year class composition for most fish species analyzed.

Smallmouth bass in Cranberry Lake indicate an upward trend in mercury uptake from age IV to age IX fish. The discrepancy observed in age III samllmouth bass is attributed to the mercury analysis of only one fish. Samllmouth bass in Stillwater Reservoir display a slightly greater increase in mercury accumulation with age when compared to Cranberry Lake. More mercury may be available to the fish in Stillwater Reservoir via food composition or increased mercury availability in the environment.

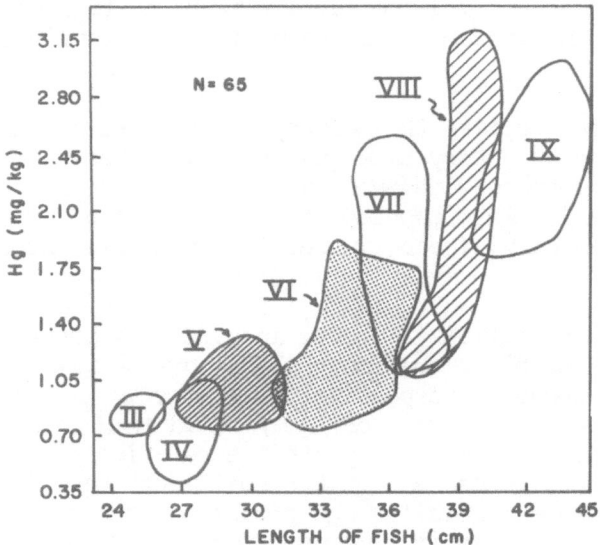

Figure 11. Scatter diagram of mercury concentration versus
length and age for smallmouth bass from Cranberry
Lake. "V" represents a five-year old individual.

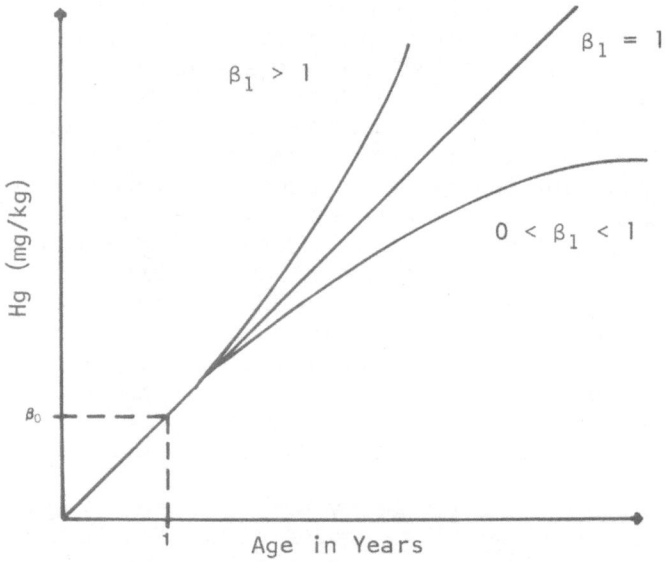

Figure 12. Plot of the empirical relationship between fish
mercury concentration and age for various values
of β_1 from Equation 1.

Table VIII. Mercury Cumulative Uptake Rates for Fish
in Cranberry Lake and Stillwater Reservoir
(in mg Hg/kg body weight per year)

Cranberry Lake

Smallmouth Bass
   IX (age class)   0.26 (3)
   VIII   0.25 (6)
   VII   0.23 (19)
   VI   0.21 (17)
   V   0.19 (14)
   IV   0.20 (5)
   III   0.28 (1)
   average   0.23 (65)
Yellow Perch, Age II 0.16 (2)
Rock Bass, Ages IV-V 0.06 (2)
Pumpkinseed, Age III 0.04 (1)
Golden Shiner, Ages
   III-IV   0.05 (4)
White Sucker, Ages
   I-IV   0.05 (5)

Stillwater Reservoir

Splake, Ages III-IV 0.51 (7)
Smallmouth Bass
   V   0.32 (4)
   IV   0.27 (10)
   III   0.21 (17)
   II   0.19 (8)
   average   0.25 (39)
Yellow Perch,
   Ages V-VII   0.20 (7)

Trophic Level ───→ (vertical axis between columns)

( ) = Number of Fish

Since the cumulative mercury uptake rate damps out the actual trend in mercury accumulation from year to year, an instantaneous mercury uptake rate for a given age class of fish can be calculated assuming the following functional form between mercury content (Hg, in mg/kg body weight) and age (A, in years):

$$Hg = \beta_0 \, A^{\beta_1} \qquad\qquad (EQ\text{-}1)$$

The parameter β_0 is numerically equivalent to the mercury content of a one-year old fish, and the parameter β_1 is a power constant. If $\beta_1 < 0$ then the mercury content decreases as the fish ages; if $\beta_1 = 0$, there is no trend with age; and if $\beta_1 > 1$, mercury increases at a fixed rate per year.

If β_1 is significantly greater than unity, the rate of uptake of mercury increases as the fish ages, possibly indicating a change in food preference of less efficient conversion of food to biomass with a fixed proportion of mercury being retained from food (Figure 12). An instantaneous rate (in mg Hg/kg-yr) between two year classes can be calculated by evaluating the integral of the differential form of Equation 1 and dividing by the change in age:

ϕ = Instantaneous Uptake Rate   (EQ-2)

$$\phi \left.\right|_{a_1}^{a_2} = \frac{1}{a_2 - a_1} \int_{a_1}^{a_2} \beta_0 \beta_1, A^{\beta i - 1} \, dA$$

which reduces to evaluating Equation 1 for successive year classes and subtracting the results:

$$\phi \left.\right|_{a_1}^{a_2} = \frac{\beta_0}{a_2 - a_1} (a_2^{\beta_1} - a_1^{\beta_1})$$ (EQ-3)

Using regression analysis, β_0 and β_i can be evaluated for smallmouth bass:

1) Cranberry Lake $\beta_0 = 0.13$, $\beta_1 = 1.28$, $r^2 = 0.60$, $n = 65$.
2) Stillwater Reservoir $\beta_0 = 0.12$, $\beta_1 = 1.58$, $r^2 = 0.83$, $n = 39$.

This indicates that one year old smallmouth bass from each water body have essentially the same mercury content. However, instantaneous uptake rates increase as the fish ages, since $\beta_1 > 1.0$. We believe that the change in uptake rate results from a change in food preference from zooplankton and benthic invertebrates to forage fish. Since the latter prey has been in the lake for a longer period of time than the former, its mercury content should be higher. Consequently, fish species of lower trophic levels (white sucker) would, after reaching maturity, be expected to accumulate mercury at a somewhat constant rate ($\beta_1 \leq 1.0$), which is evidenced by its low mercury uptake rates (Figure 9 and Table VIII). Conversely large, relatively fast growing fish of high trophic status, (splake) exhibit high mercury uptake rates ($\beta_1 \gg 1.0$) associated with a high metabolic turnover and food preference (eg. fish).

Mercury Budget

Data presented in the Results section allow calculation of a very rough mercury budget for Cranberry Lake and its watershed. The mercury budget is based on the average hydrologic budget presented in Figure 13 (NYSDEC, 1978). The mercury budget was calculated by multiplying the annual water flows by average mercury concentrations. There are obvious errors inherent in this type of calculation, the most severe being that the average mercury concentrations measured in the field differ from the actual central tendency. Only additional field sampling can resolve this issue.

The annual mercury budget in Cranberry Lake is presented

CHANGE IN LAKE VOLUME	+ 6.49
OUTFLOW FROM LAKE	321.45
EVAPORATION FROM LAKE	21.11
PRECIPITATION ON LAKE	32.07
PRECIPITATION ON WATERSHED	391.65
INFLOW TO LAKE	316.98
CHANGE IN WATER TABLE @ BRASHER FALLS, NY	+ 0.70
"EVAPOTRANSPIRATION FROM WATERSHED"*	73.97

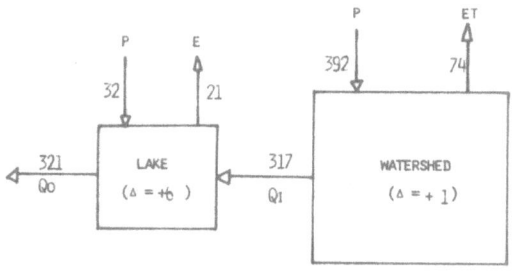

* - THIS IS BY DIFFERENCE

AVERAGE VOLUME =	52.24
HYDRAULIC RETENTION TIME =	0.15 YR (54 DAYS)

Figure 13.   Annual Cranberry Lake Hydrologic Budget (NYSDEC, 1978), in $10^6 m^3$ water.

OUTFLOW FROM LAKE	9.34
PRECIPITATION ON LAKE	0.42
PRECIPITATION ON WATERSHED	7.06
INFLOW TO LAKE	10.03
AMOUNT IN LAKE	1.52
AMOUNT IN TOP 5 CM OF ALKE SEDIMENT	27.60
AMOUNT IN FISH BIOMASS*	≈ .70
NET ACCUMULATION BY FISH PER YEAR*	≈ .003

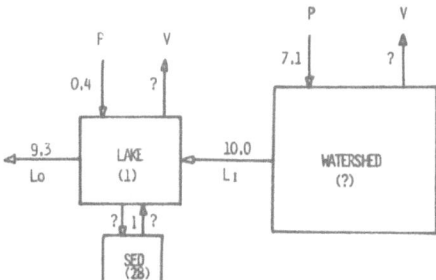

* BASED UPON A FISH BIOMASS OF 11 KG/HA AND AN ANNUAL PRODUCTION OF 2 KG/HA-YR.

Figure 14.   Annual Cranberry Lake Mercury Budget, in Kilograms Mercury

in Figure 14 with annual fluxes presented in kg Hg/yr. Additionally, the amount of mercury (kg) in the lake water and the top 5
cm of lake bottom sediments are shown in parentheses.

The major input of mercury to the lake and watershed is from
atmospheric fallout and is 7.5 kg Hg/yr. This compares with the
desktop work on Cranberry Lake by Eisenbud et al., (1978) of 70
kg Hg/yr. and the work of Harris (NYSDEC, 1978) which yields an
atmospheric deposition rate at Cranberry Lake during 1973-74 of
17.6 kg Hg/yr.

Surface runoff contributed 4.2 kg Hg/yr to Cranberry Lake,
according to Eisenbud,with a range of 0.8 to 9.8 kg Hg/yr. We
estimate 10.0 kg Hg/yr. If the watershed retention coefficient
(ρ) is defined as

$$\rho = \frac{\text{annual input of Hg to watershed from atmosphere}}{\text{annual outflow of Hg from watershed in surface water}}$$

our study yields $\rho = 0.70$, indicating that there is a positive net
contribution of mercury from the Cranberry Lake watershed.
Eisenbud's work yields $\rho = 16.7$ indicating the watershed would
retain most of the mercury contributed by atmospheric fallout.

No information is available on volatilization of mercury
from either the lake or watershed. There is also no estimate
of exchange between lake water and sediment. Work is in progress
to evaluate these fluxes. Since Cranberry Lake's seasonal thermo-
cline is quite deep (10m), most of its bottom waters are rarely
anoxic. (LaRow, unpublished data). Thus, considerable methylation of mercury should be occurring in the bottom sediments.
Samples collected during Fall, 1978 indicate little difference
between the surface and bottom water mercury concentration.
For comparison, the mercury in the top 5 cm of bottom sediments
exceeds the amount of mercury in the lake water by a factor of
18.

LaRow's unpublished work on Cranberry Lake (1975-1978) shows a
lake of low productivity. The lake is occasionally thermally statified.
rarely with anoxic hypolimnetic waters. During the summer (June-
September) chlorophyll a ranges from 2-8 µg/l, soluble reactive
phosphorus ranges from 2-5 µg/l and Secchi disc depths average
2.7 m. Using the empirically derived Morphoedephic Index (MEI) of
Ryder et al., (1974) with a total dissolved solids content of about
8 mg/l and a mean depth of 1.9 m, the MEI is 4.3, indicating a
potential fish production of 2 kg/ha-yr. Fish biomass was estimated to be 11 kg/ha yr. Given these estimates, the fish play a
numerically insignificant role in the Cranberry Lake mercury budget (Figure 13).

SUMMARY AND CONCLUSIONS

Despite ten years of study, the source of mercury and the mechanism responsible for mercury bio-availability in the Cranberry Lake watershed remains unresolved.

Although similar studies have linked mercury levels in fish to acid precipitation in Canada (Brouzes et al., 1977) and Scandanavia (Landner and Larsson, 1972), the mechanism remains obscure. The problem seems to be relating field data to laboratory bioassay studies.

In addition to mercury budget studies initiated at Cranberry Lake during 1978, more detailed work will be conducted during the next two years related to 1) establishing the mercury history of the lake through collection of more sediment cores, 2) collecting and analyzing fish from lakes proximate to Cranberry Lake and 3) analyzing other biological material from Cranberry Lake, including plankton, benthic invertebrates and fish gut contents, for mercury.

It cannot be concluded that direct atmospheric sources are responsible for elevated mercury levels in either lake water or fish flesh. However, it seems likely that the acidity of precipitation is somehow modifying the chemical nature of mercury present in Cranberry Lake water, availing the mercury to the food web.

ACKNOWLEDGMENT

The authors wish to thank Mr. Earl Harris, formerly of the New York State Department of Environmental Conservation, for providing much of the early data on mercury in Cranberry Lake rainfall and fish. We also wish to thank Mr. Ralph Karcher and Dr. Ronald Sloan for mercury analysis of fish and Mr. Robert Weinbloom for the mercury analysis of sediments. We are also indebted to Dr. Edward Ketchledge of Syracuse University, School of Environmental Sciences and Forestry for making equipment and facilities available to us. Mr. Patrick Hanley and Mr. Jeff Davis did the graphics and Miss Arlene Crawford, Miss Diane Myers, and Ms. Marilyn Munger typed the manuscript.

REFERENCES

American Public Health Association, 1976, Standard Methods for the Examination of Water and Wastewater, Fourteenth Edition, Washington, D.C., 1193 pp.

Bailey, E.H., Snavely, P.D., Jr. and D.E. White, 1961, Chemical Analyses of Brines and Crude Oil, Cymric Field, Kern County, California: U.S.G.S. Prof. Paper 424-D, p. D306-D309.

Bisogni, James J. and Alonzo Lawrence, 1975, Kinetics of Mercury Methylation in Aerobic and Anaerobic Aquatic Environments. Journal of the Water Pollution Control Federation, Vol. 47, pp. 135-152.

Boulton, P.J. and L.J. Hetling, 1972, A Statistical Analysis of the Mercury Content of Freshwater Fish in New York State, New York State Department of Environmental Conservation Technical Report No. 19, Albany, New York, 16 pp.

Brouzes R.J.P., R.A.N. McLean, G.H. Tomlinson, 1977, The Link Between pH of Natural Waters and the Mercury Content of Fish. Domtar Research Center Report, Senneville, Quebec, Canada, 37 pp.

Brown, C.E., 1970, A Sphalerite Vein and Associated Geochemical Anomalies in St. Lawrence County, New York: U.S.G.S. Prof. Paper, 700-D, pp. 162-168.

Buddington, A.F. and B.F. Leonard, 1962, Regional Geology of the St. Lawrence County Magnetite District, Northwest Adirondacks, New York: U.S.G.S. Prof. Paper 376, 145 p.

Bulkin, G.A., 1962, The Geochemistry of Mercury in the Crimean Highlands: Geokhimiza, p. 1079-1087; translation in Geochemistry, p. 1219-1230, 1962.

Buller, W., 1972, Natural Background Concentration of Mercury in Surface Water of the Adirondack Region, New York, United States Geological Survey Professional Paper 800c, pp. C-233 to C-238.

Cember, H., E.H. Curtis, 1978, Mercury Bioconcentration in Fish: Temperature and Concentration Effects, Environ. Pollut. 17, pp. 311-319.

Cogbill, C.V. and G.E. Likens, 1974, Acid Precipitation in the Northeastern United States: Water Resources Research, Vol. 10, p. 1133.

Cronan, C.S. and C.L. Schofield, 1979. Aluminum Leaching Response to Acid Precipitation: Effects on High Elevation Watersheds in the Northeast: Science 204, pp. 304-305.

deFreitas, A.S.W., B. Case, J.S. Hart, P. Clay, 1972, Distribution and Transport of Persistent Chemicals in Flowing Water Ecosystems, Ottawa, River Programme, National Research Council of Canada, Ottawa, Interim Report No. 1, Report 13.

Eisenbud, M., D. Axelrod, G.C. Butler, T. Clarkson, T. Hullar, E.J. Massaro, G. Reader, C. Schofield and P. Whitney, 1978, The Health Implications of Methyl Mercury in Adirondack Lakes, New York State Department of Health, 41 pp.

Fagerström, T., B. Asell, 1973, Ambio, 2, 5, 164.

Fagerström, T., B. Asell, A. Jernelov, 1974, Oikos, 25, 14.

Farrah, H., Pickering, W.F., 1977, The Sorptions of Mercury Species by Clay Minerals. Water, Air and Soil Pollution, Vol. 9, pp. 23-31.

Fleischer, Michael, 1970, Summary of the Literature on the Inorganic Geochemistry of Mercury, In Mercury in the Environment, U.S.G.S. Prof. Paper 713, pp. 6-13.

Furukawa, K., and K. Tonomura, 1971, Enzyme Systems Involved in the Decomposition of Phenyl Mercuric Acetate by Mercury-Resistant Pseudomonas, Agricultural and Biological Chemistry 35, pp. 604-610.

Furukawa, K., and K. Tonomura, 1972, Metallic Mercury Releasing Enzyme in Mercury-Resistant Pseudomonas, Agricultural and Biological Chemistry 36, pp. 217-226.

Galloway, S.N., Likens, G.E., and E.S. Edgerton, 1976, Acid Precipitation, pH and Acidity, Science, Vol. 194, p. 722.

Gardner D., 1978, Dissolved Mercury in the Lake Windermere Catchment Area, Water Research, V12, pp. 573-575.

Gladney, E.S. and G.E. Gordon, 1978, Coal Combustion, Source of Toxic Elements in Urban Air? J. Environ. Sci. Health, Al (7), pp. 481-491.

Glooschenko, W.A., 1969, Accumulation of ^{203}Hg by the marine diatom Chaetoceros costatum, Journal of Phycology 5:224-226.

Gregory, K.J. and D.E. Walling, 1973, Drainage Basin Form and Process, John Wiley and Sons, New York, New York, 456 pp.

Hamilton, A.L., 1971, Accumulation of Mercury in Fish Food Organisms. pp. 73-90, Proceedings of the Symposium on Mercury in Man's Environment, 15-16 February 1971, Ottawa, Canada, Royal Society of Canada.

Hannerz, L., 1968, Experimental Investigations on the Accumulation of Mercury in Water Organisms. Report on the Institute of Freshwater Research Drottningholm 48:120-176.

Harris, E.J., 1971, Unpublished data, New York State Department of Environmental Conservation, Albany, New York.

Harris, E.J., R.W. Karcher, Jr., and J. Praznik, 1970, The Determination of Microgram Quantities of Mercury in Fish by Atomic Absorption, New York State Department of Environmental Conservation Technical Report, Albany, New York, 13 pp.

Hem, J.D., 1970, Chemical Behavior of Mercury in Aqueous Media, in Mercury in the Environment, U.S.G.S. Survey Prof. Paper 713, pp. 19-22.

Järvenpää, T., M. Tillander, J.K. Miettinen, 1970, Suomen Kemistilehu, B 43, 439.

Jensen, S., A. Jernelov, 1969, Biological Methylation of Mercury in Aquatic Organisms, Nature 223, pp. 753-754.

Jernelov, A., 1972, Mercury and Food Chains, pp. 174-177, Environmental Mercury Contamination, edited by R. Hartung and B.D. Dinman, Ann Arbor, Mich., Ann Arbor Science Publishers.

Jernelov, A., H. Lann, 1971, Mercury Accumulation in Food Chains, Oikos 22:403-406.

Jernelov, A., L. Landner, T. Larsson, 1975, Swedish Perspective on
    Mercury Pollution, Journal Water Pollution Control Federation,
    Vol. 47, n. 4, pp. 810-822.
Joensuu, O., 1971, Fossil Fuel as a Source of Mercury Pollution,
    Science, Vol. 172, pp. 1027-1028.
Khalid, R.A., R.P. Gambull, W.H. Patrick, Jr., 1975, Sorption and
    Release of Mercury by Mississippi River Sediment as Affected
    by pH and Redox Potential, ERDA Symposium on Biological
    Implications of Metals in the Environment, pp. 297-314.
Krishnamurty, K.V., M.M. Reddy, 1975, The Chemical Analysis of Water
    and Sediments in the Genesee River Watershed Study, New York
    State Department of Health, Albany, New York, 688 pp.
Kudo, A., J.S. Hart, 1974, Uptake of Inorganic Mercury by Bed
    Sediments, J. Environ. Quality, Vol. 3, No. 3, pp. 273-278.
Kudo, A., D.C. Mortimer, J.S. Hart, 1975, Factors Influencing
    Desorption of Mercury from Bed Sediments, Can. J. Earth Sci.,
    Vol. 12, pp. 1036-1040.
Lagler, K.F., 1956, Freshwater Fishery Biology, Second edition, Wm.
    C. Brown Co., Dubuque, Ia., 421 pp.
Landner, L., P.O. Larsson, 1972, Biological Effects of Mercury Fallout
    into Lakes from the Atmosphere, Report on the Swedish Institute
    for Water and Air Research (IUL), Stockholm, Sweden, 18 pp.
Lange, A.D., G.M. Forbes, eds., Handbook of Chemistry, 10th Ed.,
    N.Y. McGraw-Hill, 2,000 pp.
LaRow, E., 1977, Unpublished data, Cranberry Lake Biological Field
    Station, School of Forestry and Environmental Sciences,
    Syracuse University, Syracuse, New York.
Likens, G.E., F.H. Bormann, N.M. Johnson, 1972, Acid Rain,
    Environment, 14, p. 33.
Likens, G.E., F.H. Bormann, 1974, Acid Rain, A Serious Regional
    Environmental Problem, Science, Vol. 184, pp. 1176-1179.
MacLeod, J.C., E. Pessah, 1973. Temperature Effects on Mercury
    Accumulation, Toxicity and Metabolic Rate in Rainbow Trout
    (Salmo gairdneri), Jour. Fish. Res. Bd., Canada, 30, 485.
Mitchell, D.G., K.M. Aldous, F. Ryan, W.M. Mills, in preparation,
    The Determination of Mercury in Whole Blood and Urine by Atomic
    Fluorescence Spectrometry, New York State Department of Health,
    11 pp.
New York State Department of Environmental Conservation, 1978,
    Cooperative Agreement for State Toxic Substances Control
    Projects, Proposal to United States Environmental Protection
    Agency, 112 pp.
Olsson, M., 1976, Mercury Level as a Function of Size and Age in
    Northern Pike, One and Five Years After the Mercury Ban in
    Sweden, Ambio 5, 2.
Panov, B.S., 1959, Mercury in Rocks of the Southwestern District of
    the Donets Basin, Donets Ind. Inst. Trudy 37, p. 149-152,
    (Russian), Chem. Abs. 55, p. 9192, 1961.

Reimers, R.S., P.A. Krenkel, 1974, Kinetics of Mercury Adsorption and Desorption in Sediments, Water Pollution Control Federation, Vol. 46, No. 2, pp. 352-365.

Ryder, R.A., S.R. Kerr, K.H. Loftus and H.A. Regier, 1974, The Morphoedaphic Index, a Fish Yield Estimator, Review and Evaluation, J. Fish. Res. Bd. Canada, V31, pp. 663-688.

Schofield, C. L., 1973, The Ecological Significance of Air Pollution induced Changes In the water Quality of Dilute Lake Districts in the Northeast: Trans. N.E. Fish and Wildlife Conf., 1972, Ellenville, N.Y., p. 98-112.

Schofield, C.L., 1975, Lake Acidification in the Adirondack Mountains of New York: Causes and Consequences: Proc. First International Symp. Acid Precipitation and the Forest Ecosystem: USDA Forest Services General Technical Paper NE-23.

Schofield, Carl L., 1977, Acid Snow-melt Effects on Water Quality and Fish Survival in the Adirondack Mountains of New York State: Research Project Completion Report, Project No. A-072-NY, OWRT, 27 p.

Schottel, J., A. Mondal, D. Clark, and S. Silver, 1974. Volatilization of Mercury and Organomercurials Determined by Inducible R-Factor Systems in Entire Bacteria: Nature 251, pp. 335-337.

Shan-Ching Tsai, 1975. Importance of Water pH in Accumulation of Inorganic Mercury in Fish. B. Envir. Contam. and Toxc. 13(2): 188.

Suggs, J.D., D.H. Petersen, and James B. Middlebrook, J., 1972, Mercury Pollution Control in Stream and Lake Sediments: Water Pollution Control Research Series, U.S. EPA, 39 p.

Summers, A.O., and S. Silver, 1972. Mercury Resistance in a Placmid-Bearing Strain of Escherichia coli: Journal of Bacteriology, 112, pp. 1228-1236.

Summers, A.O. and L.I. Sugarman, 1974. Cell-free Mercury Strain of Escherichia coli: Journal of Bacteriology, 119, pp. 242-249.

Tezuka, T. and K. Tonomura, 1976. Purification and Properties of an Enzyme Catalyzing the Splitting of Carbon-Mercury Linkages from Mercury-Resistant Pseudomonas K-62 Strain: Japanese Journal of Biochemistry 80, pp. 79-87.

United States Environmental Protection Agency, 1978, Methods for Metals in Drinking Water, Methods for Chemical Analysis of Water and Wastes, Cincinnati, Ohio.

Wollast, R., G. Billen and F.T. Mackenzie, 1975. Behavior of Mercury in Natural Systems and its Global Cycle, in Effects of Heavy Metals and Organolalogen Compounds, pp. 145-166; Ecological Toxicology Research, Proceedings of a NATO Science Comm. Conf., A.D. McIntyre and C.F. Mills, eds., New York, Plenum Press.

Wood, J.M., 1974. Biological Cycles for Toxic Elements in the Environment: Science 183, pp. 1049-1052.

DISCUSSION

TOMLINSON: One thing that I think is interesting is that
Cranberry Lake is a humic lake, a brown lake. In Quebec and
some of the Ontario lakes, it appears that when this brown
color of the water is combined with a low pH, fish show higher
levels of mercury than in clear lakes of similar pH. When Dr.
Jernelov's co-worker's first reported the combination of
elevated mercury in fish with low pH of water, it was noted
that the lakes were brown in color.[*] It could be very well
that the mercury coming in via the rain is methylating more
rapidly in these waters and the methylmercury becomes available
to fish. So when you are checking these other lakes, you want
to look at the color. The Stillwater Reservoir is quite clear,
is it not?

[*]Lardner, L., Larsson, P.O., IVL Report B-115, Swedish
Institute for Water and Air Pollution Research, Stockholm, 1972.

BLOOMFIELD: Essentially the same. One interesting thing about
Cranberry Lake is that its large tributary, the Oswegatchie
River, has a fairly flat, boggy type of watershed. The other
tributaries we sampled have fairly steep watersheds and the
streams are less brown in color. Yet there is no difference in
the stream mercury levels that we can detect.

What we may be seeing here is mercury as an indicator of
how fast acid precipitation is affecting the fishery of the
lake. Toxicity of mercury is probably not a factor, but it's a
time-averaged indicator when you find mercury in comparable
species at higher levels in one lake than the other.
Information from our fish management unit this year is that
when they went to collect their 1979 samples from Cranberry and
Stillwater Reservoir to determine mercury in smallmouth bass,
they could not find any smallmouth bass in Stillwater
Reservoir. It was not possible to secure any, although this
lake had been quite adequate in producing fish in the past
years. We also found this year that the pH of Cranberry Lake,
despite it's large size, was top to bottom pH 4.5 during
snow-melt conditions, exactly the same as the influent water.
At the same time, precipitation measured on site had a pH of
3.9.

We may be seeing a trend in the Adirondacks, from short
retention time, small ponds losing their fisheries to larger
flow-through reservoirs. I think that these reservoirs should
be the next place we look for changes in fisheries. So we'll
be watching that very carefully next year and using mercury as
an indicator. I think it's a better indicator than actually

measuring the pH, because the fish mercury content is an
average of several growing seasons. The pH is usually measured
in summer in most of these places, and it seems that the
fishery is lost when the pH is year around below pH 4.5.
Although the pH of lakes sampled in New York State may be in
the rante from 6 to 7 during the summer months, during
snow-melt periods the pH may drop considerably. So the
critical factor is the shape of that pH curve versus time,
which may probably start off with a median near pH 5.8. This
curve then translates downwards, until you get to a point where
the pH is continually below 4.5 or 4.

JERNELÖV: You said, to look at the instant accumulation rate
you divide the concentration of mercury in the fish by the
age. To get a very similar type of information, we have been
measuring it as milligrams per gram in square (mg/gm^2). That
means measuring the concentration as a function of the weight
instead of the age. Earlier you discussed this uncertainty in
age determination which you do not have in weight
determination. Why did you choose to use age rather than
weight?

BLOOMFIELD: Well, for a preference, I would choose length as
the best predictor of mercury level for a given species, not
comparing Northern Pike to smallmouth bass, obviously. There
is some inherent geometry between the size of the bolus the
fish consumes and its length. I think length would probably be
the best indicator, if one assumes the major uptake of mercury
is from the consumption of prey.

     Why did we use age? Because one of the areas that some
people are trying to develop is a bioassay apparatus. What we
were trying to do by dividing the mercury content by age is to
come up with an uptake rate that could be compared to bioassay
work.

JERNELÖV: Let me rephrase the question somewhat. The reason
why we took it as nanogram per gram squared was based on a
similar argument; only we wanted to have some measurement of
the type to relate to bioassays. We also thought that it was
good to have the squared function as it resembles an
acceleration. We've been doing this for about 10 years. You
do something else. Have you thought of a good argument to
change this system?

BLOOMFIELD: Well, the only problem is that the one variable in
fish that conceivably can come to a dead halt is the weight,
and in fact the weight can change negatively. So if you're
looking for an uptake over time, weight or weight squared,

although that's a way of normalizing the numbers, it may not be
a true uptake rate when comparing a number of different lakes
with various weight, age, length relationships. Furthermore,
sampling smallmouth bass in the Spring and Fall, and
introducing the sex of the fish as another factor may give you
a different weight versus length relationship.

BERG: In Lake Ontario some years ago we made a small
collection in which we tested mercury in these fish against
both weight and age. The data came out rather neatly, that for
a cohort of the same age, there was a straightforward dilution
effect of growing weight. As you went to bigger fish they had
less per gram and more per body. But I would consider a
reference to weight to require modeling two opposite
contributions of the weight factor: one, the trophic level
contribution, and the other, the dilution by growth which work
in opposite directions.

McLEAN: The most detailed study on the relationship of mercury
uptake to the various growth factors (by Scott[+] in Clay Lake)
showed that in fact there is a clear, second-order effect that
as the fish grows older the relative uptake increases, i.e. the
older fish have _more_ mercury per gram, not less.[*]

BERG: I was speaking of changes within the same age class, as
opposed to progression from one age to the other. From one age
class to the other, taking average weight, there was an
increase, just as Dr. McLean says. Within each age class,
however, there was a dilution effect.

[+]Scott, D.P., "Mercury Concentration of White Muscle in
Relation to Age, Growth and Condition in Four Species of Fishes
from Clay Lake, Ontario." J. Fish. Can. 31, 1723 (1974).

[*]N.B. This contradicts Dr. Berg's point.

THE EFFECTS OF ACIDITY

ON THE UPTAKE OF MERCURY IN FISH

Arne Jernelöv

Swedish Water and Air Pollution Research Institute

Gothenburg, Sweden

ABSTRACT

The effect of pH on the transformation, transport, and
bioaccumulation of mercury are discussed. Through its effect
on mercury retention in the water-body, and on the biomass and
growth rate of aquatic organisms, low pH leads to higher
mercury levels in fish. For a better understanding of mercury
turnover and transport, however, greater knowledge concerning
re-emission of mercury from both land and water is required.

Acidity can affect the uptake of mercury in fish through
its influence on two levels of organization: the individual
(physiological processes) and the ecological levels.
Physiologically, pH may affect the penetrability of cell
membranes, leading to either higher or lower transfers of
methylmercury through these membranes. However, no such
specific studies have been found in the literature, nor does
field data suggest that this type of effect would be of major
importance.

Another effect to consider is that, in natural waters, low
pH occurs in water with a low salt content, as these waters are
generally more susceptible to acidification. A low salt
content in turn means that a fish living in that water will
drink more water, taking up the salts to maintain its salt
content. This will increase the amount of mercury ingested
with the water. However, mercury uptake from water is so
heavily dominated by the water passing the gills for oxygen
uptake, that an increase in the amount of swallowed water will

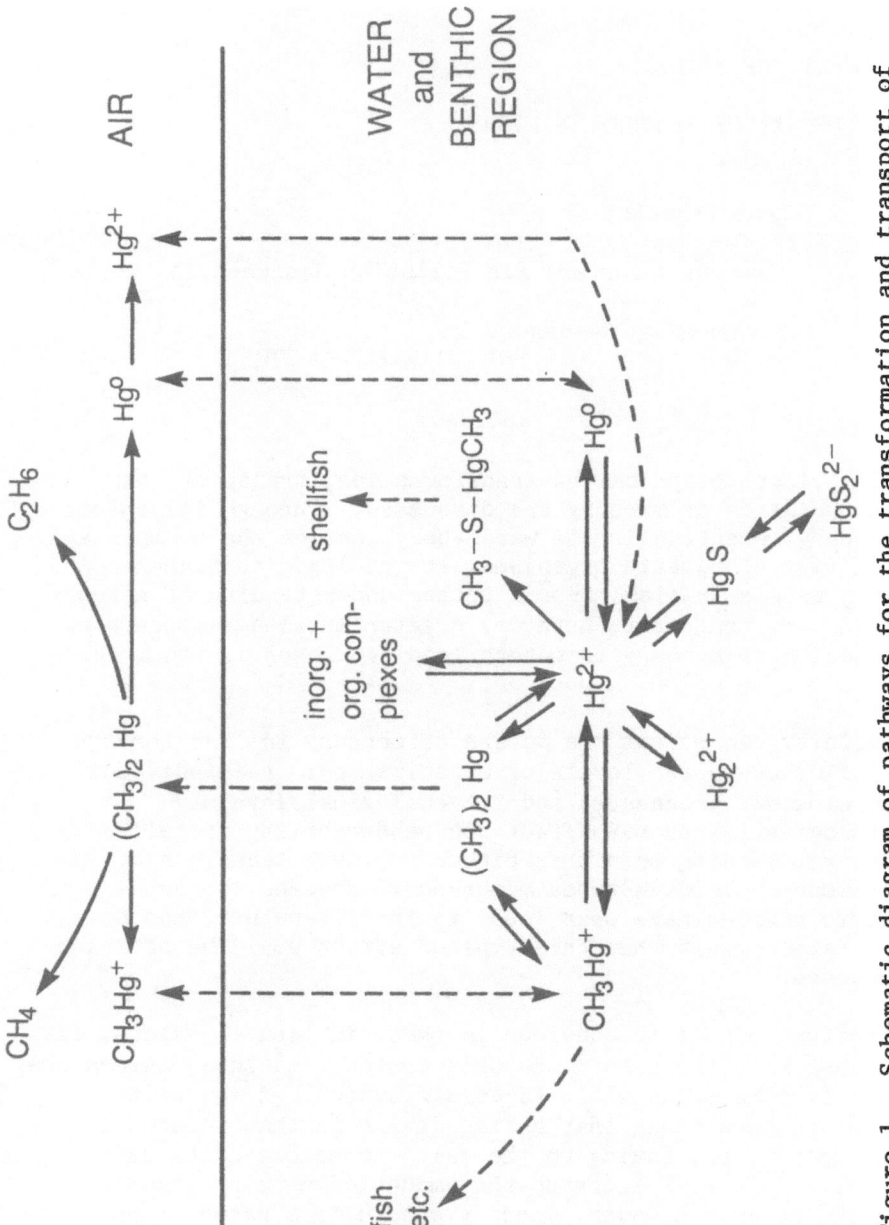

Figure 1. Schematic diagram of pathways for the transformation and transport of mercury in air, water and sediment.

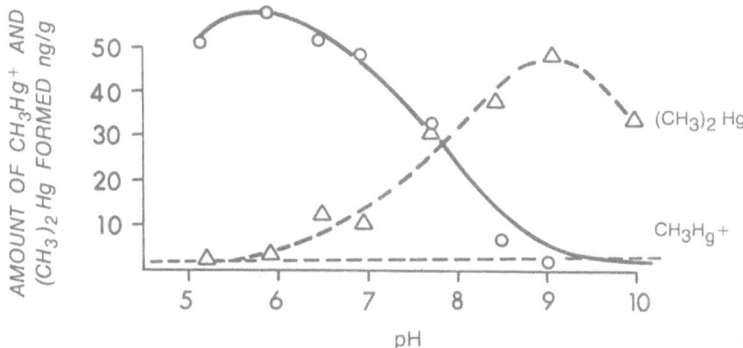

Figure 2.   Formation of mono- and dimethylmercury as a function of
pH in the medium.

have a very marginal effect on the total uptake of mercury. It
thus can be seen that the effects of acidity on the uptake of
mercury in fish at the physiological level are of minor
importance. The important effects will be seen to be at the
ecological level.

For some time now, both the chemical form of the mercury
and the principles governing how water pH affects the
solubility of various mercury levels have been known. The
quantitative relationships also have gradually been better
understood. Important work in this context has been
accomplished by Sillen, among others.

The instability of dimethylmercury at low pH values has
long been known. The environmental consequences of this were
discussed at an early stage (IVA Annual Report, 1967). The
final outcome of biological methylation of mercury was found to
be either dimethyl- or monomethylmercury (Jensen and Jernelov,
1967, 1968). The ratio of these products was found to be a
function of the pH value of the media in question (Jernelov and
Fagerstrom, 1971). Through its general effect on biological
processes, including growth rate and reproductive capacitiy,
the pH value affects a number of processes within the chain
that extends from the formation of methylmercury by
microorganisms to the accumulation levels of mercury by
predatory fish (Jernelov, 1968). Laboratory experiments show
that, for a given amount of total mercury in an aquatic
ecosystem, higher levels of mercury are found in fish at low pH
values than at high pH values. The biomass available for

diluting the methylmercury formed is decreased along with the
retarded growth rate of the fish (in certain stages of the
acidification process).

    In an in situ study of three lakes in western Sweden, the
amount of mercury reaching the respective lakes was found to be
approximately equal, as was the total amount of mercury found
in the lakes. The varying levels of mercury in pike from the
three lakes were to a decisive extent determined by the
nutrition and acidification status of the lake, and by the flow
of mercury between the trophic levels (Grahn, Hultberg, and
Jernelöv). This data further substantiated the data of Landner
and Larsson in which they reported a correlation between pH
values and mercury levels in the fish of the lakes in Dalsland,
western Sweden.

    For a long time, the significance of airborne mercury to
(methyl) mercury levels in fish has been the object of
differing opinions. In global perspective, the discussion has
been influenced by reports concerning the mercury level of the
Greenland ice. However, the very same scientists who had
studied the Greenland ice sheet have lately had reason to
revise their reports. The basis for a great many hypotheses
and calculations has thus disappeared. Most of the models
created in order to explain the observations were impaired by a
logical flaw that deserves some attention. The assumption had
been made that man's contribution could be calculated by
relating the total deposition during one year to anthropogenic
discharges during the same period. This, however, did not
allow for the possibility of mercury re-emission. Previous
calculations will be entirely wrong if the mercury is actually
deposited, re-emitted, deposited again, etc. The existence of
re-emission will also affect the conclusions drawn from
investigations of moss and snow samples taken in the vicinity
of a chloralkali plant (Jernelöv, Wallin, and Skärby). These
analyses measured the net, not the total deposition. The
conclusion that only a few percent of the total mercury
discharge would be deposited within a limited vicinity (radius
of 5 Km) must accordingly be changed into the following: only a
few percent [of the total mercury discharges] will remain in
the area in question during the observation period. When
formulated like this, the resemblance is quite strong between
the curves showing the correlation between mercury
level/distance from a source obtained through a) analyses of
moss and snow samples, and b) direct measurements of air
levels (Svedung).

In attempting to estimate the effect of the air emitted mercury from Skoghall, a chloralkali plant in western Sweden, the mercury level of pike was related to the distance from the source in two directions. This was done for the main wind direction, and for the transsect (Hasselroth and Björklund). The authors found that higher mercury levels in fish were found downwind, and thus concluded that local atmospheric emission was important.

Since it was found that the pH values of the lakes in the varying directions differed, and that pH values alone were the dominant statistical explanation for the measured and estimated variation of mercury levels in pike of 1 kilogram weight, the conclusions had to be revised (Jernelöv, Linse, and Hansson). In a further development of these studies, the mercury level of the 1-kilo-pike was related to the pH color, and conductivity of the water in a number of areas. Certain comparisons were also made with lakes in southeast Sweden (Hultberg and Jernelöv). These studies again show an overall strong correlation between mercury levels of the fish and pH values of the lakes. Statistically, with a multiple correlation analysis, pH explains about 50% of the variation in mercury levels of 1-kilo-pike. Keeping the effects of pH in mind, the study also shows a differentiation among the various areas. Along with pH, the study indicates that the fish mercury level is higher at high color values and lower at high conductivity. These tendencies are in accordance with earlier laboratory experiments, though not entirely significant when analyzing the material from the in situ study.

The significance of re-emission is illustrated by various calculations and measurements. Based on the measurement by Svedung of the air levels surrounding a chloralkali plant (and a series of plausible assumptions), it has been estimated that approximately 50% of the gross mercury deposition is re-emitted and that 20% is primarily deposited within the first few kilometers from the source (Högström and Enger). In another study, done 8 months after the temporary closing of a chloralkali plant in Colombia, the re-emission from the plant area has been estimated at approximately 1 ton/year (Eriksson and Jernelöv). Estimates have also indicated that a very large part of the mercury brought to a lake by rainfall is immediately re-emitted to the atmosphere (Hultberg and Svedung). One possible explanation states that, due to its low pH values, the rain is capable of washing out more mercury than the lake water is capable of retaining. The tendency of a lake to retain mercury would increase at very low pH values with additional retention when high levels of organic material are

present because of the complexation of the mercury. The water
pH interaction with the mercury cycle thus allows for the fact
that, given the same amount of mercury cycled in the
environment, acidified lakes will not only take up a greater
net total of mercury than lakes with a higher pH; they will
also transform a greater amount into monomethylmercury,
resulting in higher methylmercury levels in fish.

Applicable to Sweden in general is that extremely few lakes
with pH values below 5 have mercury levels in 1-kilo-pike of
less than 1 mg/kg. At a pH value of 6, for instance, the
normal level for the same pike would be approximately 0.6
mg/kg. In the interval of pH values between 4.5 - 6.5, a 0.5
unit change in the pH value corresponds to a change in the fish
mercury level of approximately 0.25 ppm. There is a tendency
for the pH-effect to be greater at low pH values, indicating
that the correlation is not linear when acidification is
measured with pH as a logarithmic unit.

It might finally be said that research in the last few
years shows the mobility and possibility of air transport for
mercury through re-emission to be greater than expected. This
fact, coupled with the effect that low pH has on retention of
mercury (with its subsequent effect on fish) could mean that,
if acidification continues at the present rate in Sweden,
thousands of additional lakes could be blacklisted within the
next decade.

REFERENCES

1) IVA Annual Report 1967 Swedish Academic of Ingenieuring
   Sciences, Stockholm, Dec. 1967.

2) Jensen and Jernelöv 1967 Biocidinformation 10:4
   Nordforsk, April 1967.

3) Jensen and Jernelöv 1968 Biocidinformation 14:3
   Nordforsk, Feb. 1968.

4) Jernelöv and Fagerström 1971 Water Research 5, 121, 1971.

5) Grahn, Hultberg and Jernelöv 1976 IVL B 291 Report from
   Swedish Inst. for Water and Air Pollution Research.

6) Landner and Larsson 1972 IVL B 115, Swedish Inst. for
   Water and Air Pollution Research.

7) Jernelöv, Wallin and Skärby 1971 Atmosphere Environment
   1971.

8) Svedung pers. comm.

9) Hasselrof and Björklund 1974 SNV report series, Swedish
   Environment Protection Board.

10) Jernelöv, Linse and Hansson IVL B publ 245, Swedish
    Institute for Water and Air Pollution Research.

11) Hultberg and Jernelöv 1976 IVL B Publ 316, Swedish
    Institute for Water and Air Pollution Research.

12) Högström and Enger 1978 Report for Swedish Energy
    Commission.

13) Eriksson and Jernelöv 1978 FAO report on pollution of
    Cartagena Bay, FAO, Rome, Nov. 1978.

14) Hultberg and Svedung Report from Swedish Institute for
    Water and Air Pollution Research.

DISCUSSION

BLOOMFIELD: In 1973 and 1974, there was a magnetite sintering operation about 30 km to the Northwest of Cranberry Lake. We were measuring total fallout with bulk precipitation collectors. Right on that site when the thing was operating and producing a fairly large particulate emission, there was about a 270 micrograms per square meter per year mercury deposition rate. About a kilometer or two away, the level was down to about 35. Near Cranberry Lake, which is about 30 km away, the level was about the same number. We thought initially that site, since it was so close to many of the lakes in the Northwest Adirondacks, was probably contributing mercury and that might be a cause of the problem. We were never able to demonstrate that the sintering operation produced anything but a localized effect.

The plant has been shut down since 1975. At Cranberry Lake we're still getting approximately the same contribution of mercury in atmospheric fallout. So I don't know whether that may indicate that the plant had no effect except locally, or that the inactive site is still contributing mercury at some level. It would be of interest to see if that site is contaminated and giving off mercury to the air.

TOMLINSON: In Quebec City last week there was a paper given by Mr. Bachand on the Le Grand River basin lakes. This is the area that's being flooded by Hydro-Quebec's great Northern Quebec James Bay Project. Data which he found were very similar to those which you showed today. Similar data was found in Ontario. So local fallout may have been a factor but probably not the main factor. Can you comment on that?

JERNELÖV: Not very much.

McLEAN: Were your re-emission studies based on studies using transplanted mosses?

JERNELÖV: We started with that. But the problem with the mosses is that they keep some of the mercury but not all of it. It is very hard to measure how much they actually keep and when and how and for how long. We started doing that because it worked so well with a lot of other heavy metals. When we got some slightly confusing results, we started to transplant the mosses from the contaminated site back to background. And it turned out that they very quickly got down to background values. So we had to assume a much higher loss from them.

SCRUDATO: You talk about mercury inputs to various aquatic systems, let's say coming in from the atmosphere. But you didn't mention what, if any of the mercury, is going to be bound to organic or inorganic particulates, and how that adsorption-desorption processes would affect re-emission. The mercury, whether dimethylmercury or methylmercury, has a fair affinity for various types of particulate matter, I would imagine. How have you accounted for resorption-desorption in your consideration of re-emission of mercury back in the atmosphere?

JERNELÖV: That takes a long discussion to answer. We have looked at them for dimethylmercury. The only form of attachment we find is direct solution. Monomethylmercury will bind to various groups in organic particles. It will bind to sulfur groups to some extent, like it does in fish proteins. It will bind to nitrogen groups and occasionally also to carbon. It also, to some extent, dissolves in fatty substances.

But the interesting thing is that we seem to have the type of very quick equilibrium. If we measure the methylmercury present in an ionic form in the water itself, and that attached to some particulate matter, then we will have some certain distribution here which depends on concentrations and type of organics. In an example where we have 2% in this form and 98% in that, let us put in some fish. We find that the fish starts accumulating this mercury. But if we continue to measure what relative proportion we have here versus here, we find that they are roughly constant, despite the fact that the fish is taking mercury very quickly. Obviously this equilibrium is fast.

The same goes if we use an enclosed atmosphere in a closed system, and add some methylmercury in the gaseous state. We will have some methylmercury dissolved in water and some attached to particles. Of course, we can change this equilibrium, and addition of more organics will get some of that from the atmosphere down in the waters and further onto the fish. But if we flush away the methylmercury in the atmosphere, we get the same quick equilibrium. In these closed laboratory systems, it is comparatively quickly that we regain equilibrium.

To what extent it would be different out in the environment, I don't know. If we compare these results with inorganic matter, it seems as if the binding to particles is substantially stronger and equilibrium is not attained as fast.

EVANS: It seems to me it's very difficult for you to separate out the effects of acidity and methylmercury. Do you think you

have a system in which you could judge the relative amounts of
each of these? You seem to have the compounding effect of
acidity with mercury levels. In nature, do you think you can
separate out the effects of one versus the other?

JERNELÖV: We tried this in the three lakes I talked about that
are situated quite close to one another, with the same fallout
and roughly the same amount of mercury that we can find in
water and sediment and so on. The total amount of mercury
varied. But, the fraction of mercury in the biota is very much
higher in the more acid ones. We also have done this in lab
systems where it is much easier because we had given the amount
of mercury and the given amount of acid and so on. But in most
natural cases that is a problem.

I would like to underline here once again what I think is
important. And that is, when we talk about mercury in the
lake, we should remember that a typical picture for Swedish
situations depends on the depth considered. We would find that
in the sediments we have 95% or so of the mercury in the
system. Sometimes up to 99%. In the water, we have something
between roughly 1 and 5%. In the organisms, in the form of
methylmercury, we have a very tiny fraction which is frequently
not more than .1%. This is for total mercury. If we are
considering only methylmercury, we would have the dominating
part being methylmercury. This means that in practically any
lake there is enough mercury to get the fish levels high, if
all the biological and biochemical conditions are in favor.

TOMLINSON: You showed in the lower section of your chart the
benthic and water communities as alternatives. Your original
work dealt with sediments alone. But presumably if the mercury
enters the stream from the rain or some other source, then it
can be methylated quite rapidly in the water column, perhaps
piggyback on the microbes. Can you comment on the relative
contributions to methylation in the benthos and in the water?

JERNELÖV: I think you're quite right in pointing out that when
we started, we thought of the direct discharge of mercury. We
had the pulp industry and the chloralkali industry, so we
started with the sediments. What is obvious when it comes to
the biological methylation is that we need some
microorganisms. Of course, we have such in the water, but they
are more frequently in the very top layer of the sediment. On
the other hand, they are more active on the suspended
particles. I think this will to a very large extent depend on
the type of water we have. There obviously are many waters
where the sediments in themselves are perhaps not so important,
but where organics suspended in the water may be of larger

importance. Then, the third factor that also comes in is the question of the monomethylmercury concentrations in the atmosphere. How high are they?

There are some reports, we have some results at IVL that so far I am reluctant to trust, that say the percentage of methylmercury out of total mercury in the atmosphere should be 1, or 2 or 3%. If that's the case, then it could well be that the atmosphere is not predominantly a sink of methylmercury but rather the source for it from the lake's point of view.

McLEAN: I believe that you can modify the argument that most of the mercury deposited from the atmosphere is rapidly re-emitted. It is now agreed that 90-95% of the mercury in the atmosphere exists in the vapor phase and this will be the major contributor to the deposition - re-emission process. However, most of the mercury coming out from the atmosphere in precipitation is associated with particulate matter, and this mercury can be removed by filtration through a 0.45 μ filter. In addition, it often takes two or three days in an acid oxidizing solution (2% v/v H_2SO_4 + 0.1% $K_2Cr_2O_7$) before the mercury is extracted from the particulate matter to be determined in solution using the standard analytical method. In other words the mercury in precipitation appears to be fairly tightly bound to particulate matter.

JERNELÖV: I think if you use the reducing media instead, you would convert it to elementary mercury quite quickly, as you do in your analytical procedure.

McLEAN: However, mercury bound to particulate matter is not easily removed under mild reducing conditions used in analytical procedures.

JERNELÖV: You're saying that you have your mercury bound to organic or inorganic particles. But this is mercury 2+. If it's re-emitted, it's quite probable that it's in that form also. Which means even if it's a little hard to extract this in an oxidizing medium, you should put a reducing agent on instead.

McLEAN: Yes, that is true. However, I would be surprised if such a reduction of mercury associated with particulate matter took place very rapidly in the natural environment.

JERNELÖV: Oh, if you you a redox potential that will put electrons in.

McLEAN: However, is it not more likely that the mercury which
is being rapidly re-emitted is that which has been deposited
from the vapor phase and which has not been so strongly bound?

JERNELÖV: I don't exclude it. It's not the way I pictured it.

BLOOMFIELD: It's an interesting thought experiment that just
occurred to me. Your hypothesis is that perhaps it's the ratio
of the total amount of methylmercury in a body of water to the
total amount of biota, biomass, that is important, and the
limiting factor might be the amount of methylmercury in the
water. Therefore, if you have less biota you can get more
methylmercury in that biota on a concentration basis.

    One interesting experiment which could be used to test that
hypothesis is to find a body of water and continually stock it
with a large amount of forage base every year, for example,
such as Golden Shiner or whatever fits into that body of
water. And then to continue to monitor Northern Pike to see if
you've diluted out the mercury by adding that biomass.

JERNELÖV: You would have to design this so that you don't
affect the growth rate of the pike too much, when you either
provide it with more food or take some away from it.

SCHOFIELD: You've noted changes in the fish populations. Have
you looked at plankton?

JERNELÖV: As a function of acidity, do you mean? Yes.

SCHOFIELD: And the mercury problem?

JERNELÖV: I don't think mercury in itself has that much affect
on the populations of fish. There are very elaborate studies
on the effect of acidity on plankton and whatever you find in
the water. Basically the conclusion is that you have a clear
decrease in productivity at most stages.

THE ROLE OF pH AND OXIDATION-REDUCTION POTENTIALS IN THE MOBILIZATION OF HEAVY METALS

John M. Wood

University of Minnesota
Gray Freshwater Biological Institute
Box 100
Navarre, MN 55392

ABSTRACT

The effect of pH and standard reduction potential ($E°$) is discussed in terms of the kinetics and mechanisms for the biomethylation of a number of toxic metals. In the aerobic environment, conditions have been delineated to explain the methylation of Hg^{II}, Pb^{IV} and Se^{VI} compounds.

The methylation and mobilization of mercury is especially important in lakes which are susceptible to acid precipitation. The accumulation of methylmercury in fish taken from acid-sensitive oligotrophic lakes is rationalized in terms of the chemistry and biochemistry of mercury compounds.

INTRODUCTION

At the turn of the century Gosio observed that an unusual gas was evolved from arsenical wallpaper containing the pigments Scheele's green and Paris green.[1] Thirty years later Challenger showed that Gosio's gas was pure trimethylarsine.[2] Challenger's discovery provided us with the first example of how biological systems possess the capability for synthesizing very toxic organo-arsenic compounds from less toxic inorganic substrates.[3,4]

In the last decade the case of methylmercury pollution has demonstrated the profound importance of understanding biologically mediated transformation reactions that produce organometallic compounds with a high potential for bioaccumulation and toxicity.

Most metal-alkyls are poisonous to the central nervous systems of
higher organisms, and these compounds often bioaccumulate in cells.
Such reactions are of considerable environmental significance since
in general the methylated organo-metallic derivatives are more toxic
to higher organisms than their inorganic precursors. In order to
understand the movement of toxic elements in the biosphere, it is
necessary to examine several factors. The most crucial factors
have been postulated recently by Jernelöv.[5] Jernelöv presents a
comprehensive set of environmental parameters that, when determined,
permit an evaluation of potential undesirable inputs for toxic
elements: a) production and emission in relation to natural flux;
b) residence time in various reservoirs; c) bioaccumulation, both
"passive" and active transport; d) physical and chemical properties
relating to dispersion (e.g., volatility, adsorption, dissociation-
association reactions, oxidation-reduction reactions, formation of
insoluble precipitates, etc.); e) toxicity to aquatic organisms;
f) toxicity to man and other mammals; g) long-term biological
effects on ecosystem metabolism; and h) transformation reactions
by organisms. The latter point on transformation by organisms is
of interest because both microorganisms and higher organisms are
capable of synthesizing toxic products from innocuous substrates.

In 1968 we discovered that methylcobalamin was capable of
transferring a methyl-carbanion to mercuric salts in aqueous
solutions.[6] This initial discovery opened the door to a study of
reactions between methylcobalamin and a number of metal and
metalloid ions. Methyl-transfer in biological systems naturally
depends on the co-enzymes which are available to perform this
function. Three co-enzymes have been found which are capable of
the transfer of methyl groups: (1) methylcorrinoid derivatives,
(2) S-adenosylmethionine, and (3) N^5-methyltetrahydrofolate
derivatives.

For these three co-enzymes only methylcorrinoid derivatives
are capable of transferring the methyl-group as a carbanion (CH_3^-).
Clearly, on the basis of charge, CH_3^- is most likely to react with
positively charged metal ions (M^{n+}). S-adenosylmethionine and
N^5-methyltetrahydrofolate derivatives transfer methyl-groups as
carbonium ions (CH_3^+), and so it is unlikely that this CH_3^+ species
would be involved in transfer to a positively charged metal ion.

Once synthesized, these methylated metals are invariably more
toxic than their inorganic substrates. This toxicity is probably
due to the non-polar nature of many organo-metallic compounds
which allows them to diffuse rapidly into and through cell membranes. Dynamic aspects of these methylation reactions are of
critical importance, because even though most methylated metals
are thermodynamically unstable in water, many of them are kinetically
stable. In fact, it is well known that metals which are lower in
their periodic groups form metal-alkyls which are kinetically more

stable. For example, mercury, platinum and possibly lead offer
potentially stable systems, whereas palladium, chromium and cadmium
do not.

Alternate Mechanisms for Co-C Bond Cleavage

Methyl-transfer from methylcobalamin requires cleavage of the
Co-C bond. This bond can break under different conditions to give
a carbanion (CH_3^-), a radical ($CH_3\cdot$) or a carbonium ion (CH_3^+).
Figure 1 presents six alternate reaction pathways which lead to
the transfer of the methyl-group by each alternative mechanism.[7]

Reaction 1. This pathway involves a single electron oxidation
of methylcobalamin by "outer sphere" electron acceptors. The
oxidized methylcobalamin product is extremely labile and produces
a methyl radical and aquocobalamin.

Reaction 2. This reaction involves heterolytic cleavage of
the Co-C bond with the transfer of a carbanion to the attacking
metal ion.

Reaction 3. This reaction has been described as a "Redox-
Switch" mechanism. The metal ion in its lower oxidation state
first forms an "outer sphere" complex with the corrin macrocycle,
followed by a two electron transfer of the complexed metal ion to
a two electron acceptor. Oxidation of the complexed metal ion
facilitates carbanion transfer from the cobalt to the complexed
metal.

Reaction 4. This reaction pathway involves homolytic cleavage
of the Co-C bond by "inner sphere" free radical attack.

Reaction 5. In this mechanism the reagent transfers a single
electron to methylcobalamin by an "outer sphere" mechanism. The
product of this reaction is very labile, producing a methyl radical
and Cob(I)alamin (B_{12-s}).

Reaction 6. This reaction involves heterolytic cleavage of
the Co-C bond by an attacking nucleophile which displaces a carbonium
ion to produce Cob(I)alamin (B_{12s}).

In our studies, the transfer of CH_3^- or $CH_3\cdot$ has been found
to be the most predominant reaction mechanism for a number of
metals and metalloids. Methylcobalamin has been found to be very
stable to nucleophilic attack, but very susceptible to electrophilic
attack and free radical attack.

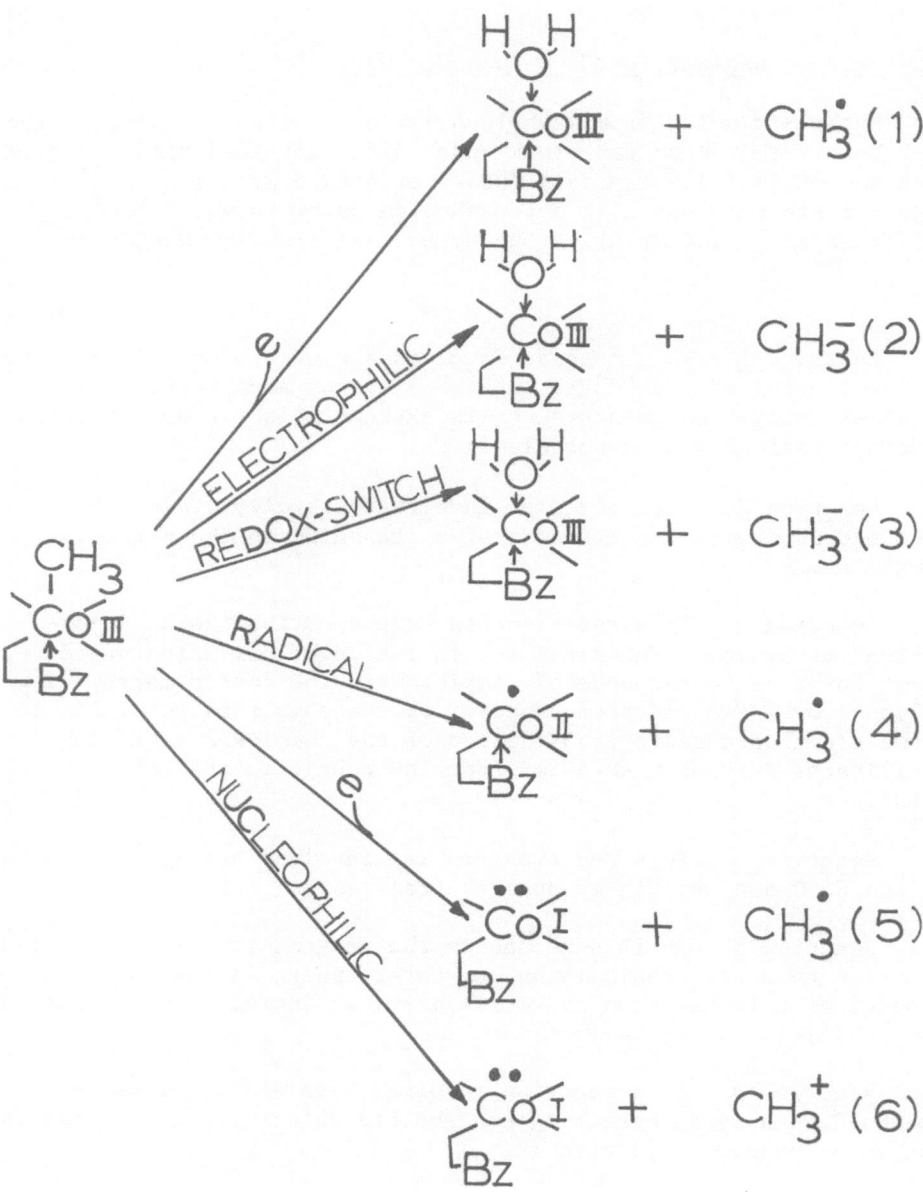

Figure 1.   Alternate pathways for transfer of methyl-group.

Electrophilic Attack

The reaction between mercuric ion and methyl corrinoids is an example of carbanion methyl-transfer (Figure 2). Because mercuric ion is a good electrophile, it also coordinates to the nitrogen of the 5,6-dimethylbenzimidazole base to give a mixture of "base off" and "base on" methylcobalamin. The "base on" species reacts 1000 times faster than the "base off" species to give methylmercury as the product.[6] Other metals which are known to react with methyl-cobalamin by a similar mechanism to mercuric ion are lead (Pb^{IV}), thallium (Tl^{III}) and palladium (Pd^{II}).[9-11]

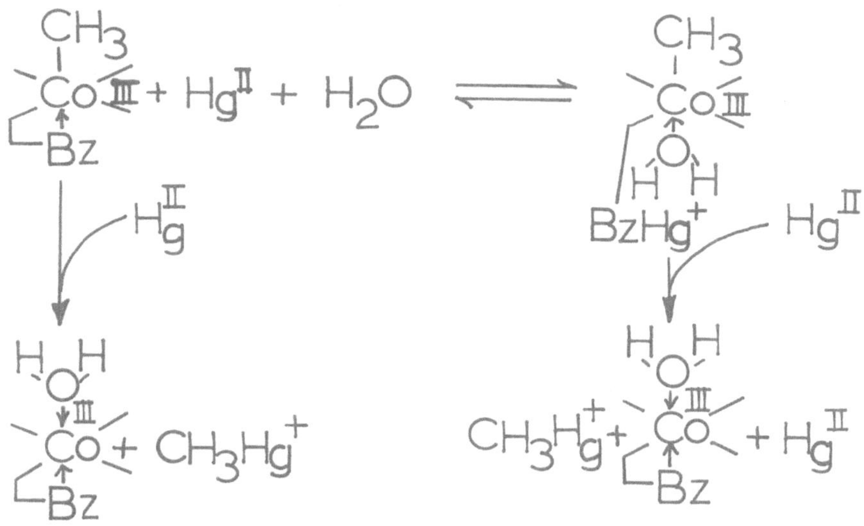

Figure 2. Methylcobalamin-dependent methylation of Hg^{II} salts.

The reactions described above all involve the displacement of a carbanion from the cobalt atom of methylcobalamin. These reactions occur under aerobic conditions with rate constants in the order of milliseconds. It is apparent that metals which react by electro-philic attack on the Co-C bond (SE$_2$ mechanism) occur with the more oxidized state of the metal, i.e., Pb^{IV}, Tl^{III}, Hg^{II}, Pd^{II}, which have standard reduction potentials greater than 0.8 volts. Because of the "base on - base off" equilibrium these reactions are pH dependent.[12]

Free Radical Attack

Homolytic cleavage of the Co–C bond of methylcobalamin leads to methyl-radical transfer. For this reaction to occur, the attacking species must be a free radical, and so the generation of such a radical intermediate is necessary either by the one equivalent oxidation of a metal ion in the reduced state of a redox couple (e.g., $Sn^{II} \xrightarrow{\; - e- \;} Sn^{III}$), or by a one equivalent reduction of a metal ion in the oxidized state (e.g., $Au^{III} \xrightarrow{\; +e- \;} Au^{II}$). The attacking free radical has been shown to require a standard reduction potential lower than 0.5 volts.

Methylcobalamin has been shown to react with a transient Sn^{III} radical intermediate to give methyl-Sn^{IV} and Cob(II)alamin (B_{12-r}) as the products [13] (Figure 3).

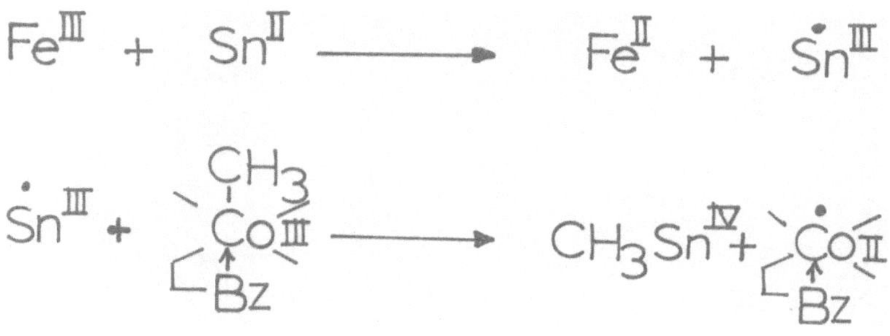

$$Fe^{III} + Sn^{II} \longrightarrow Fe^{II} + \dot{S}n^{III}$$

Figure 3. Methylcobalamin-dependent methylation of Sn^{II} salts.

The biomethylation of tin is of environmental significance because methyl-tin compounds have been detected in blood and urine samples as well as in aquatic food chains.

In the case of gold we have shown that homolytic cleavage of the Co–C bond occurs with an Au^{II} radical intermediate which can be generated by single electron reduction of Au^{III} (Figure 4). This reaction is initiated by pre-incubating catalytic amounts of $Au^{III}Cl_4^-$ with Cob(II)alamin (B_{12-r}) or Fe^{II}. A similar mechanism to that demonstrated with Sn^{III} and Au^{II} has been reported for Cr^{II} [11] and radicals produced by a one equivalent oxidation of thiols.[12] Contrary to reactions on the Co–C bond with electrophiles, free radical attack appears to be pH independent; no appreciable difference in reaction rates are observed for methyl-transfer from "base on" versus "base off" methylcobalamin. These methyl radical transfer reactions can be viewed as "inner-sphere" one equivalent

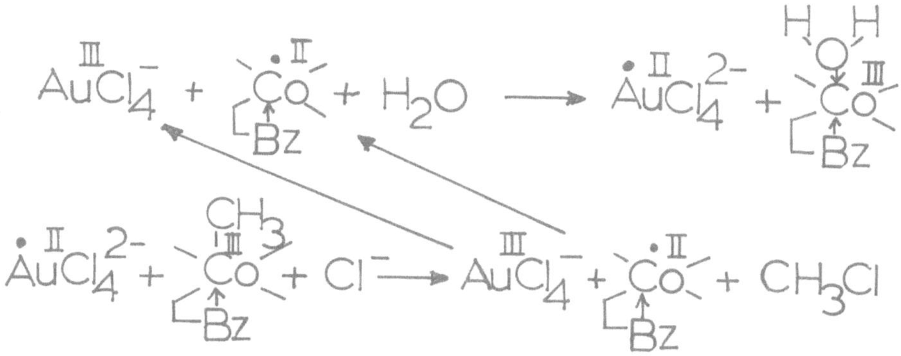

Figure 4.   Demethylation of methylcobalamin by Au^{III} salts.

oxidative-addition with the methyl group acting as a bridging ligand.[12]

Standard Reduction Potential and Biomethylation

The connection between the standard reduction potential and the mechanism for biomethylation seems highly rational, because E° describes the relative thermodynamic tendency for the metals involved to accept or donate electrons. The electrophilic mechanism occurs with inorganic compounds with a standard reduction potential (E°) of +0.85 volts or higher (type 1) while the free radical mechanism occurs with inorganic compounds having E° of +0.50 volts or lower (Type 2) (Table 1).

Table 1.   Relationship Between Standard Reduction Potential (E°) and the Mechanism of Methylation for Selected Elements

Redox Couple	E° (volts)	Mechanism of Methylation
Pb(IV)/Pb(II)	+1.46	Type 1
Tl(III)/Tl(I)	+1.26	Type 1
Se(VI)/Se(IV) acid	+1.15	
Pd(II)/Pd(0)	+0.987	Type 1
Hg(II)/Hg(0)	+0.854	Type 1
Pt(IV)/Pt(II)	+0.760	Redox Switch
As(V)/As(III) acid	+0.559	
Au(III)/Au(II)	+0.50[a]	Type 2
Sn(IV)/Sn(II)	+0.154	Type 2

Se(VI)/Se(IV) base	+0.05	
Cys-S-S-Cys/2Cys-SH	−0.22	Type 2
Cr(III)/Cr(II)	−0.41	Type 2
As(V)/As(III) base	−0.67	

[a]The Au(III)/Au(II) couple is estimated.[14]

The Type 1 reactions occur under aerobic conditions and are pH-dependent. Therefore, in the aerobic environment and under acidic conditions we should be concerned with B_{12}-dependent methylation of Pb^{IV}, Tl^{III}, Se^{VI}, Pd^{II} and Hg^{II}. These methyl-transfer reactions to a variety of metals and metalloids are not only of environmental significance, but the B_{12}-reactions reported here appear to be a good model for reactions which occur in complex biological media such as lake sediments.

Predictions for B_{12}-dependent Methylation in Acid Susceptible Lakes

The biochemistry reported above outlines the necessary physical parameters for the transfer of CH_3^- to positively charged metal ions in the aqueous environment. Under aerobic conditions it is clear that the heavy metals Hg^{II}, Pb^{IV} and the metalloid Se^{VI} are more soluble and available to react in an acidic environment. These toxic elements are biomethylated under acidic conditions to give methylated products which accumulate in aquatic predators. The process of methylmercury uptake by lake biota is extremely rapid. We have shown that methylmercuric chloride diffuses through a 40 Å membrane in 20×10^{-9} seconds.[15] This means that one can predict elevated concentrations of methylmercury in biota from oligotrophic lakes for three principle reasons. 1) In acid lakes there will be more inorganic mercuric ion in solution, and therefore the rate of synthesis of methylmercury will be enhanced. 2) The low pH favors methylmercury rather than dimethylmercury synthesis. 3) The uptake of methylmercury by lake biota should be rapid, and in acid sensitive lakes there will be less biota to accommodate the total methylmercury available. This will lead to an increased burden of methylmercury in predators.

The increase in the concentration of methylmercury in predators is apparently not offset by the synthesis of hydrogen sulfide from sulfate by Desulfovibrio species in the sediments. Sulfide precipitates all mercuric salts and H_2S has been shown to disproportionate methylmercury to dimethylmercury. However, dimethylmercury is hydrolyzed back to methylmercury in acidic conditions.[16]

The dangers of acid precipitation, coupled with heavy metal fallout from particulates emitted from smelters and coal-fired

power plants, should not be underestimated. Significant damage to
both Canadian and Scandinavian lakes has already occurred. Meanwhile pollution control strategies have been slow to develop in both
North America and in Europe. What is the price of 300 lakes in
North America? What is the expected damage level to one third of
the North American continent? These are difficult questions to
answer, but it does appear that the checks and balances between
the acquisition of energy and the protection of the environment
are not effective in the "so-called" advanced industrial society.

Acknowledgements

I wish to thank my students R.E. DeSimone, W.P. Ridley,
L.J. Dizikes and Y.-T. Fanchiang for their contribution. The
Northwest Area Foundation and NIH AM 18101 supported this research.

References

1. B. Gosio, An unusual gas formed from arsenical wallpaper, Arch.
    Ital. Biol. 35:201 (1901).
2. F. Challenger, C. Higginbottom, L.J. Ellis, The synthesis of
    trimethylarsine by the bread mold Scopulariopsis brevicaulis,
    J. Chem. Soc., p. 95 (1933).
3. F. Challenger, Biological methylation of arsenic and selenium
    salts, Chem. Reviews 36:326.
4. F. Challenger, Biosynthesis of organometallic and organometal-
    loidal compounds, Organometals and Organometalloids,
    Occurrence and Fate in the Environment, ACS Symposium Series
    82, eds., F.E. Brinckman and J.M. Bellama, p. 1-23 (1978).
5. A. Jernelöv, Heavy Metals, Metalloids and Synthetic Organics,
    The Sea, ed., E. Goldberg, Wiley Interscience, N.Y. 5:799
    (1974).
6. J.M. Wood, F.S. Kennedy and C.G. Rosen, Methylmercury synthesis
    by extracts of a methanogenic bacterium, Nature (London)
    220:173 (1968).
7. J.M. Wood and Y.-T. Fanchiang, Mechanisms for B_{12}-dependent
    methylation, Third European Symposium on Vitamin B_{12} and
    Intrinsic Factor, ed., B. Zagalak, Zurich, Switzerland
    (1979)(in press).
8. R.E. DeSimone, M.W. Penley, L. Charbonneau, S.G. Smith, J.M.
    Wood, H.A.O. Hill, J.M. Pratt, S. Ridsdale and R.J.P.
    Williams, The kinetics and mechanism of methyl and ethyl
    transfer to mercuric ion, Biochim. Biophys. Acta 304:851
    (1973).
9. J.M. Wood, Y.-T. Fanchiang and W.P. Ridley, The biochemistry
    of toxic elements, Quarterly Reviews of Biophysics 9:2
    000 (1979)(in press).

10. G. Agnes, H.A.O. Hill, J.M. Ridsdale, F.S. Kennedy and R.J.P. Williams, B_{12}-dependent methyl-transfer to metals, <u>Biochim. Biophys. Acta</u> 252:207 (1971).

11. W.H. Scovell, The mechanism for B_{12}-dependent methyl-transfer to palladium, <u>J. Am. Chem. Soc.</u> 96:3541 (1974).

12. Y.-T. Fanchiang, W.P. Ridley and J.M. Wood, Kinetic and mechanistic studies on B_{12}-dependent methyl-transfer to certain toxic metal ions, Organometals and Organometalloids, ACS Symposium Series 82, eds., F.E. Brinckman and J.M. Bellama, p. 54-64 (1978).

13. L.J. Dizikes, W.P. Ridley and J.M. Wood, A mechanism for the biomethylation of tin by reductive Co-C bond cleavage in alkylcobalamins, <u>J. Am. Chem. Soc.</u> 100:1010 (1978).

14. R.L. Rich and H. Taube, Aqueous ion chemistry of gold salts, <u>J. Phys. Chem.</u> 58:1,6 (1954).

15. J.M. Wood, A. Cheh, L.J. Dizikes, W.P. Ridley, S. Rakow and J.R. Lakowicz, Biomethylation of toxic elements, <u>Fed. Proc.</u> 37:No. 1, 16 (1978)

16. P.J. Craig, Metals and biological methylation, Handbook of Environmental Chemistry, ed., O. Hutzinger, Springer-Verlag, Heidelberg (1979)(in press).

DISCUSSION

CLARKSON: That thing about methylmercury chloride and its rapid penetration across membrane bothers me. It is not at all clear that methylmercury chloride will be present in the waters in their natural condition. Methylmercury chloride is a very lipid soluble compound of methylmercury; we convert methylmercury to methylmercury chloride which is extracted into benzene for gas chromatography. And it is not at all clear that the conclusions with methylmercury chloride would apply to the forms of methylmercury which naturally exist under natural conditions and which I would imagine would be bound to a sulfhydryl group in an amino acid or protein. This would be water soluble, not lipid soluble as methylmercury chloride is.

WOOD: I would argue that methylmercury chloride is an important species in the environment for the following reasons:

1.   It is the principle species of methylmercury in seawater.

2.   Because of the [Cl$^-$] concentration, it is clearly present in blood.

3.   I don't know what its like in Rochester, but in Minnesota we apply large quantities of salt to our roads in winter (thousands of tons); and therefore [Cl$^-$] enrichment of freshwater lakes is quite significant in the USA and Canada.

Recently, we have shown precisely by using fluorescence polarization and 270 MHz NMR methods, that methylmercury chloride is not particularly lipid soluble. In fact it has a partition coefficient lipid/water of about 2, but a translational diffusion rate of 20 x 10^{-9} seconds (1,2). In other words it diffuses rapidly, but it does not partition very well in membranes.

1.   Wood et al, Fed. Proc. Fed. Am. Soc. Exp. Biol. 37:1, 16-21 (1978).

2.   Anderson and Lakowicz, J. Biol. Chem. (in press) (1979).

McLEAN: Methylmerury chloride is quite soluble in water (.1 g/liter), so you could expect to find it in natural waters, particularly sea water. However, I agree; it seems much more likely that methylmercury would be associated with thiol groups in water.

WOOD: Although the stability constants for the interaction of thiols with methylmercury certainly favor complexation, it should be pointed out that thiols are oxidized by molecular oxygen, and so the available [RS⁻] will depend on how anaerobic the system is.

TOMLINSON: Well, I understood from a discussion with Dr. Wood some time ago that one of the reasons for the rapid transfer through the gills was that the cysteine groups in the hemoglobin in the blood transferred it very rapidly, so it was taken away from the side of the gill very quickly. These sulfhydryls pick up mercury very rapidly. I have the impression that the ionic form of methylmercury is picked up by the thiol groups in hemoglobin and eventually transferred to methionine at the blood-brain barrier, and for that reason it is much more of a problem than dimethylmercury which is fat soluble, but which does not react with thiol groups and therefore does not accumulate in the system. I also thought that methylated arsenic, which you said was very poisonous was somewhat less poisonous than arsenic itself. Can you comment on that, Ron?

McLEAN: I believe that trimethyl arsine is somewhat less toxic than dimethyl and monomethyl arsine and inorganic arsenic (III) species.

JERNELÖV: I thinks it's likely--or that dimethylmercury or trimethyl arsine, as such, are probably not very toxic. But they are not very stable, which means that they will be in the body degraded to monomethylmercury and to dimethyl and monomethyl arsine. Those would be toxic agents.

McLEAN: In this case, arsenic and mercury are not analogous since trimethyl arsine can be converted to arsenic (V) species, e.g., trimethyl arsine sulfide, which is much less toxic. For dimethylmercury, the only transformation is demethylation.

TOMLINSON: I've been under the impression that methylmercury, which could be normally a chloride or in the ionic form, is particularly a great problem because of the strong binding of mercury to thiol groups which can occur in sediments and through the whole environmental system. Of course, tetraethyl lead, which has been used in gasoline and still is, can't be that much of a problem. So that methylated and ethylated metals are not necessarily that great a problem as compared with monomethylmercury which is so serious if present in large enough quantities.

TORIBARA: In various conferences I have been to it's been
obvious that the form of the metal is going to be increasingly
the subject of investigation. Most methods so far measure
total amounts, but I notice that many methods are being
developed on the proportion of the various species of that
metal so that more precise correlations with the toxicity may
be made.

DOHERTY: You mentioned as fish grew larger there would be
growth dilution in terms of concentration of mercury. That
seems illogical to me if there is a constant caloric
requirement per unit increment in body weight. How do you
explain the dilution in terms of amount? How much mercury is
entering the fish by diffusion, in terms of methylmercury? How
much methylmercury is being formed? What are the upper limits
in both of these by chemical conversion?

JERNELÖV: To the first question, any organism will have
basically three uses of energy: first, it has a type of basic
metabolism to maintain life functions of various sorts and
replace enzymes and proteins and whatever is lost; then it has
a need for energy to move around, and then thirdly, if it still
has an excess of energy, it can transfer these into fat and
muscles to gain weight and size. This means if you have a very
low input, it is a small fraction of this that can be used to
grow. If you have a higher input, it's a higher fraction of
the total that's used to grow. And this means that if they
grow fast, they'll basically have a better conversion rate and
that means they need less food.

To the second question, which involves chemical alkylation,
there are at least three or four different aspects to it. One
type of chemical alkylation is a transalkylation from other
metals, from methyl lead or methyl tin. Then the extent of it
is only dependent on the amount of other alkylated metals.

Next there is a type of chemical alkylation which is
difficult to classify as purely chemical, and that is when
substances as humic acids and humic substances are involved.
There are some groups, methyl groups, that are sticking out,
pictorially speaking, to which mercury will attach. So a humic
acid with a methyl group with the mercury on it, may be
considered methylmercury or not. But when it is extracted for
analysis, it will become methylmercury. No doubt about it.
This type of available methyl group will increase when the
bacteria attack the humic acids, and in the gradual breakdown
of humic acids more and more of it is produced. I don't know
whether this a chemical alkylation. It is biochemical in a way
because it occurs after the extracellular enzymes attack the

humic acids. On the other hand, it doesn't occur when fed to
fish, for example. The fish does not accumulate this
methylmercury attached in this form. If fed to an organism
with low pH in the stomach, the same reactions as in your
chemical analysis will occur. I don't know the total
theoretically possible extent of this humic acid type of
chemical alkylation.

The third type you can find may be considered an
atmospheric gas reaction with elementary mercury to form
methylmercury. And this process, as far as I know, is only
known as a possible principle. So I'm afraid in most cases
there is no quantitative answer.

McLEAN: I think you're referring to the work of Rogers[+] on
the methylation of mercury by soil extracts containing
fulvates. That is the only conclusion that one can make if
this does take place, and you seem to cast some doubt on it.

[+]Rogers, R.D., "Abiological Methylation of Mercury in Soil"
J. Environ. Qual. 6, 463 (1977).

JERNELÖV: No, it does take place. It's just a question of
when you start calling it methylmercury.

McLEAN: The conclusion is that there is methylation of mercury
(0) by a carbonium ion (CH_3^+) rather than methylation of
mercury (II) by carbanion (CH_3^-) and therefore, in this
case the generally accepted mechanism involving vitamin B_{12}
cannot work.

WOOD: Humic substances provide a wealth of free radical
chemistry, and so the transfer of $CH_3\cdot$ (methyl-radicals)
should not be ruled out.

SESSION III:
EFFECTS ON PLANTS

FOLIAR RESPONSES THAT MAY DETERMINE PLANT INJURY BY SIMULATED

ACID RAIN

LANCE S. EVANS
LABORATORY OF PLANT MORPHOGENESIS, MANHATTAN COLLEGE
BRONX, NY 10471 and
DEPARTMENT OF ENERGY AND ENVIRONMENT, BROOKHAVEN
NATIONAL LABORATORY, UPTON, NY 11973

ABSTRACT

Experiments were performed to categorize the responses of
foliage of several plant species after exposure to simulated acid
rain in order to predict the relative sensitivities of plants to
acid precipitation in nature. The present investigations were
performed to (1) identify leaf indumentum responses, (2) determine histological responses, and (3) determine whole plant and
individual leaf responses that may be used to diagnose acid rain
injury. Plants were exposed to simulated rain at pH levels of
5.7, 3.4, 3.1, 2.9, 2.7, 2.5, and 2.3. Sporophyte leaves of
bracken fern (P. aquilinum) and foliage of pinto bean, soybeans,
and sunflower were most sensitive to simulated acid rain among
the species tested. About 5% of the surface area of older leaves
of sunflower, soybeans, and pinto beans was injured after exposure to 4 rainfalls at pH 2.5 (a single-six min rainfall every
four days). Foliage of pin oak (Q. palustris) exhibited less than
one-percent leaf area injury after exposure to simulated rain at
pH 2.5 after 10 rainfalls (one-twenty min rainfall daily). The

Research supported in part by United States Department of
Energy Contract #EY-76-C-02-0016 and in part by Associated
Universities Contract #469167. By acceptance of this article,
the publisher and/or recipient acknowledges the United States
Government's right to retain a non-exclusive, royalty-free
license in and to any copyright covering this paper.

responses of poplar (<u>Populus</u> sp.) and spiderwort (<u>Tradescantia</u>
sp.) were intermediate between these two extremes. Histological
observations show that lesion development results in collapsed
leaf tissue in most sensitive species. Gall formation that resulted from both cell hypertrophy (abnormal cell enlargement) and
hyperplasia (abnormal cell proliferation) occurred in lesions of
spiderwort, poplar, and oak. Limited hyperplastic and hypertrophic reactions occurred in soybean foliage after exposure to simulated acid rain but no leaf galls resulted. Sporophyte foliage
of <u>P</u>. <u>aquilinum</u> and leaves of pinto bean and sunflower exhibited
neither hyperplasia nor hypertrophy after exposure to simulated
acid rain. Injury occurred most frequently near vascular tissues
and trichomes in all species. In general, plant species that show
cell hyperplasia and hypertrophy of leaf tissues after exposure
to simulated acid rain are injured less than species that do not
show these responses.

INTRODUCTION

Sulfur and nitrogen oxides emitted into the atmosphere may
react with atmospheric moisture to form acidic solutions. Natural
rain water, in equilibrium with ambient CO_2 concentrations has an
approximate pH of 5.6. At the present time precipitation in the
northeastern United States has an average annual pH of approximately 4.4, with individual rainfalls as low as pH 2.1 (Likens
et al., 1972). This increased acidity in precipitation, attributed to oxidation and hydration of oxides of sulfur and nitrogen
from anthropogenic sources, may have detrimental effects upon
plant foliage (Jonsonn and Sundberg, 1972; Wood and Bormann, 1974;
Ferenbaugh, 1976; Shriner, 1976; Evans, Gmur, and Da Costa, 1977,
1978; Evans and Curry, in press).

Pathological changes that can alter the metabolism of an
organism may be manifested on the cellular and tissue levels.
Alterations at these hierarchial levels in one species may be
used as tools to predict the influences that these malfunctions
may have on other animals and plants. It would be advantageous
if the cell and tissue responses could be used as early indicators of a pathological condition. If the responses of many
organisms may be categorized then a list of symptoms can be used
to mark the progressive stages of the pathological condition.

Not all organisms may respond in the same manner or to the
same extent to a particular level of an abiotic or biotic influence. Because of innate differences between various animal or
plant species, some organisms may be less sensitive than others
to an environmental stress or pathogen. These differences in
relative sensitivity may be reflected by cell or tissue responses.

If the differences in responses of cells or tissues can be systematically characterized among various organisms, maybe they could
be used to help predict the relative responses or extent of injury of other species not exposed to the environmental stress or
pathogen.

The aim of the research presented here is to try to use visual, scanning electron micrographs, and histological preparations
as tools to predict the relative sensitivity of various plant
species to simulated acid rain. It is hoped that these results
might enable a prediction of the relative sensitivities of major
plant groups of economic and aesthetic interest to air pollutants.

MATERIALS AND METHODS

Plant exposures -- Rhizomes of Pteridium aquilinum were pur-
chased from Panfield Nurseries Inc. (Huntington, NY 11743). Clo-
nal plants of Quercus palustris cv. Sovereign were obtained from
Princeton Nurseries (Princeton, NJ 08540). Three clones of
Tradescantia species were obtained from the late Dr. Arnold Spar-
row (Biology Department, Brookhaven National Labs, Upton, NY
11973). Stem cuttings of six clones of hybrid poplar (Populus
sp.) were obtained from Dr. K. Jensen (United States Department
of Agriculture, Forest Service, Delaware, OH 43015; see Evans,
Gmur, and Da Costa, 1978 for information about clone numbers and
hybrids). Seeds of pinto bean (Phaseolus vulgaris, Univ. of
Idaho, Asgrow Seed Co., Kalamazoo, MI), sunflower (Helianthus
annuus, cv. Teddy Bear, Asgrow Seed Co., Kalamazoo, MI), and soybean (Glycine max "Amsoy 71" Asgrow Seed Co., Kalamazoo, MI) were
planted in 20 cm plastic pots and were allowed to germinate for
several days.

Plants were propagated in a closed greenhouse equipped with
activated charcoal air-filters. The air temperature was maintained between 23-27 C. Plants were transferred to a controlled
environment chamber with activated charcoal air-filters at least
3 days prior to each experiment. Plants were kept at 23 ± 1 C,
60-70% relative humidity, with an 18 hr light period daily. The
light intensity was 90 microeinsteins m^{-2} sec^{-1} at plant height
as measured with an LI-COR Quantom Sensor (LI-190S, LI-COR
Instrument Corp. Lincoln, NB 68504) that measures only the Photosynthetically Active Range (PAR) of irradiation.

Plants were transferred to a special rain chamber and exposed to a simulated sulfate acid rain solution for one, 20 min
period daily (except Tradescantia, 2 rainfalls daily). Descript-
ions of the chamber, the simulated rain solution, and the rain
drop size distributions are given elsewhere (Evans, Gmur, and Da
Costa, 1977). The pH levels of the simulated rain solutions were
5.7, 3.4, 3.1, 2.9, 2.7, 2.5, and 2.3. A plant was exposed to

only one pH level throughout the experiments. Between each rainfall, plants were returned to the controlled environment chamber.

   Histological preparations -- Leaf samples were fixed in acrolein, and dehydrated through a series of alcohols (Feder and
O'Brien, 1968). After dehydration, leaves were infiltrated and
embedded with paraffin and sections were stained (Shellhorn and
Hull, 1960).

   Scanning electron microscopy -- Leaf samples for scanning
electron microscopy (SEM) were fixed in acrolein and dehydrated
in ethanol. Samples were critical-point dried (Evans, Gmur, and
Da Costa, 1977) and a layer of carbon or silver was applied to
each sample in a vacuum evaporator with a rotating and precess-
ing stage. The scanning electron microscope (Model 700, Materials Analysis Co., Palo Alto, CA 94303) was operated at 5 kV or
10 kV.

   Density of trichomes and stomata -- Densities of trichomes
and stomata were determined from positive leaf impressions by
the silicone rubber technique of Sampson (1961). A total of
eight areas were sampled randomly on each of four leaf replicas.
No estimate of the functional capacity of stomata was obtained.

   Determinations of percent leaf area injured -- Lesions were
scored for percent leaf area injured. Topographical assessments
were made using a dissecting microscope. Lesions were categorized into four groups dependent upon the diameter of the injured
area (less than 0.25 mm, 0.25 to 1.00 mm, 1.00 to 2.00 mm, and
greater than 2.00 mm). The number of lesions per group per field
was tabulated to give the percentage of the leaf surface injured.

RESULTS

   Relative sensitivities among plant species -- The relative
sensitivities of foliage of several plant species are shown in
Table 1. Oak is least sensitive. Less than one percent of the
leaf is injured after 10 rainfalls of simulated acid rain at pH
2.5. Spiderwort and poplar exhibited intermediate responses.
Only 3.7 and 4.6 of the leaf area of spiderwort and poplar was
injured after numerous exposures to pH levels of 2.3 and 2.7, respectively. The value for poplar is high because the percentage
reflects a significant contribution by galls (see below). Soybeans, sunflower, pinto bean, and bracken fern exhibited the most
injury. About five to seven percent leaf area was injured in
pinto bean, sunflower, and bracken fern after exposure to simulated rain of pH 2.5. About six percent of the foliage of soybeans
was injured after six rainfalls of pH 2.7.

Table 1. Relative sensitivities of various plant species to simulated acid rain

Plant species	Rain pH	Number of rainfalls[a]	Duration of experiment (days)	Percent of leaf area with lesions
Oak (Quercus palustrus)	2.5	10	10	0.6
	2.7	"	"	0.1
	3.1	"	"	0.1
Spiderwort (Tradescantia sp.)	2.3	16	9	3.7
Poplar (Populus hybrids)	2.3	6	6	8.4[b]
	2.5	"	" •	5.9[b]
	2.7	"	".	4.6[b]
	2.9	"	"	4.6[b]
	3.1	"	"	2.2[b]
Soybeans (Glycine max)	2.7	6	6	5.7
	2.9	"	"	4.9
	3.1	"	"	2.8
	3.4	"	"	1.4
Sunflower (Helianthus annuus)	2.3	4	12	10
	2.5	"	"	5
	2.7	"	"	1
	2.9	"	"	1
	3.1	"	"	0
Pinto bean (Phaseolus vulgaris)	2.3	"	"	10
	2.5	"	"	5
	2.7	"	"	5
	2.9	"	"	1
	3.1	"	"	1
Bracken fern (Pteridium aquilinum)	2.5	"	"	7.4
	2.7	"	"	6.1
	3.1	"	"	0.9
	3.4	"	"	0.8

[a] Rainfalls were always 20 min in duration daily except for sunflower and pinto bean (3 min in duration). For spiderwort, two rainfalls were given daily except the first and last day. For sunflower and pinto beans, one rainfall was given every 3 days.

[b] These values include galls and lesions. Histological preparations of galls on poplar leaves suggest that a photosynthetically active mesophyll remains intact.

Leaflet sensitivity vs development -- In experiments to view
relative sensitivity versus age of first trifoliate leaflets of
P. vulgaris, plants were exposed to three daily, 6-min exposures
to sulfate acid rain, pH 2.7 (Fig. 1). The density of lesions is
low in very young leaves (before the maximum rate of leaf expansion). The curve of lesion density is very similar to the curve
of total leaflet area. Densities of stomata and trichomes were
very high in young leaflets (10-day-old plants) but decreased to
very low levels before the maximum rate of leaflet expansion (14-
day-old plants). The leaflet stage with the greatest density of
both stomata and trichomes was not the age of maximum sensitivity
to simulated acid rain. A difference in sensitivity during leaf
development is also present in soybeans. Foliage of soybeans exhibited lesions on about 6 to 7% of the leaf surface after 6,
daily exposures to simulated rain of pH 2.7 (Fig. 2). Exposure
to rainfalls of pH 2.9 produced lesions on about 5% of the leaf
surface. In general, younger soybean leaves were more sensitive
than fully expanded leaves. Percent leaf area injured was greater
after four rainfalls than six rainfalls because the leaves became
less responsive to acid rain as leaf expansion continued. In
this manner, the number of lesions per leaflet did not increase
significantly after day 4 but leaflet area increased markedly.

Lesion localization vs leaf surface structures -- In most
of the plant species studied about 75% of all lesions developed
near veins (vascular tissues). In Figures 3 and 8 a stoma is
located among 3-5-flaccid epidermal cells injured after exposure
to low pH rain. In another case two stomata are shown among
several epidermal cells (Fig. 4). A slightly larger lesion in
Fig. 5 shows a stoma and a remnant of a trichome among injured
epidermal cells. Glandular (hydathodes) and spike trichomes
(Fig. 7) were both affected by simulated acid rain. Initial lesions were usually in epidermal cells adjacent to trichomes and
stomata. In our experiments lesions originated in epidermal cells
adjacent to trichomes: 75%, stomata:20%, and areas not associated
with either trichomes or stomata: 5%, respectively. No attempt
was made to separate percentages of lesions near hydathodes from
lesions near spiked trichomes.

Comparative leaf histological responses -- Lesion development in fronds of P. aquilinum and leaves of P. vulgaris and H.
annuus was very similar. In these three species initial injury
was characterized by collapse of one to six cells on the adaxial
epidermis. Normal turgid cells which comprised a continuous epidermis became flaccid with an initial increase in cellular stain
intensity. This tissue collapse resulted in a depression on the
leaf surface. Underlying mesophyll tissues showed little or no
distortion at this stage.

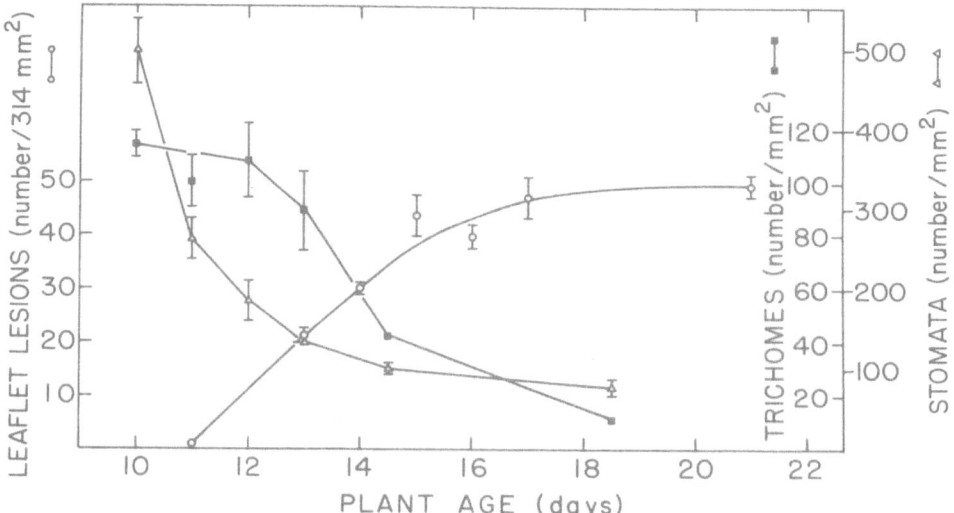

Fig. 1.   Relationships between density of middle leaflet lesions,
          density of stomata, and density of trichomes on first tri-
          foliate leaflets with plant age (Phaseolus vulgaris). Den-
          sities of stomata and trichomes are high in unexpanded
          leaves, which are quite insensitive to simulated acid rain
          than more expanded leaves that have lower densities. The
          leaflet areas for 12, 14, 16, and 18-day-old seedlings are
          1.3, 3.0, 3.6, and 3.8 x 10^{-3} mm^2, respectively.

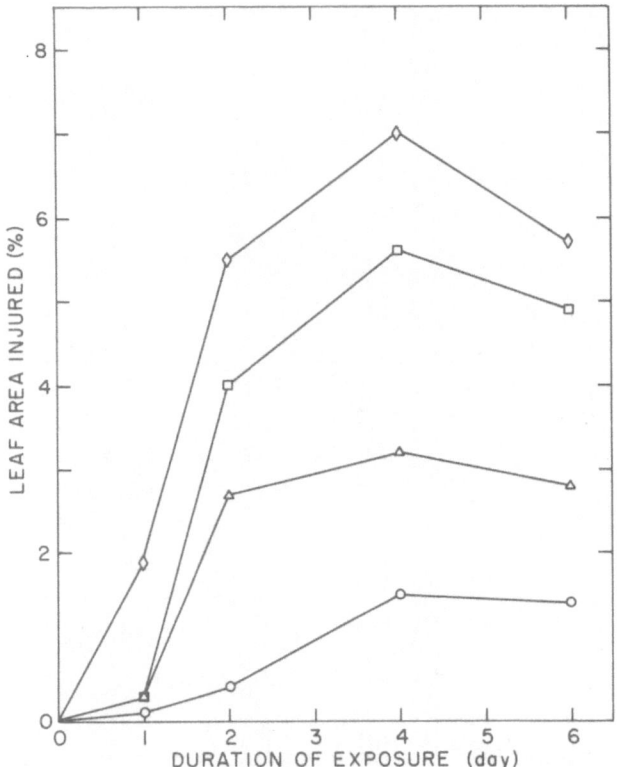

Fig. 2. Relationship between percent leaf area injured and
number of daily exposures (20 min each) of simulated rain
on middle leaflets of first trifoliate leaves of soybean
(<u>Glycine</u> <u>max</u>). Diamonds, squares, triangles, and circles
represent injury after exposure to simulated rain of pH
2.7, 2.9, 3.1, and 3.4, respectively.

At a later stage of injury palisade parenchyma cells collapsed adjacent to collapsed epidermal cells. The area of injured
epidermis in most lesions increased in diameter so that ten to
fifteen epidermal cells comprised the lesion when palisade cells

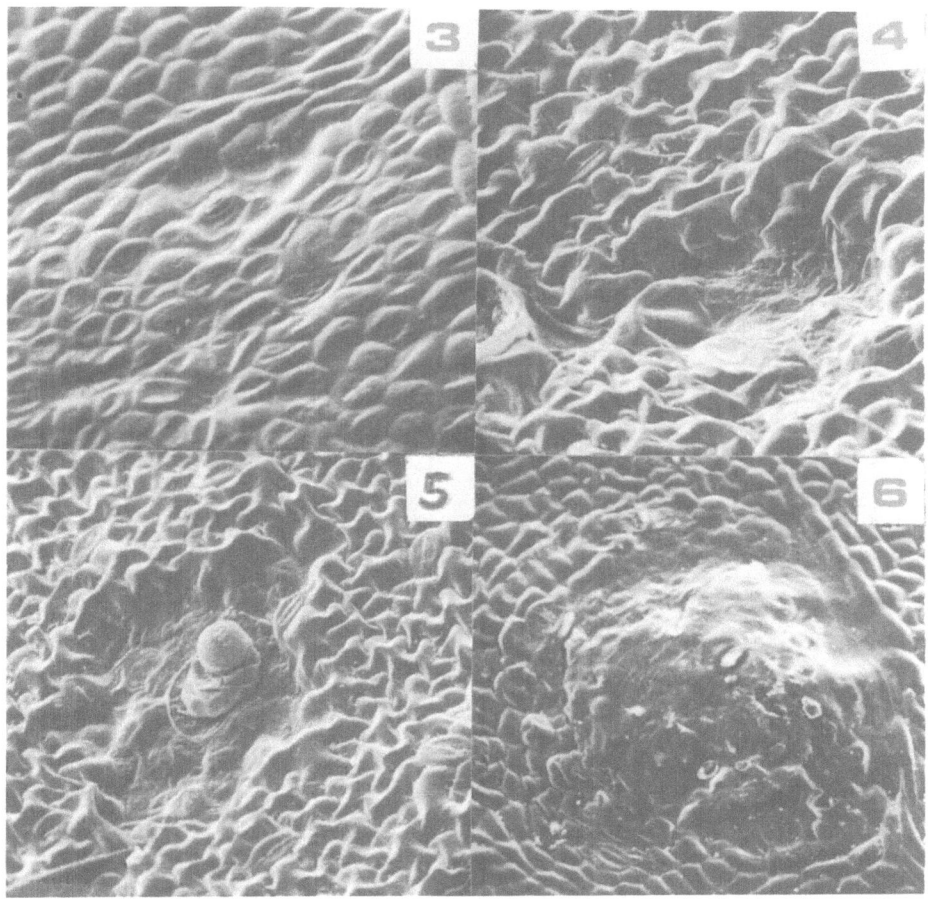

Fig. 3-6. Scanning electron micrographs of adaxial leaf surfaces of <u>Helianthus annuus</u>, <u>Phaseolus vulgaris</u>, and hybrid clones of <u>Populus</u> exposed to simulated acid rain pH 2.7 or pH 3.0 from one to several rainfalls. 3. Several flaccid subsidiary cells surround the guard cells of the stoma of <u>Helianthus annuus</u>. 4. Epidermal cells that surround the stomata are injured. The injured cells are surrounded by normal, turgid epidermal cells of <u>P. vulgaris</u>. 5. Collapse of several epidermal cells and a trichome hydathode on the leaf surface of <u>P. vulgaris</u>. 6. A gall on the leaf of <u>Populus</u>. The diameter of this large gall is about 0.3 mm.

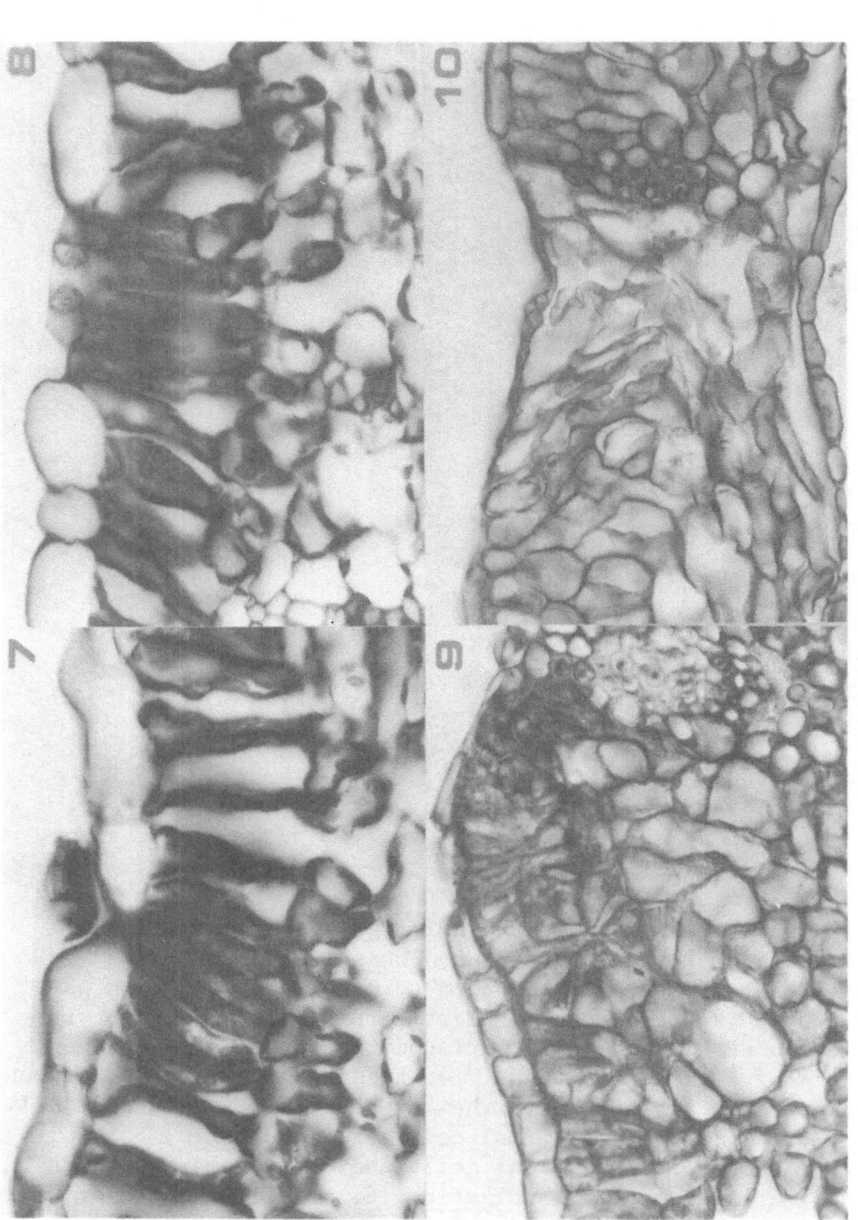

Figs. 7-10.

began to collapse. Cells in the first layer of the palisade
parenchyma tissue collapsed periclinally while those in the second palisade layer (if present) collapsed anticlinally. Supportive cells also collapsed when injury was present near vascular
tissues.

The third stage of lesion development was characterized by
collapse of spongy parenchyma cells adjacent to epidermal and
palisade parenchyma cells (Fig. 9). Initially, spongy mesophyll
cells exhibited cell wall distortions. Eventually they collapsed
completely. In the last stage of injury, all tissues except vascular tissues were completely collapsed.

The histological stages of lesion development in leaves of
G. max were similar to those in P. aquilinum, P. vulgaris, and H.
annuus. Lesions were initiated by collapse of several cells on
the adaxial epidermis. Epidermal cell collapse was followed by a
distortion of palisade parenchyma cells. This mesophyll tissue
layer exhibited extensive hyperplasia. Occasionally, slightly
enlarged cells were observed concomitant with this hyperplasia.
Hyperplasia and hypertrophy occurred prior to cell collapse.
These limited cellular responses were also present in paraveinal
and spongy mesophyll cells. Although cell hypertrophy was greatest in spongy mesophyll cells, these cells eventually collapsed
with continued exposure to simulated acid rain.

Foliage of six hybrid clones of poplar (Populus species)
were exposed to simulated rain at various pH levels. The responses of two clones (8 and 327) of poplar were markedly different
from those of the other four clones used. In clones 8 and 327,
exposure to simulated acid rain resulted in gall-like outgrowths
after one or two, 6-min daily exposures (Fig. 6 and 10). The
epidermis of the gall and the tissues that surround the gall col-

---

Fig. 7-10. Leaf cross sections of Phaseolus vulgaris, hybrids of Populus sp., and Quercus palustris after one to
several rainfalls of simulated acid rain of low pH. Fig. 7.
Initial injury of epidermal cells near a trichome hydathode.
Note injury is greatest at the anticlinal walls of epidermal
cells at the base of a hydathode on P. vulgaris. Fig. 8.
Initial injury of subsidiary cells near the guard cells of a
stoma. The cells of the palisade parenchyma appear normal
on P. vulgaris. Fig. 9. A large lesion with an injured adaxial epidermis, palisade parenchyma, and spongy parenchyma
near vascular tissues in poplar. Fig. 10. A cross-section
of a gall on a leaf of Q. palustris. Note both collapsed
epidermal and palisade cells. Hypertrophy and hyperplasia
of spongy mesophyll cells are also evident.

lapsed. The collapse of epidermal cells around the gall formed a
depression on the leaf surface. This depression was present in
interveinal tissues but not if galls were near or within vascular
tissues. On occasion, galls were observed on clones No. 207 and
No. 211 but none was present on clones No. 43 and No. 44.

In a later stage of gall formation, cells of both the palisade and spongy parenchyma areas exhibited hypertrophy and hyperplasia in response to simulated acid rain. These cell abnormalities generally caused elevations on both adaxial and abaxial
leaf surfaces.

The responses of clones No. 207 and No. 211 were intermediate
between: (i) those of both No. 43 and No. 44, and (ii) those of
clones No. 8 and No. 327. Clones No. 8 and No. 327 exhibited
galls with abnormal cell enlargement and cell proliferation, but
clones No. 43 and No. 44 did not produce either abnormal cell enlargement or abnormal cell proliferation.

Within cross-sections of _Tradescantia_ there is a regular
distribution of both trilaminar epidermal complexes and vascular
bundles within the leaf mesophyll. The leaf mesophyll consists
of parenchyma cells which are uniform in size and shape. Initial
exposure to simulated acid rain of pH 2.7 resulted in the collapse
of outer epidermal cells in some areas. Distortion of underlying
epidermal cells in the trilaminar sites and disruption of upper
mesophyll cells also occurred. The next stage of lesion development varied slightly among lesions. In most lesions, collapse of
mesophyll cells near the adaxial epidermis was evident. However,
cell hypertrophy was also observed in about 50% of all lesions.
Hypertrophic cells attained a cell volume three times that of unaffected cells. Supportive cells near vascular bundles were usually partially collapsed at this stage.

At a later stage a complete breakdown of cellular organization from the adaxial epidermis to the lower mesophyll layer
occurred. A depression of the adaxial leaf surface was evident
upon microscopic examination. Necrotic cells appeared flaccid
and stained heavily. Cells in lesions containing zones of hypertrophy exhibited an increase in cross-sectional area due mostly
to an elevation of the adaxial leaf surface. Most affected mesophyll cells exhibited hypertrophy after collapse of epidermal and
supportive cells. Hyperplasia of mesophyll cells was not present
with hypertrophy in _Tradescantia_. Epidermal and vascular cells
did not exhibit any hypertrophy. In addition, it appeared that
vascular tissues were unaffected by acid rain exposure.

Each stage of lesion development showed a progressive increase in cell deterioration from the leaf exterior toward the
interior. Lesions were observed on the abaxial leaf surface at

the apical end (tip) of the leaves. This was attributed to the curvature of the leaf along the margin which allowed exposure of the lower surface to rainfall.

Leaves of pin oak showed a different sequence of steps in leaf injury from that seen in other plant species. Initially, injury began with collapse of adaxial epidermal cells. Initial lesions consisted of approximately one to six collapsed epidermal cells which resulted in a slight depression on the leaf surface. At this stage, all cells in the mesophyll layers, and the abaxial epidermis were unaffected. After several more simulated acid rainfalls larger lesions, with increased surface area and depth, developed from smaller lesions. These lesions encompassed five to fifteen collapsed epidermal cells. Penetration of acidic solutions into the mesophyll tissues resulted in collapse of epidermal and underlying palisade parenchyma cells.

The third stage of injury produced lesions with mesophyll tissue abnormalities. Epidermal cell collapse occurred concomitant with collapse of palisade parenchyma cells. Hyperplastic and hypertrophic responses of spongy mesophyll gave rise to a gall (Fig. 10). As cell enlargement and an increase in cell numbers occurred, the depression in the leaf surface became less evident and eventually gave rise to a convex protrusion on the leaf surface. The fourth stage of injury which was normally characterized by exhibited necrosis of all tissue layers, did not occur in oak leaves as it did in P. aquilinum. As a result, lesions of oak leaves were not observed past the third stage.

DISCUSSION

Lesion development after acid rain exposure is localized. Lesions were localized initially near trichomes and stomata on adaxial leaf surfaces. Scanning electron micrographs showed that 75% of all lesions began at the bases of spiked and glandular trichomes while 20% originated at or near stomata. Very few (5%) lesions began in areas unassociated with these structures.

Our results indicate that once a lesion develops, injury to adjacent cells occurs only in a localized fashion, also. Increase in lesion dimensions both in surface area and in internal leaf structures affected after intermittent exposures to acid rain. Once a lesion is formed in several plant species, it serves as a depression for collection of subsequent rain. Enlargement of these small lesions is probably accelerated by this preferential pooling of these latter rainfalls.

Injury to foliage of many species is related to leaf development. In general, leaves that are expanding rapidly or re-

cently expanded are most sensitive to simulated acid rain. Very
small, immature and older leaves are more insensitive (Wood and
Bormann, 1974; Evans et al., 1977; 1978; Evans and Curry, in
press). The reason(s) for this age-dependent relationship remains unknown.

Reports by Wood and Bormann (1977), and Wood and Pennypacker
(1976) show that pine needles are rather insensitive to simulated
acid rain. When these results are compared with those for foliage
of birch (Wood and Bormann, 1974), maple (Wood and Bormann, 1975),
poplar and oak, the foliage from broadleaved trees is injured to
a greater extent. These results suggest that the deciduous forests
of eastern North America would be more sensitive to acid precipitation than the predominantly coniferous forests of Scandinavia.
More experimentation is necessary to make more definitive comparisons between deciduous and coniferous plant species.

Foliage of herbaceous species such as bracken fern, soybean,
pinto bean, and sunflower are very sensitive to simulated acid
rain. Foliage of Tradescantia and poplar exhibited less visual
injury compared with these four species while pin oak exhibited
the least amount of visible injury among these species. From
these experimental results, it may be suggested that foliage of
broadleaved herbaceous plants is more sensitive than foliage of
woody broadleaved plants. More plant species should be tested
to see if this relationship is correct.

Oak leaves are quite insensitive to simulated acid rain. Oak
leaves may exhibit little foliar injury because hyperplastic and
hypertrophic reactions occur in the spongy mesophyll when the
overlaying palisade cells collapse. These rapid cell division
and cell enlargement responses produce a gall that elevates the
leaf tissues above the level of the preceding epidermis. Similar
responses are exhibited by some clones of poplar and Tradescantia.
In pinto bean, sunflower, and fern sporophyte foliage, these
hyperplastic and hypertrophic responses are not apparent. The
depression of the leaf surface in these species may provide locations for pooling of water from subsequent rainfalls. This
preferential pooling may accelerate lesion development.

The experimental results demonstrate that initial injury to
foliage by simulated acid rain preferentially affects the leaf
indumentum near trichomes and vascular tissues. Characteristics
of the adaxial leaf surface are probably important after one or
two rainfalls. However, after many rainfalls, in which epidermal
cells are injured, other characteristics of the leaf may become
important. In general, foliage of plant species that exhibits
extensive hyperplastic and hypertrophic responses has less injury
than foliage of species that have these two responses. In this
manner, the hyperplastic and hypertrophic responses of foliage
may alleviate extensive foliar injury.

REFERENCES

Evans, L. S. and Curry, T. M. (in press) Differential responses
    of plant foliage to simulated acid rain. Amer. J. Bot.

Evans, L. S., Gmur, N. F. and Da Costa, F. (1977) Leaf surface
    and histological perturbations of leaves of Phaseolus vulgaris
    and Helianthus annuus after exposure to simulated acid rain.
    Amer. J. Bot. 64: 903-913.

Evans, L. S., Gmur, N. F. and Da Costa, F. · (1978) Foliar re-
    sponse of six clones of hybrid poplar to simulated acid rain.
    Phytopathology 68: 847-856.

Feder, N., and O'Brien, T. P. (1968) Plant microtechnique:
    Some principles and new methods. Amer. J. Bot. 55: 123-142.

Ferenbaugh, R. W. (1976) Effects of simulated acid rain on
    Phaseolus vulgaris L. (Fabaceae). Amer. J. Bot. 63: 283-288.

Jonsonn, B. and Sundberg, R. (1972) Has the acidification by
    atmospheric pollution caused a growth reduction in Swedish
    forests? A comparison of growth between regions with differ-
    ent soil properties. In "Supporting studies to air pollution
    across national boundaries. The impact on the environment of
    sulfur in air and precipitation. Sweden's Case study for the
    United Nations Conference on The Human Environment". B. Bolin
    (ed.). Royal Ministry of Foreign Affairs, Royal Ministry of
    Agriculture, Stockholm.

Likens, G., Bormann, F. H. and Johnson, N. M. (1972) Acid rain.
    Environment 14: 33-44.

Sampson, J. (1961) A method of replicating dry or moist surfaces
    for examination by light microscopy. Nature 191: 932-933.

Shellhorn, S. J., and Hull, H. M. (1961) A six-dye staining
    schedule for sections of mesquite and other desert plants.
    Stain Technol. 36: 69-71.

Shriner, D. S. (1978) Effects of simulated acidic rain on host-
    parasite interactions in plant diseases. Phytopathology 68:
    213-218.

Wood, F. A. and Pennypacker, S. P. (1976) Etiological study of
    the short-needle disease of pine. Page 977-978. In "Proceed-
    ings of the First International Symposium on Acid Precipita-
    tion and the Forest Ecosystem." U.S. Forest Service General
    Technical Report NE-23. U.S. Forest Service, Upper Darby, PA.

Wood, T. and Bormann, F. H.   (1974)   The effects of an artificial
    acid mist upon the growth of Betula alleghaniensis Britt.
    Environmental Pollut. 7: 259-268.

Wood, T. and Bormann, F. H.   (1975)   Increases in foliar leaching
    caused by acidification of an artificial mist. Ambio 4: 169-171.

Wood, T. and Bormann, F. H.   (1977) Short-term effects of a simu-
    lated acid rain upon the growth and nutrient relations of
    Pinus strobus L.   Water, Air, and Soil Pollut. 7: 479-488.

DISCUSSION

ELIAS: How did you determine that the epidermal cells
collapsed before the palisade cells expanded?

EVANS: By taking histological samples at different time
periods after the initial rain treatment. On poplar, for
example, we sectioned over 700 leaf pieces for those particular
studies. So we're very confident that there is some cell wall
deformation of epidermal cells before palisade cell enlargement.

McFEE: How soon after the acid precipitation did these lesions
appear?

EVANS: We found galls and lesions one day after exposure to
one-20 min rainfall of simulated rain of pH 2.9.

McFEE: When do the cells around these stomata collapse?

EVANS: Within one day after exposure to simulated acid rain.

LILJESTRAND: I'm interested in the leaf-area injured
versus-age relationship you showed. Do you think there's
repair by the ongoing cells or is there just a change in the
type of cells that are on the surface as the plant grows?

EVANS: The second part of your question is difficult to
answer. There is an increased enlargement of the leaf over the
time period shown. The percent leaf area that was actually
injured was reduced since the number of lesions remained the
same.

ROBERTSON: Why do you think the monocots are less sensitive
than the dicots?

EVANS: We think that there's much more runoff on monocot
leaves such as Tradescantia because they're usually in a much
more vertical position. When a leaf, such as on Trad, droops,
rainwater starts to drip right away. In expanded leaves of
broad-leaved-plants, the leaves are more horizontal and the
rain could pool in more areas. We've also found that the
cuticle is fairly thick on Tradescantia.

SHRINER: Just to expand on that a little more, patterns of
venation on leaves seem to contribute to this pooling which
results in the effects. With monocots, of course, you have a
parallel venation pattern which tends to drain the leaves
rather than to collect surface moisture.

KADLECEK: With respect to conifers, when the leaf starts to
fold back from an emerging fascicle or at least a pack of
needles, it seems to me you've got the pocket you need to hold
the water. Since that's also the growing tissue, would that
not seem to be a fairly sensitive area and explain the
sometimes- seen necrotic bands that seem to develop outwardly
with time?

EVANS: Yes, there has been some work done in that area by
Clancy Gordon* at the University of Montana (Gordon, 1972). He
suggests that cells in the fascile are particularly sensitive
to acid rain. However, others have also done those
experiments** (Wood and Pennypacker, 1976; Jacobson, personal
communication) and not come up with quite the sensitivity that
Dr. Gordon finds. I haven't worked with it myself directly but
suspect that needle enlargement or elongation would be a very
sensitive stage. To the actual extent that is occurring in the
field or in the laboratory, I don't really know.

*Gordon, C.C. 1972. Short-long conifer needle syndrome.
Interim Report to the Environmental Protection Agency.

**Wood, F.A. and S.P. Pennypacker 1976. Etiological study of
the short needle disease of pine. pages 977-978. In
Proceedings of the First International Symposium on Acid
Precipitation and the Forest Ecosystem. USDA Forest Service
General Technical Report NE-23. Upper Darlay, Pennsylvania.

Miller: Could you comment on the thickness of the cuticle and
the possibility that the location of the lesions could be
related to cracks in cuticles?

EVANS: Yes, several of my cohorts here have that particular
philosophy. From our experiments it looks as if simulated acid
rain may injure the epicuticular waxes that are on the leaf
surface. I think Dr. David Shriner*** (1974) has shown that.
I don't think, however, that the acidic solutions that we're
using will significantly degrade the cuticles themselves,
because isolated cuticles are not degraded by very strong
sulphuric acid with a little zinc chloride. The chemical
solution eats away everything except the cuticle. So I really
don't think that the solutions we're using are degrading the
cuticle appreciably.

***Shriner, D.S., 1974. Effects of simulated rain acidified
with sulfuric acid on host-parasite interactions. Ph.D.
Thesis. North Carolina State University, Raleigh, 79 pages.

MILLER: But what about cuticular cracks already existing?

EVANS: Some people don't believe that there are cracks in the cuticle ****(Martin and Juniper, 1970). Of course, since acid rain seems to affect particular locations of foliage I would suggest that there is evidently preferential penetration or action of aqueous solutions in particular locations on plant surfaces. And tricomes have been shown to be involved in this to a much greater extent than other epidermal cells *****(Evans and Curry, 1979).

****Martin, J.J. and B.F. Juniper 1970. The cuticles of plants. St. Martin's Press, New York.

*****Evans, L.S. and T.M. Curry 1979. Differential responses of plant foliage to simulated acid rain. Amer. J. Bot. 66: 953-962.

VERMEULEN: What's the sensitivity for tomato plants?

EVANS: I have no idea.

JACOBSON: We did some work with tomatoes in the greenhouse and obtained foliar lesions at acidities between pH 2 and 3 by repeated exposure. They're not as susceptible as bean, sunflower, spinach or lettuce.

VEGETATION SURFACES: A PLATFORM FOR

POLLUTANT/PARASITE INTERACTIONS[1]

D. S. Shriner

Environmental Sciences Division
Oak Ridge National Laboratory
Oak Ridge, Tennessee 37830

ABSTRACT

Above-ground surfaces of higher plants are the site of deposition of a variety of airborne biological materials and atmospheric pollutants. These biological materials (pollen, fungal spores, bacterial cells, waste from insect feeding) and the pollutants (as aerosols, gases, or dissolved in rain) may interact to influence the host plant, the pathogen, or the host/pathogen association. The conditions of the plant's surface environments can affect the behavior of a pathogen and alter the chemical characteristics of throughfall or stemflow originating at those surfaces. Host-pathogen interactions represent a sensitive indicator of plant stress, since the changes in balance between host and parasite may reveal a subtle abiotic stress long before the response would be detectable in the healthy plant alone. The impact of wet and dry deposition of pollutant materials on the plant surface alone, and on potentially pathogenic organisms, is necessary in order to fully interpret many plant:parasite interactions. This paper discusses examples of pollutant-parasite interactions at the leaf surface to illustrate the role that pollutants can play in the relationship of susceptible host and virulent pathogen.

---

[1]Research sponsored by the Office of Health and Environmental Research, U.S. Department of Energy, under contract W-7405-eng-26 with Union Carbide Corporation. Publication No. 1373, Environmental Sciences Division, ORNL.

259

Introduction

The above-ground portions of plants represent a large mass of tissue which is potentially vulnerable to impact from atmospheric contaminants. In terrestrial ecosystems, leaf area is commonly used as an index of photosynthetic biomass. Depending on vegetation type, structure, and maturity, the leaf surface area exposed to light may vary from as little as double to as much as 10 to 12 times the ground surface area. Photosynthetic biomass is generally regarded in plant systems as the most sensitive tissue to atmospheric contaminants. The responses of leaf surfaces are of importance in evaluation of the impact of specific forms of atmospheric contaminants on vegetation because of their sensitivity to stress, physiological importance to the plant, and large surface area.

Leaf surfaces are continually changing. Their physical and chemical characteristics can vary significantly, even on a single plant, depending on age and environmental variables. The degree to which a vegetation surface is an efficient scavenger of suspended material in the surrounding air mass is, to a large degree, a function of the same variables.

The air masses to which leaves of plant canopies are continually exposed are also dynamic in nature. For the purposes of discussion, let us consider a few of the types of biotic and abiotic contaminants most commonly impacting vegetation:

(1) Dry particulate matter – dust, fly ash, organic residues from insect feeding, etc.
(2) Aerosols – both neutralized and acidic.
(3) Gases – a wide variety, both inorganic and organic.
(4) Biotic propagules – pollen grains, fungus spores, bacterial cells, virus particles.
(5) Precipitation – in the form of rain, mist, snow, or fog.

The objective of this paper will be to review and illustrate some of the more important factors involved in the interaction of plant (foliage) surfaces with pollutants and plant parasites.

Deposition to the Plant Surface

Deposition and retention of particulate matter by leaf surfaces may be approximated by leaf washing to remove those particulates lodged on the leaf surfaces.[13,25] Using such a technique, estimates of leaf surface loading of airborne fly ash have ranged as high as 100 $\mu g/cm^2$ of leaf area (unpublished data, Shriner).

Particulate matter may physically become a factor in leaf physiological function if deposition is heavy enough to block light (critical to the photosynthetic machinery of the leaf) or if stomata become occluded by the particles. In the latter case, both uptake of carbon dioxide for photosynthesis and water loss by the plant could be reduced. Particulate matter also has a potential to chemically alter the micro-environment of the leaf surface. Trace concentrations of heavy metals associated with deposited particles, as well as strong acid anions, may significantly modify the chemical nature of water drops or films of water in contact with plant surfaces.

For a majority of fungal and bacterial plant pathogens, infection occurs from water droplets or films of free moisture on the plant surface, or in an atmosphere of greater than 95% relative humidity. For many of these pathogens, primary and secondary dispersion occurs in rain splash or wind-blown rain, as well.

Changing the character of the leaf surface may be the most significant pathway through which plant host-parasite relationships are modified by pollutants in wet and dry deposition. Several mechanisms are potentially involved, and will be discussed in order:

(1)   Direct chemical effects of polluted rain on the leaf epidermal tissues.
(2)   Direct chemical effects of polluted rain on the pathogen.
(3)   Combined physical/chemical degradation of protective waxes on the leaf surface by polluted rain.
(4)   Indirect effects on leaf surface microflora resulting from shifts in pH or nutrient leaching, and causing shifts in competitive microflora.
(5)   Pollutant-induced changes in the response of the plant to other pollutants, resulting in alteration of plant tissue susceptibility to infection.
(6)   Pathogen-induced changes in resistance or sensitivity to pollutant injury.

Direct Chemical Effects of Precipitation on Leaf Epidermal Tissues

Precipitation incident to a plant canopy can be relatively well characterized chemically. Table I shows the composition of "unpolluted" rain, based on analytical data for each of the constituents in precipitation and assuming equilibrium with atmospheric concentrations of carbon dioxide.[25] In this system, the solution pH would be expected to be approximately pH 5.65. The major strong acid anions which commonly occur in precipitation as a result of atmospheric pollution, and which contribute to the

Table I. Typical composition of rain water of pH 5.6

Ion	mg/liter	μeq/liter
CA^{++}	0.22	10.5
Mg^{++}	0.06	4.9
Na^{+}	0.12	5.1
K^{+}	0.08	2.3
NH_4^{+}	0.22	12.2
$SO_4^{=}$	0.53	11.0
NO_3^{-}	0.74	12.0
Cl^{-}	0.42	12.0
H^{+}	0.002	2.2
HCO_3^{-}	0.13	2.2

Source: Ref. 25.

formation of so-called "acid rain" (precipitation which has a pH lower than pH 5.6) are sulfate ($SO_4^{=}$), nitrate (NO_3^{-}), and chloride (Cl^{-}). The relative importance of these anions at a given site can be strongly influenced by local sources of pollutants, as well as long-range transport of the pollutants, depending on whether the sulfur and nitrogen oxides are incorporated in the precipitation through cloud-forming processes, or are "washed" out of the atmosphere by rain falling through a polluted air mass. At the present, most of the land area of the eastern United States is exposed to annual average rainfall acidity of pH 4.5 or lower.

The direct effects of acidic precipitation on vegetation have been characterized by a number of researchers.[29,11,3,23] Visible injury typically has been described as small necrotic lesions. The significance of these small lesions to host-parasite relationships has been discussed by Shriner.[24] Disease incidence was shown to increase as a function of plant exposure to simulated acidic rain of pH 3.2, especially in the case of facultative parasites, which are most successful as pathogens when using breeches in host defense as

a route of penetration. The formation of pollutant-induced necrotic
lesions should be expected to favor many plant pathogens which enter
the plant by colonization of dead tissue and advancement to living
tissues.

Considerable research has documented the importance of leaf
exudates, rich in mineral nutrients and carbohydrates, in providing
a source of nutrients for growth of pathogenic fungi in infection
drops.[1,16,21,6] These and other authors also have shown the
presence of pathogen-inhibiting substances in these leaf
exudates.[5,18] To the extent that the acidity of water droplets
leaching leaf tissues can alter the rate of extraction and/or the
composition of nutrients diffused out of the tissues, this more
indirect pathway may also affect host-parasite relationships.

Direct Chemical Effect of Polluted Rain on the Pathogen

A second major area where acidic precipitation can potentially
affect host-parasite relationships is that of direct chemical
effects on the microorganism. During dispersion, transport, and
germination of a fungus spore in a rain droplet, a period of several
hours could typically elapse during which the spore and thinner-
walled germ tube were in direct contact with acidic water. Data of
Leben,[12] evaluating the effectiveness of acidic buffer sprays
ranging from pH 3.2 to 6.0, suggested that growth and lesion production by each of 15 plant pathogenic fungi tested were significantly
reduced by the acidic conditions (pH 3.2).

Studies with simulated acidic rain found growth of the bacterial plant pathogen Pseudomonas phaseolicola totally inhibited by a
solution of pH 3.2.[24] When these solutions were used for
inoculation of Phaseolus vulgaris 'Red Kidney' test plants, no
infection or disease development was observed. Control plants
inoculated with solutions at pH 5.6 developed normal disease patterns. Additional studies with other host-parasite pairs are summarized in Table II.

Effect of Combined Physical/Chemical Weathering
of Protective Leaf Surface Waxes

The outer-most layer of leaf tissue exposed to pollutants is
the epicuticular wax layer. Purnell and Preece[17] and Martin and
Juniper[15] described by scanning electron microscopy the changes
in structure of epicuticular waxes on the surfaces of plants as a
result of weathering agents (rain, wind, dust, abrasion with foreign
objects). Rentschler[20] reported that the degree to which leaves
are coated with epicuticular waxes is important in the deposition
and retention of solid and liquid aerosols on leaf surfaces, and in

Table II.   Response of host/pathogen pairs to simulated acidic rain

Host/pathogen	Treatment pH	Treatment caused injury to host	Treatment inhibited growth of pathogen
Quercus phellos/ Cronartium fusiforme	3.2	Yes	Unknown
Zea mays/Helminthosporium maydis	3.5	No	No
Phaseolus vulgaris/ Meloidogyne hapla	3.2	No	Yes
Phaseolus vulgaris/ Uromyces phaseoli	3.2	No	No
Phaseolus vulgaris/ Pseudomonas phaseolicola	3.2	Yes	Yes

the absorption of gaseous pollutants. Shriner[22] found that in
addition to the physical aspect of raindrop weathering of these
waxes, the epicuticular wax of willow oak (Quercus phellos) appeared
to be more readily weathered by simulated rain solutions of pH 3.2
than by control solutions of pH 5.6. Irving and Miller[10] reported
decreased weight of cuticle from soybean (Glycine max) following
treatment by simulated acidic rain, also suggesting increased
weathering as a function of rain pH.

The increased erosion (weathering) of plant cuticles by simulated "rain" acidified with sulfuric acid may disrupt processes
that occur at the leaf surface. The most important function of
the cuticle is to reduce water (transpiration) loss.[8,15] Removal
of leaf surface waxes has been reported to increase wettability of
those surfaces.[8,15,20] Leaves which had been wetted and dried
are subsequently wetted more easily.[14] The water repellency of
a plant surface is important in preventing the establishment of
infections from water-borne inocula.[7,15] Dickinson[2] proposed
that the thigmotropic response of germ tubes of certain fungi is a
function of the structure and thickness of the cuticle. Both of
these features of cuticular waxes appear to be altered by simulated
rain acidified with sulfuric acid. The apparent loss of surface

waxes as a result of washing by simulated rain supports similar observations by Purnell and Preece,[17] and complements the data of Holloway[8] relating the loss of the hydrophobic characteristics of leaves to the loss of surface waxes.

The cuticle also has an important function in leaf-surface ion-exchange. Leaf surfaces of plants become more susceptible to leaching as wettability increases,[27] or if injured.[28] The buffeting action of rain may increase leaching.[20] Tukey[27] reports that leaching of cations involves exchange reactions on the leaf surfaces in which cations on exchange sites of the cuticle are exchanged by hydrogen from leaching solutions. Cations and organic metabolites may also move directly from the translocation stream within the leaf into the leachate by diffusion and mass flow through areas devoid of cuticle. Wood and Bormann[29] have reported a significant increase in the leaching of potassium, magnesium, and calcium from pinto bean and sugar maple leaves exposed to mists acidified to pH 2.3 with sulfuric acid. Fairfax and Lepp[4] also have reported increases in foliar leaching of calcium from tobacco leaves exposed to a simulated rain of pH 3.0.

The cuticle also provides a barrier to penetration by bacteria, viruses, and many fungi.[15] Enhanced development of a bacterial blight of bean plants previously exposed to simulated acidic rain was thought to have resulted from the creation of infection courts by damage to leaf surfaces by the rainfall, as mentioned earlier.

Other evidence suggests that superficial waxes may play a part in the repellency of some plants to insects.[15] These waxes also may be involved in the control of leaf temperature and the protection of leaf tissues from ultra-violet radiation.[15]

Because of the differences in temperature, light, relative humidity, soil moisture, and other factors, the cuticular development of plants grown in greenhouses is different from plants grown out-of-doors.[15] Cuticle thickness of leaves outdoors was greater than for plants grown in a greenhouse.[9] Since the observations of Shriner[22] dealt entirely with greenhouse-grown plants, one can only speculate whether similar effects would occur in plants grown out-of-doors under natural rainfall of equivalent acidity (pH 3.2).

The evidence above suggests that another of the means by which plants may be affected by acid rains in nature is through effects on the submicroscopic structure of the epicuticular wax layer. Such effects could in turn influence functions of the wax relating to plant water relations, protection against stress and radiation balance.

Indirect Effects on Leaf Surface Microflora

Pollutants in rain may alter the pH of the leaf surface. Shifts
in leaf surface pH may result in changes in microfloral populations.
As discussed earlier, certain classes of microorganisms, notably
bacteria, are quite sensitive to pH, and others, quite tolerant.
Although at the present time no research of this type has been conducted, the exposure of leaf surface microbial populations to acidic
rain could conceivably favor acid-tolerant microorganisms over less
acid-tolerant leaf surface microorganisms. Since phylloplane microorganisms are predominantly bacteria and yeast-like fungi,[26] it
would appear reasonable to assume a relatively high potential for
this type of interaction.

Pollutant-Induced Changes in the Response
of the Plant to Other Pollutants

Any exposure to pollutants which results in increased tissue
injury increases the potential for invasion by pathogens. When two
or more pollutants interact, increased tissue injury frequently
occurs. Here again, little exists in the way of direct evidence
for this type of effect involving acidic rain. Interactions between
sulfur dioxide, ozone, and nitrogen dioxide have been shown to
result in additive, greater than additive, or less than additive
increases in tissue injury.[19] Such increases in tissue injury
clearly increase the potential for invasion of the tissues by
saprophytes or facultative parasites.

Preliminary research in our laboratory on the effects of wet
or dry foliage on injury response of Phaseolus vulgaris 'Bush Blue
Lake 274' to ozone has shown that the rainfall pH (4.0 or 5.6) did
not affect ozone injury response following subsequent exposure to
0.15 ppm ozone for three hours. Wet foliage, regardless of pH,
however, did show up to 5% greater injury. When ozone concentration
was increased to 0.25 ppm, wet foliage experienced up to 13% greater
injury.

Pathogen-Induced Changes in Resistance
or Sensitivity to Pollutant Injury

Biotic agents may act (1) as pathogens, through their own
synthesis or through their direction of host synthesis of growth
regulatory substances, to significantly alter plant response to
pollutants; (2) as pathogens, to interfere directly with stomatal
action and/or membrane integrity; or (3) as symbionts, through
mycorrhizal or root nodule associations, to affect mineral nutrition
and water relations.

Diseased tissues may be more prone to leaching by acid rains than are healthy tissues, much as mechanically wounded tissues have been shown to be.[28]

Conclusions

Leaf surfaces of plants are not passive receptors, but are in a dynamic relationship with the atmosphere. Wet and dry deposition of atmospheric contaminants may alter a plant's response to parasitic microorganisms in a variety of ways, depending on the sensitivity of the plant and the microorganism to the contaminant(s) in question.

The potential would appear to be great for particles lodged on leaf surfaces to interact chemically with either ambient gaseous pollutants or precipitation. Such particles could effectively increase the surface area of leaves available for absorption of gaseous pollutants and, depending on their source, either neutralize or further acidify rain water contacting the leaf surface. Such interactions certainly occur in nature, and may be important in the ultimate response of a plant to complex mixtures of biotic and abiotic stress factors.

Literature Cited

1. J. P. Blakeman, The chemical environment of the leaf surface in relation to growth of pathogenic fungi, in: "Ecology of Leaf Surface Microorganisms," pp. 255-268, Academic Press, New York (1971).

2. S. Dickinson, The nature of thigmotropic stimuli affecting uredospore germ tubes of Puccinia spp., British Mycol. Soc. Trans. 47:300 (1964).

3. L. S. Evans, N. F. Gmur, and F. DaCosta, Leaf surface and histological pertubations of leaves of Phaseolus vulgaris and Helianthus annuus after exposure to simulated acid rain, Amer. J. Bot. 64:903-913 (1977).

4. J. A. W. Fairfax, and N. W. Lepp, Effect of simulated "acid rain" on cation loss from leaves, Nature 255:324-325 (1975).

5. A. K. Fraser, Growth restriction of pathogenic fungi on the leaf surface, in: "Ecology of Leaf Surface Microorganisms," pp. 529-535, Academic Press, New York (1971).

6. B. E. Godfrey, Leachates from aerial parts of plants and their
    relation to plant surface microbial populations, in: "Micro-
    biology of Aerial Plant Surfaces," pp. 433-439, Academic Press,
    New York (1976).

7. P. H. Gregory, "The Microbiology of the Atmosphere," Leonard
    Hill, London, p. 152, (1961).

8. P. J. Holloway, The chemical and physical characteristics of
    leaf surfaces, in: "Ecology of Leaf Surface Microorganisms,"
    T. F. Preece and C. H. Dickinson, eds., Academic Press, Inc.,
    New York, pp. 39-53 (1971).

9. H. M. Hull, The effect of day and night temperature on growth,
    foliar wax content, and cuticle development of velvet mesquite,
    Weeds 6:133-142 (1958).

10. P. M. Irving, and J. E. Miller, Response of soybeans to acid
    precipitation alone and in combination with sulfur dioxide.
    Radiological and Environmental Research Division Annual Report,
    ANL-77-65, Part III, pp. 21-27 (1977).

11. J. S. Jacobson, and P. Van Leuken, Effects of acidic precipita-
    tion on vegetation. Proc. of the Fourth International Clean
    Air Congress, Tokyo, May 16-20, 1977, pp. 124-127 (1977).

12. C. Leben, Influence of acidic buffer sprays on infection of
    tomato leaves by Alternaria solani, Phytopath. 44:101-106
    (1954).

13. S. E. Lindberg, R. C. Harriss, R. R. Turner, D. S. Shriner,
    and D. D. Huff, Mechanisms and rates of atmospheric deposition
    of selected trace elements and sulfate to a deciduous forest
    watershed, ORNL/TM-6674, Oak Ridge National Laboratory, Oak
    Ridge, Tennessee (1979).

14. H. F. Linskens, Uber die Anderung der Benetzbarkeit von
    Blattoberflachen und deren Ursache, Planta 41:40-51 (1952).

15. J. T. Martin, and B. E. Juniper, "The Cuticles of Plants."
    St. Martin's Press, New York, 347 p. (1970)

16. G. J. F. Pugh, and N. G. Buckley, The leaf surface as a sub-
    strate for colonization by fungi, in: "Ecology of Leaf Surface
    Microorganisms," pp. 431-445, Academic Press, New York (1971).

17. T. J. Purnell, and T. F. Preece, Effects of foliar washing on
    subsequent infection of leaves of swede (Brassica napus) by
    Erysiphe cruciferarum, Phys. Plant Path. 1:123-132 (1971).

18. T. J. Purnell, Effects of pre-inoculation washing of leaves
    with water on subsequent infections by Erysiphe cruciferarum,
    in: "Ecology of Leaf Surface Microorganisms," pp. 269-275,
    Academic Press, New York (1971).

19. R. A. Reinert, A. S. Heagle, and W. W. Heck, Plant responses
    to pollutant combinations, in: "Responses of Plants to Air
    Pollution," J. B. Mudd and T. T. Kozlowski, eds., Academic
    Press, New York, 383 pp. (1975).

20. I. Rentschler, Significance of the wax structure in leaves for
    the sensitivity of plants to air pollutants, International
    Clean Air Congress, Third, Dusseldorf, Proc., pp. A139-A142,
    (1973).

21. J. K. Sharma, and S. Sinha, Effect of leaf exudates of Sorghum
    varieties varying in susceptibility and maturity on the germi-
    nation of conidia of Colletotrichum graminicola (Ces.) Wilson,
    in: "Ecology of Leaf Surface Microorganisms," pp. 597-601,
    Academic Press, New York (1971).

22. D. S. Shriner, Effects of simulated rain acidified with sul-
    furic acid on host-parasite interactions, Ph.D. Thesis, N.C.
    State University, Raleigh, 79 pp. (1974).

23. D. S. Shriner, M. E. Decot, and E. B. Cowling, Simulated acidic
    precipitation causes direct injury to vegetation, Proc. Am.
    Phytopath. Soc. 1:112 (1974).

24. D. S. Shriner, Effects of simulated acidic rain on host-parasite
    interactions in plant diseases, Phytopathology 68:213-218
    (1978).

25. D. S. Shriner, Atmospheric deposition, Chapter 11, in: "Hand-
    book of Methodology for the Assessment of Air Pollution Effects
    on Vegetation," Air Pollution Control Association, Pittsburgh
    (1979).

26. W. H. Smith, Air pollution - Effects on the structure and func-
    tion of plant-surface microbial-ecosystems, in: "Microbiology
    of Aerial Plant Surfaces," pp. 75-105, Academic Press, New
    York (1976).

27. H. B. Tukey, Jr., Leaching of substances from plants, in:
    "Ecology of Leaf Surface Microorganisms," T. F. Preece and
    C. H. Dickinson, eds., Academic Press, Inc., New York,
    pp. 67-80 (1971).

28. H. B. Tukey, Jr., and J. V. Morgan, Injury to foliage and its effect upon the leaching of nutrients from the above-ground parts of plants, Physiol. Plant. 16:557–565 (1963).

29. T. Wood, and F. H. Bormann, The effects of an artificial acid mist upon the growth of Betula alleghaniensis, Brit. Environ. Pollut. 7:259–268 (1974).

DISCUSSION

BLOOMFIELD: Did I understand you to say that bacteria are almost equally affected by acid rain?

SHRINER: No, I hope you didn't understand that. Bacteria are generally regarded much more sensitive to pH than fungi are. Many fungi actually prefer a slightly acid substrate. Interestingly enough, Curt Leben* studied buffered acidic solutions looking for a potential control solution for a variety of different plant pathogens. All of the 15 species of fungal plant pathogens tested were sensitive to acidic solutions of pH 3.2, the subsequent growth rate was affected, and the subsequent rate of lesion development on plants treated with such solutions was also affected. We did not find that in some of the work we have done on plants treated with simulated rain solutions of similar pH. For example, with Helminthosporium maydis, which is quite an aggressive pathogen, particularly on the corn lines that we used with it, there was no apparent effect of the acidity directly on the pathogen.

*Leben, C. 1954. Influence of acidic buffer sprays on infection of tomato leaves by Alternaria solani. Phytopathology 44:101-106.

EVANS: Has anyone documented a change in pathogen sensitivity to air pollution after exposure to an air pollution?

KRUPA: Yes, Manning[a] has done it; Hibben[b] has done it; and Heagle[c] has done it. Generally they become more tolerant as opposed to changing the host and indirectly getting to them.

[a]Manning, W. J., W. A. Feder, and I. Perkins, Phytopathology 60: 669-670 (1970).

[b]Hibben, C. R., Phytopathology 56: 880 (1966).

[c]Heagle, A. S., Ann. Rev. Phytopathol. 11: 365-388 (1973).

EVANS: In these experiments were the host and the pathogen exposed at the same time?

KRUPA: Yes, and also the pathogen by itself.

SHRINER: A fungal spore is a relatively resistant structure; it has a very thick wall which probably plays a role in terms of sensitivity. A fungal spore which has germinated has a thin wall germ tube which if exposed to a stress at just the right time might have more likelihood of affecting the organism.

KRUPA: That's an interesting point though. The
Helminthosporium type organisms are thin-walled. For example,
in cases where the ozone has altered the leaf surface microbial
population, a chain of succession of organisms occurs on the
leaf surface. Some of the organisms that control the nutrient
status of the leaf may undergo a population shift.

EVANS: In some of the coniferous forests in the Northwest,
there are nitrogen-fixing organisms which play a very important
role in the nitrogen cycling of those ecosystems which are
canopy organisms[+] (Jones, 1976). It would be interesting to
look at this sort of a problem in some of the tropical areas
where you get thick crusts of nitrogen-fixing bacteria on leaf
surfaces.

[+]Jones, K. 1976. Nitrogen fixing bacteria in the canopy of
conifers in a temperate forest. In: Dickinson, C.H. and T.F.
Preece (eds.) Microbiology of Aerial Plant Surfaces. Academic
Press, New York.

RESPONSES OF PLANTS TO SUBMICRON ACID AEROSOLS

D. S. Lang, D. Herzfeld, S. V. Krupa

Department of Plant Pathology
University of Minnesota
St. Paul, MN 55108

ABSTRACT

A continuous flow system for exposing vegetation to submicron sulfuric acid aerosols was developed. The aerosols were mechanically produced using a nebulization-impaction-neutralization process. Typical aerosol distributions had a mass mean diameter of about 0.5 μm and a standard geometric deviation near 1.7. Plants were injured during acute exposures to sulfuric acid aerosols. However, the aerosol concentrations required to produce visible injury were significantly higher than the measured ambient concentrations. All species exposed to the acid aerosols exhibited marginal and tip necrosis or interveinal necrosis. However, plant responses to sulfuric acid aerosol did not follow proportional relations to pollutant dose (concentration X time). Lack of proportional dosage relations suggested that active rather than passive mechanisms may be controlling the response. In the exposed tissue no increase in the leaf surface sulfur concentrations could be detected. On the contrary, there was an increase in the internal tissue sulfur levels.

Environmental Impacts of Acidification

Environmental acidification due to the deposition of atmospheric pollutants is becoming a significant public concern. While this problem has been considered to some extent by several investigators, it has not been intensively studied in the United States compared to other major air pollutants. Acidification poses a threat to certain regions with limited substrate buffering capacity[9]. Aquatic systems are often the first to be affected by acidification[22]. In areas where snow accumulates this problem may be intensified when acidic material is released during the snowmelt, but in general this is not the case. However, severe reductions in fish population and diversity have been associated even with relatively small changes in pH. Some fish will not tolerate acidification below pH 4.5. As a consequence of acidification many lakes in Northern Europe and Northeastern United States may no longer support game fish populations.

In contrast, the effects of acidification on plant components of ecosystems are generally subtle. Significant effects on vegetation such as in the Sudbury region of Ontario is exceptional and difficult to partition between SO_2 and acidification[16]. Direct observations of acute responses of vegetation exposed to high concentrations of acidic pollutants for relatively short periods have been reported under simulated conditions in laboratory studies, but have not been documented as a natural occurrence[10,28].

Most sulfate pollution is derived from man-made sources of SO_2 emissions. Sulfate aerosols result from SO_2 oxidation which is, at least in part, photochemically derived[4,11]. Since the greatest increase in pollutant output from these sources have occurred during the past several decades, relatively little time for biological adaptation may have been allowed. Controlled adaptation of crop plants is a principal objective of agriculture and the long term effects of pollutant stress on agricultural systems is, consequently, difficult to assess. Natural ecosystems with longer generation times provide the basis needed for assessment of chronic pollution problems such as the acidification. Effects which have been observed in some areas and attributed to acidification include shifts in plant communities toward acid pollutant tolerant species and overall productivity declines[22].

Additionally, changes in nutrient cycling, uptake, and availability have been proposed in conjunction with acidification based on a few carefully controlled studies[7] but the acidification mechanism has not been fully understood. Subtle changes, such as these, which occur over plant generations are difficult to measure and require long term intensive studies for appropriate assessment.

Occurrence and Deposition of Acidic Substances

Acidic substances occur as fine particulates which are suspended in the atmosphere and are consequently transported under the influence of prevailing meteorological conditions.

Small particles ranging in size from a few hundredths of a micron
to several tenths of a micron in diameter have a long residence
time as a function of their size and can thereby be transported
over long distances of hundreds and possibly thousands of
kilometers prior to deposition[3]. Basically two mechanisms are
responsible for the deposition of acidic pollutants carried in
the atmosphere. Final deposition of the acidic aerosol may occur
either directly via sedimentation and impaction (dry deposition)
or indirectly as an acidic solution (wet deposition) which is
formed during condensation of atmospheric water vapor prior to,
or collected during, rain events[29]. Segregation of the relative
contribution of wet and dry deposition processes is difficult and
must be carefully assessed[6].

      Global sulfur deposition rates have been estimated to be in
the range of 1 to 9 grams per square meter per year[18].
Atmospheric concentrations of sulfate sulfur appear to be in the
range of 10^{-1} (remote areas) to 10^{2} (polluted areas) $\mu g \ m^{-3}$[33].
Studies relating sulfate aerosol deposition to accumulation in
terrestrial sinks are being conducted at this time. It seems
reasonable to expect that significant portions of sulfur compounds
are being deposited directly as sulfate aerosols. Sulfate aerosol
deposition velocities of 0.1 cm sec^{-1} have been observed for
grass surfaces[13] which are thought to be typical of most
vegetation. This value compares with an estimated average value
of 0.8 cm sec^{-1} for SO_2[13] and may be influenced by several
factors[32]. Predicted annual sulfate depositions of about 6
kg/hec/yr are in proportion to this deposition velocity in areas
with annual ambient sulfate concentrations of 20 $\mu g \ m^{-3}$.
Garland[13] has estimated that about 50% of the atmospheric sulfur
is deposited as SO_2 and related compounds (dry deposition) with
an associated residence time of 2 days and that the remaining 50%
which occurs as sulfate particulates is collected by impaction or
absorption on falling rain droplets (washout) or by condensation
of sulfate containing nuclei (rainout). Although sulfate aerosol
scavenging efficiencies in excess of 50% have been suggested the
mechanism is complex and difficult to predict[15]. However, the
resulting acidity of rain developed during these processes can be
dramatic, perhaps, providing adequate testimony to either the
efficiency of collection or the concentration of acid material in
the atmosphere or both. Rains with pH values near 3.0 have been
observed in northern Europe, Canada, and Northeastern United
States[14,16,22].

Effects of Acidification on Vegetation

      While the wet deposition of acidic pollutants may produce
dramatic changes in the characteristics of the physical environment and suggest associated biological effects, it may be

considered basically as localized, short duration stress. Laboratory experiments using simulated acid rains have shown acute plant responses (necrotic islands at the point of droplet contact) but these symptoms have not been found in nature[10]. Lack of agreement between laboratory and field observations regarding the effects of acidic precipitation on vegetation suggest that acute plant responses incited by high intensity – short term stress may be relatively unimportant in predicting overall impact of acidic pollutants on ecosystems. What is likely to be important, relative to observed productivity declines of natural systems, is the overall effect associated with the cumulative deposition of acidic substances. Chemical mass balances have been developed in conjunction with forest ecosystem models, and the measured annual sulfate deposition rates predict excess sulfate accumulation in regions with high sulfate loading by as much as a factor of 2 or 3 in some cases. Studies performed in Sweden suggest forest productivity declines on the order of 0.13 to 0.46 percent per year and have been attributed to ecosystem acidification[9].

Annual sulfate and other pollutant deposition needs to be proportioned on the basis of that attributable to wet and dry processes. While the long term effects of these different mechanisms may be similarly expressed as excess environmental loading by sulfate, the short term effects are known to be different. It is likely that in large areas of northeastern United States vegetation is continuously exposed to sulfate aerosol concentrations that are 3 to 5 times those found in remote regions of the U.S.

Ecological studies are needed at this time to develop present-day baseline data on which to evaluate potential future effects in areas subject to acidic pollution loading. Other forms of environmental toxification are known to occur in conjunction with aerosol scavenging via rain and investigation of this complex problem is also needed[2,16].

Our research group has recently conducted studies to ascertain the direct short term effects of submicron aerosols of sulfuric acid on representative crop and forest species[19,20]. Studies are currently being performed to examine plant responses to longer term exposures of low aerosol concentrations. We have found that plant responses to submicron aerosols of sulfuric acid are subject to influence by biological as well as physical factors and further influenced in type of expression by aerosol concentration and exposure duration. These findings suggest, on a preliminary basis, the complex nature of ecosystem acidification and the need for intensive long term studies relating to this potentially serious problem.

Methods for Exposing Vegetation to Aerosols

Design Criteria. Perhaps the most important aspect to be considered relative to biological studies with aerosols is that the aerosol produced be representative in size of the fine particulates , in nature. Additionally, aerosol chemistry and concentrations which are found naturally should be considered. Only recently have these conditions been given proper attention[5,17,19,21,23,31]. Exposure chambers should be non-corrosive to minimize uncontrolled and undesirable surface reactions and also provide for natural light intensity and quality. Further the exposure environment should be uniform in pollutant concentration and preferably of a continuous flow nature.

An aerosol exposure facility incorporating several of these important design criteria has been described recently[19]. Five areas were addressed in the design of this system:

1. Submicron aerosol size – efforts were made to assure that a substantial part (more than 80%) of the aerosol mass be submicron in size and thereby comparable to naturally occurring acid aerosol.

2. Simultaneous plant exposure to the aerosols and control environment – efforts were made to minimize individual plant differences which may be apparent even in clonally propagated species[8]. Chambers were designed to allow simultaneous, yet separate, exposure of different branches of the same plant to aerosol and control environments; where appropriate, this system also allowed whole plant exposure.

3. Natural/external lighting – chambers were constructed of clear cast acrylic plastic and lined with a clear Teflon film to utilize natural light regimes whenever possible. The exposure facility was located in a greenhouse and supplementary external lighting was provided by ultraviolet enhancing fluorescent lights to maintain photoperiods.

4. Corrosion resistance – all surfaces in contact with sulfuric acid aerosol were composed of either Teflon or stainless steel to minimize corrosion and prevent undesirable reactions that might confound results.

5. Continuous (single pass) flow with turbulent mixing – to minimize problems with varied biological effects due to localized static aerosol concentrations, baffles and flow rates were developed to produce a well mixed, evenly distributed aerosol mass within the chambers.

Chamber Construction and Operation

     Six exposure chambers (2 aerosol and 4 control) were constructed from clear cast acrylic tubes 30.5 cm in diameter and 132 cm in length. The tubes were lined with clear Teflon film 0.076 mm thick and fitted with Teflon ports 6.4 cm in diameter to allow the insertion of plant branches. Plant branches were held securely in position by wrapping a small sheet of Teflon film over a plug of expanded polyethylene which was slit to accommodate the branch and inserted as assembled into the Teflon port. The chambers were also used for exposure in which the entire plant was placed in the chamber. This practice was found to be acceptable for exposures of small plants because the biological effects of 'within chamber' and 'through port' exposures were found to be similar. Chambers were fitted with aerosol and accessory inlets and exhaust ports. The exhaust system was developed with a blower controlled by a variable power supply and flow valve to maintain a slight negative pressure within the chamber. Pressure gauges were attached to the chambers and used to balance total flow among the 6 chambers.

Aerosol Generation and Characterization

     A commercially available aerosol generator (Environmental Research Corp. model 7330) was modified to reduce the corrosive effects of sulfuric acid on critical orifices and used successfully in this exposure system. The principal components of this generator are constructed of stainless steel and are susceptible to corrosion by sulfuric acid. The corrosive effects of H_2SO_4 on stainless steel increase when the concentrated acid is diluted and are most apparent prior to the impaction stage which removes large particulates. For this reason, the Collison nebulizer stage of the generator was replaced with a unit constructed entirely of Teflon. The performance of the Teflon nebulizer was found to be similar to that of the stainless unit with 85%–90% of the aerosols, by mass in the sub-micron range. The generator impaction stage was also modified to incorporate a single orifice Teflon jet plate which reduced aerosol volumes for low concentration exposures.

     Several methods were used to characterize aerosol distributions within the exposure chambers. The principal method employed a rectangular jet impactor which was inserted through a port in the chamber and used to inertially classify particles larger than 0.3 µm in diameter. As an additional index, total sulfuric acid mass use per total mass flow was calculated to give an approximation of H_2SO_4 mass per volume. Since a substantial part of the aerosol distribution in this system was below inertial impaction limits, an electrical mobility size analyzer (Thermo-

systems model 3030) was employed to characterize particulates
between 0.01 μm and 1 μm in diameter.

Performance

Initial exposure indicated that static areas existed in the
chambers as indexed by varied symptom expression as a function of
chamber location. Teflon baffles were installed in the chambers
to improve turbulent mixing. Flow rates were adjusted such that
the chamber volumes were replaced once every two minutes. This
flow rate did not cause perceptible movement of foliage in the
chamber. Subsequent studies showed that symptom development on
the indicator species was uniform throughout the chamber after
this modification. Foliar sulfur accumulation was also independent of chamber location after these revisions.

Since branches of large plants or entire small plants can be
placed in the chamber, this system displays large chamber
advantages without concomitant increased costs. Desired aerosol
distributions can be developed rapidly and can be kept sufficiently stable. For continuous operations needed for chronic
exposures this system has been equipped with an air compression
unit fitted with appropriate air dryers and pollutant filters to
drive the aerosol generator.

Effects of Submicron Sulfuric Acid Aerosols on Vegetation

Symptoms of acute sulfuric acid aerosol injury. Typical symptoms of short term, single exposure sulfuric acid aerosol injury
(about 100 mg m^{-3} x 8 hours) is very similar to that caused by
gaseous fluoride on broad leafed plants and consists of marginal
and tip necrosis (Figures 1A and 1B). All plant species examined
developed similar symptoms and they appeared to vary, only in
degree, based upon species and plant sensitivity. Microscopic
injury from sulfuric acid aerosol was found to be similar to that
caused by simulated acidic precipitation[10]. Guard cells and
epidermal cells appeared shrunken and collapsed.

Symptoms of Chronic Sulfuric Acid Aerosol Injury

Preliminary investigations of plant responses to longer
term, lower concentration exposures indicate that symptoms
associated with acute and chronic exposure may be quite
different. Pinto bean exposed to submicron sulfuric acid aerosol
at concentrations of 1-5 mg/m^3 for periods of several days
developed a bifacial, interveinal necrosis very similar to that
caused by sulfur dioxide.

Figure 1. Symptoms of acute injury caused by sub-micron sulfuric
acid aerosols. A – Unexposed (left) and exposed (right)
leaves of Bountiful bean. B – A leaf of hybrid poplar
exposed to the aerosol.

Subtle plant responses to sulfuric acid aerosol. Foliar sulfur content was assayed to determine if sulfur accumulation
during exposure (about 100 mg/m^{-3} for 8 hours) was related to
physical factors known to condition plant response to other
pollutants. Plant response was found to be influenced by what
appears to be seasonal factors. A seasonal decrease was observed
in the ability of Bountiful bean to accumulate sulfur in exposed
foliage (Figure 2). A similar response was observed in pinto
bean, soybean and green ash. Midwinter studies with pinto bean,
under artificially maintained photoperiods, showed a marked
inability of this species also to accumulate foliar sulfur during
the winter exposure (Figure 2). Plants were exposed to 175 mg/m^3
and 390 mg/m^3 in an attempt to initiate an acute plant response
where visible injury and foliar sulfur content could be
correlated. However, at that seasonal time foliar sulfur contents
were not found to be significantly altered and visible injury was
not appreciably different. The seasonal variability in foliar
sulfur content appears to be related to variability in
macroscopic symptom expression but a quantitative relationship
could not be established. Plant age (from 15-40 days for pinto
bean) does not appear to predominantly influence the degree of
visible injury or ability to accumulate sulfur. X-ray emission
spectra (Figure 3A) suggested that more sulfur accumulated within
the leaf than on the leaf surface, suggesting that active plant
processes may play a role in the overall plant response.

Changes in leaf growth rates conditioned by sulfuric acid
aerosol exposure were not pronounced. Young leaves exposed to 175
mg/m^3 for 7 hours showed a slight (P = 0.25) stimulation in
growth which decreased as the initial leaf area increased beyond
75 cm^2 (Figure 3B), which again suggests the possibility of
active, in addition to passive, mechanisms of sulfate uptake.

Factors affecting plant response. Plant response was found
to be variable between exposure and appeared to be subjected to
control by seasonal factors. Injury ranged from 0 to 25% necrosis
of the exposed foliage on sensitive species between different
exposures but with constant conditions of age, temperature, and
pollutant concentration. The relative humidity varied between 50%
and 70% but did not appear to influence plant respnse within this
range. When injury did develop it appeared to be uniform among
plants of the same species exposed at the same time.

Plant growth and development did not appear to be affected
during short term chamber residence. Foliage of plants placed in
control chambers for 24-48 hours did not differ significantly in
leaf area from comparable plants maintained in the greenhouse
adjacent to the chambers. All individual plant species exposed to
sulfuric acid aerosol developed a similar degree of injury during
a given exposure indicating that the extent of injury was not

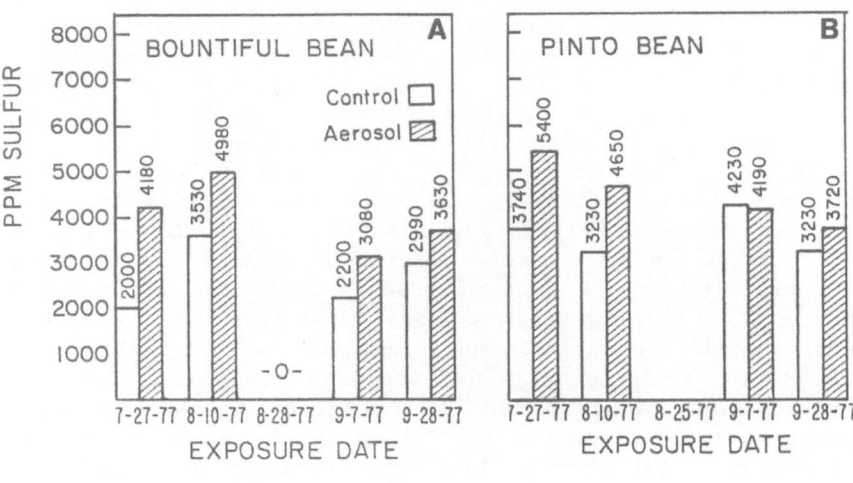

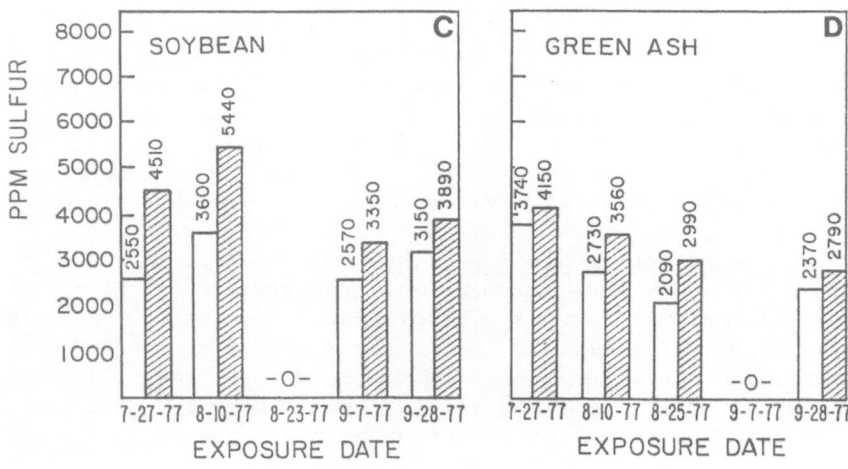

Figure 2.   Seasonal variations in the foliar accumulation of sulfur
            following acute exposure to sub-micron sulfuric acid
            aerosols. Empty bars - unexposed tissue and filled bars -
            aerosol exposed tissue.

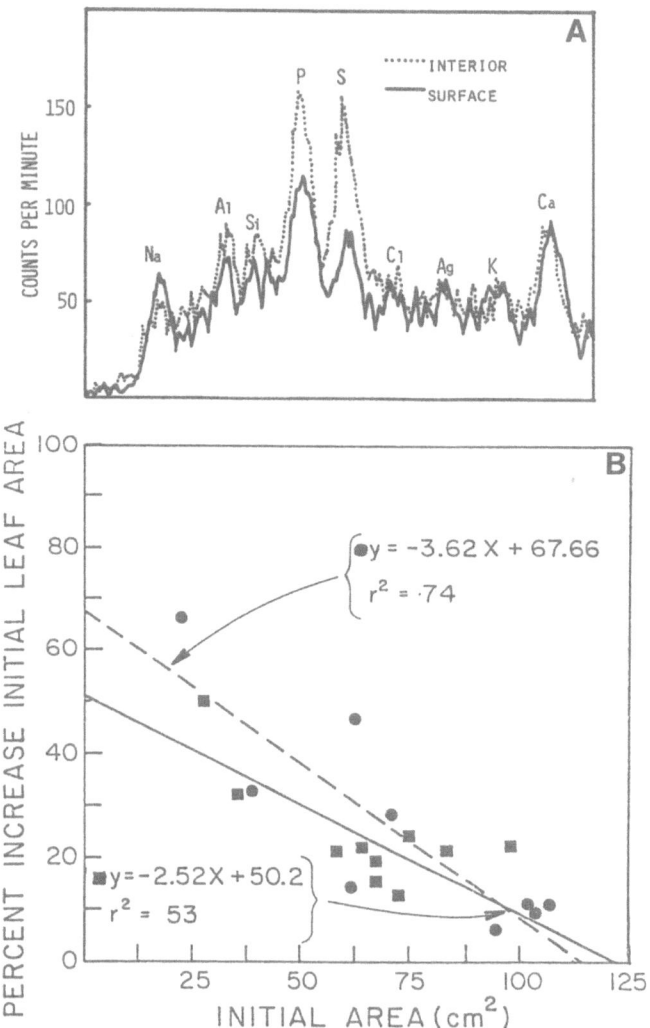

Figure 3. $\underline{A}$ - X-ray elemental microanalysis spectra of pinto bean
leaf exposed to sub-micron sulfuric acid aerosol. $\underline{Solid}$
$\underline{line}$ - surface of the leaf and $\underline{dotted}$ $\underline{line}$ - interior of
the leaf. $\underline{B}$ - Relationship between growth and initial leaf
area of aerosol exposed (dashed line) and non-exposed
(solid line) trifoliate leaves of pinto bean.

dependent upon chamber location.

Physiological Aspects of Sulfuric Acid Aerosol Injury to Vegetation

Plant responses to sulfuric acid aerosol have been studied under acute conditions designed to produce visible symptoms characteristic of this form of pollution so that other factors influencing injury could be evaluated. Exposure to 100 mg/m^3 of sulfuric acid aerosol for 8 hours are injurious to vegetation and induce both visible and subtle responses. Plants appear to vary in sensitivity to this pollutant based on phenotypic as well as genotypic characteristics. Acute plant responses to sulfuric acid aerosol appear to be influenced by factors under biological control which are expressed as seasonal differences. It has not been possible to accurately define plant responses, either visible or subtle, in terms of physical factors known to condition plant responses to other pollutants. Plants did not show predictable changes in injury related to increased pollutant concentration beyond a threshold value, duration of exposure, or plant age. Similarly, preliminary investigations on the influence of temperature, relative humidity and light did not show predictable relationships.

These observations do not suggest that physical factors are not involved in conditioning plant responses to sulfuric acid aerosol injury. It is probable that optimum temperature, concentrations, humidity and light regimes exist which will, when the plant is in a susceptible condition, combine to produce maximum injury. However, highly variable observations relating to plant responses at different times to a constant level of a given physical factor has led us to suggest that biological factors may be principally involved in controlling plant response to sulfuric acid aerosol.

The ability of plants to metabolize sulfate brings into consideration a number of factors which may complicate and interfere with sulfate aerosol exposure studies. Varied plant responses to sulfate as described in the results section may be a reflection of their effects. Sulfate uptake seems to be accomplished by active transport mediated by carrier with enzyme properties[27]. Transport is unidirectional from outside to inside the cell and is governed by pH, temperature, sulfate concentration and available metabolic energy. Free internal sulfate is also known to depress sulfate transport as is excess endogenous sulfur[27]. Possibly the growth stimulation shown in Figure 2B for young leaves exposed to sulfuric acid aerosol was due to the relief of an initial endogenous sulfur deficit. Conversely, the growth inhibition of older leaves may have been due to a net excess of endogenous sulfur due to sulfate accumulation. The foregoing discussion suggests that plant responses to acute sulfate exposure may be unpredictable, based on the limited studies performed to date.

Plant response to sulfuric acid aerosol is further regulated by stomates. It is currently thought that stomates respond first to interstomatal carbon dioxide concentration and only indirectly to other factors[26]. Stomatal function is known to be energy dependent[26]. Since sulfate activation is thought to be necessary prior to transport it is possible that the same, energy dependent, factors which control stomate opening and closure may regulate sulfate transport. Simultaneous interference with these processes would likely cause substantial changes in plant response to sulfuric acid aerosols.

A consideration of the complex interactions regulating sulfate uptake and stomatal action can help explain the varied responses which have been reported. Plants were found to be alternately sensitive and insensitive to conventional factors controlling injury. Pollutant concentration, temperature, duration of exposure and age of plants do not appear to be fundamentally involved in determining plant responses to sulfuric acid aerosols. Our experience combined with specific information from other studies[26,27] suggests that several additional factors may be involved in conditioning plant response to sulfuric acid aerosol. These include endogenous plant sulfur (especially as sulfate), plant energy status, and pH of guard cell cytoplasm which subsequently influence sulfate metabolism and stomatal diffusion resistance[26,27]. Quantification of these factors, in conjunction with aerosol exposure, will be required to clarify plant responses to sulfuric acid aerosol.

Most of our work has involved the acute exposure of plants to fine aerosols of sulfuric acid. The concentrations used were not intended to be analogous to those found in the ambient environment. They were designed to produce acute plant responses (necrosis and wilt) which could then be related to subtle plant responses (pollutant accumulation and growth). Correlation of acute and subtle plant response to pollutant stress is necessary to predict when significant injury may result from chronic exposures. Whitby[33] has reported atmospheric sulfate concentrations of 80 $\mu g/m^3$ under highly polluted conditions. This value is two to three orders of magnitude less than the concentration we found necessary to induce acute injury to vegetation by one-time, short-term exposure and is 10 to 100 times larger than the concentration which could be attributed to automobiles equipped with catalytic converters[25]. Research, to date, indicates that injury to vegetation is unlikely to result from single, few-hour exposures to sulfur acid aerosol alone at concentrations which are presently known to occur in the environment[19,20]. However, since sulfuric acid aerosols are known to occur as a component of complex pollutant mixtures in the atmosphere the question of interactive or more than additive effects also needs to be considered. Additionally, recent research has shown that sulfuric acid aerosol can be accumulated in plants, over a period of several days, to toxic concentrations (Herzfeld, unpublished). If a similar accumulation response is found during chronic exposures to sulfuric acid, the long term effects of this pollutant may be of much greater significance than those of immediate concern.

Preliminary investigations, in our laboratory, involving chronic exposures to sulfuric acid aerosols have shown that this pollutant is injurious to sensitive vegetation over at least a hundred fold range (1 to 100 mg m^{-3}) although symptoms appear to be distinctly different. Acute injury was seen as a marginal and tip necrosis similar to that caused by gaseous fluoride while chronic injury was observed to consist of an interveinal necrosis resembling that caused by sulfur dioxide. Although explanations based on aerosol deposition frequency, have been offered to account for the type of injury we have seen during acute exposures[23] we do not feel they are particularly applicable to submicron aerosols. It is our opinion, at this time, that both acute and chronic responses are subject to organismic control in terms of uptake, translocation and subsequent toxicity.

The implications of ecosystem acidification are considered to be serious and are likely to develop in a subtle manner over a period of years. The acidification trend, once initiated, has been shown,[14] at least in part, to be self sustaining and irreversible[14]. The acidification process is complex with deposition occurring in both wet and dry forms and possibly in conjunction with additional toxicants such as heavy metals[16] or radionucleides[1,12]. Effects may be mitigated to a large extent by the environment[34] or may be increased as a function of acid composition[7,24] determined by the chemical and physical characteristics of atmospheric pollutants prior to rain events. Indirect effects may also occur in conjunction with ecosystem acidification such as aerosol alteration and exclusion of incoming radiation[30]. Effects such as these which induce subtle ecosystem perturbations, have the potential of confusing observations unless they are included in overall impact assessments.

LITERATURE CITED

1. Adriano, D. C., and J. E. Pinder. 1977. Aerial deposition of plutonium in mixed forest stands from nuclear fuel reprocessing. J. Environ. Qual. 6:303-307.
2. Askne, C. and C. Brosset. 1971. Determination of strong acid

in precipitation, lake-water, and air-borne matter. IVL
Report B107. Gothenburg.

3. Altshuller, A. P. 1976. Regional transport and transformation
of sulfur dioxide to sulfates in the U.S. J. Air Poll.
Control Assoc. 26:318-3244.

4. Atkins, D. H. R., R. A. Cox, and A. E. J. Eggleton. 1972.
Photochemical ozone and sulfuric acid aerosol formation
in the atmosphere over southern England. Nature
235:372-376.

5. Carlson, R. W., F. A. Bazzaz, and J. J. Sturel. 1976.
Physiological effects, wind reentrainment, and rain wash
of Pb aerosol particulate on plant leaves. Environ. Sci.
and Tech. 12:1139-1142.

6. Clough, W. S. 1975. The deposition of particles on moss and
grass surfaces. Atmos. Environ. 9:1113-1119.

7. Cronan, C. S., W. A. Reiners, R. C. Reynolds, and G. E. Lang.
1978. Forest floor leaching: Contributions from mineral,
organic and carbonic acids in New Hampshire subalpine
forests. Science 200:309-311.

8. Dochinger, L. S. et al. 1972. Responses of hybrid poplar
trees to sulfur dioxide fumigation. J. Air Poll. Control
Assoc. 22:369-371.

9. Dochinger, L. S. and T. A. Seliga. 1975. Acidic precipitation
and the forest ecosystem. J. Air Poll. Control Assoc.
22:11.

10. Evans, L. S., N. Gmur, and F. Dacosta. 1977. Leaf surface and
histological perturbations of leaves of Phaseolus vulgaris and Helianthus annuus after exposure to simulated acid
rain. Am. J. Bot. 64:903-911.

11. Fennelly, P. F. 1976. The origin and influence of atmospheric
particulates. Amer. Scientist 64:46-56.

12. Fleisher, R. L. 1974. Aerosol particles on tobacco trichomes.
Nature 250:158-159

13. Garland, J. A. 1978. Dry and wet removal of sulfur from the
atmosphere. Atmos. Environ. 12:349-362.

14. Grahn, O. 1976. Macrophyte succession in Swedish lakes caused
by deposition of airborne acid substances. Water Air
Soil Pollut. 6:295-305.

15. Hales, J. M. 1978. Wet removal of sulfur compounds from the
atmosphere. Atmos. Environ. 12:389-399.

16. Hutchinson, T. C. and L. M. Whitby. 1977. The effects of acid
rainfall and heavy metal particulates on a boreal forest
ecosystem near the Sudbury smelting region of Canada.
Water Air Soil Pollut. 7:421-438.

17. Klepper, B. and K. K. Craig. 1975. Deposition of airborne
particulates onto plant leaves. J. Environ. Qual.
4:475-499.

18. Knabe, W. 1976. Effects of SO_2 on terrestrial vegetation.
Ambio. 5:213-218.

19. Lang, D. S., S. V. Krupa, and D. S. Shriner. 1978. Injury to

vegetation incited by sulfuric acid aerosols and acidic
rain. Proc. Air Poll. Control Assoc. 78-7.3:1-15.

20. Lang, D. S., and S. V. Krupa. 1979. Effects of sulfuric acid
    aerosols on vegetation. EPA-600/3-79-002. 81. p.

21. Lerman, S. L. and E. F. Darley. 1975 Particulates. In
    Responses of Plants to Air Pollution. Mudd, J. B. and
    T. T. Kozlowski (eds.). Academic Press, Inc., New York.
    383 p.

22. Likens, G. E. and F. H. Bormann. 1974. Acid rain: A serious
    regional environmental problem. Science 184:1176-1179.

23. Little, P. 1977. Deposition of 2.75, 5.0 and 8.5 μm particles
    to plant and soil surfaces. Environ. Poll. 12:293-315.

24. Marsh, A. R. W. 1978. Sulphur and nitrogen contributions to
    the acidity of rain. Atmos. Environ. 12:401-406.

25. Maugh, T. H. 1977. Sulfuric acid from cars: A problem that
    never materialized. Science 198:280.

26. Raschke, K. 1975. Stomatal action. Ann. Rev. Plant Phys.
    26:306-40

27. Schiff, J. A., and R. C. Hodson. 1973. The metabolism of
    sulfate. Ann. Rev. Plant Physiol. 24:381-414.

28. Shriner, D. S. 1974. Effects of simulated rain acidified with
    sulfuric acid on host-parasite interactions in
    plant disease. Ph.D. Thesis, North Carolina State Univer-
    sity 85 p.

29. Smith, W. H. 1977. Removal of atmospheric particulates by
    urban vegetation implications for human and vegetative
    health. Yale J. of Biol. and Med. 50:185-197.

30. Unsworth, M. H. and H. A. McCartney. 1975. Effect of
    atmospheric aerosols on solar radiation. Atmos. Environ.
    7:1173-1185.

31. Wedding, J. B., R. W. Carlson, J. J. Sturel, and F. A.
    Bazzaz. 1977. Aerosol deposition on plant leaves.
    Water Air and Soil Poll.7:545-550.

32. Wesley, M. L., and B. B. Hicks. 1977. Some factors that
    affect the deposition rates of SO_2 and similar gases on
    vegetation. J. Air Poll. Control Assoc. 27:1110-1116.

32. Whitby, K. T. 1978. The physical characteristics of sulfur
    aerosols. Atmos. Environ. 12:135-159.

34. Winkler, E. M. 1977. Natural dust and acid rain. Water Air
    Soil Poll. 7:295-302.

DISCUSSION

MILLER: If stomatal action is involved, then testing in light and darkness, which vary stomatal action, is one way to test your hypothesis of stomatal action as a factor in lesion development.

KRUPA: We are actually doing one better than that. That is to use S^{35} to see whether the radioactive material is being actively absorbed using autoradiography and scintillation. We do know that in darkness, the sulfate intake seems to be lower.

EVANS: Are you generating aerosols with sulfur-35?

KRUPA: I am going to.

HICKS: The migration of particles through this stomatal opening must be extremely slow. Do I assume correctly that the migration is _via_ a liquid phase?

KRUPA: I think so. Sulfate uptake is quite complex as far as the leaf is concerned. There is no direct diffusion because it is polar. I suspect once it enters the system, it moves along the translocation pathway.

HICKS: What's the relative humidity typically in that area of the leaf?

KRUPA: The relative humidity can be varied from 50-70% without changing the aerosol particle size. If you go beyond 70%, then you start getting bigger particles due to condensation. So if you're talking about 50 and 70% relative humidity, the variation doesn't seem to do anything to it.

McFEE: What's the form of sulfur in the aerosol?

KRUPA: Sulfuric acid. The way we generate sulfate aerosols is simply to take concentrated sulfuric acid and atomize it at high pressure, like 60 p.s.i. high. Then send the aerosols through an impaction stage and remove all the particles not wanted, that is particles in excess of one micron. The aerosol is sent through a Krypton 85, neutralizer, and the charge is removed.

EVANS: Have you looked at the nitrogen status of your substrate relative to the lack of correlation between sulfate uptake and injury?

KRUPA: Yes, we have pretty well standardized the potting mix
and NPK levels are kept rather constant. The thing that I want
to emphasize is that visible injury cannot be really used
because of the inconsistency in the appearance of visible
symptoms. What one has to do perhaps is something like
measurements of C-14 uptake at various times and exposures to
the aerosol.

EVANS: Why use C-14?

KRUPA: For a physiological standard. You have to use some
physiological parameter because evaluation of plant response
through visible injury is a rather crude technique and I'm sure
Jay Jacobson and others know that even with ozone, for example,
at concentrations like .1 ppm per half hour will result in a
physiological depression, even though you don't see any
corresponding visible effect.

TOMLINSON: Can you comment on the amount of SO_3, and the
distribution between SO_2 and SO_3 at various distances down
from the stack.

KRUPA: I do not know exactly how much SO_3 is in the plume,
but EPA flew the plume last summer and we also have stack
analysis on it. One way people measure sulfate is to modify a
flame photometer, one that would normally measure SO_2. An
electrical chopper is used in front of the flame photometer and
the aerosol is neutralized. Subsequently, total sulfur versus
SO_2 is measured. The SO_2 concentration is subtracted from
the total sulfur to calculate SO_4 levels. The only person I
know that might have done significant work on this is Etough at
the University of Utah. Etough has found that sulfite, up to
30% in some smelters, may exist as a stable complex with
transient metals, such as lead, copper, iron [Atmos. Environ.
12: 263-272 (1978)] and Jake Hales has done the same thing on
the MAP 3S study over Lake Michigan in precipitation. [Atmos.
Environ. 12: 389-400 (1978)]. Data of Hales indicate that
maybe 10-20% of the sulfur could be as a stable sulfite.

EVANS: How do you define a stable sulfite?

KRUPA: People used to think sulfite would be catalytically
oxidized to sulfate rather effectively and, thus, sulfite
species cannot persist in smelter plumes, etc. Recent
evidence, however, shows that stable metal sulfites can be
present in plumes as such [Atmos. Environ. 12: (1978)] .

POLLUTED RAIN AND PLANT GROWTH*

J. S. Jacobson, J. Troiano, L. J. Colavito, L. I.
Heller, and D. C. McCune
Boyce Thompson Institute
Cornell University
Ithaca, New York 14853

ABSTRACT

   Experiments were performed to determine whether the effects of
acidic precipitation on agricultural crops of the eastern U.S. are
altered by differences in the supply of nitrate and sulfate in rain
or the concentration of ozone in the atmosphere. In tests with
greenhouse-grown lettuce plants, better growth was obtained by
treatment with simulated acidic rain at pH 3.2 containing high sulfate to nitrate ratios. When acidic precipitation at pH 2.8, 3.4,
and 4.0 was applied intermittently to field-grown soybeans, the more
acidic treatments produced a shift in the partitioning of photosynthate from the vegetative (leaves and stems) to the reproductive
organs (seeds) in those plants exposed to low concentrations of
ozone (hourly average concentration did not exceed 0.03 ppm during
the growing season). No shift in photosynthate partitioning was
observed when simulated acidic rain was applied to soybeans exposed
to ozone concentrations that reached a maximum of 0.125 ppm and
significantly reduced growth and yield. These results demonstrate
that both the nutrition of plants and the atmospheric concentrations
of ozone must be controlled or carefully monitored in experiments
on the effects of acidic precipitation on agricultural crops.

---

*
This research was financed in part with Federal funds from the
Environmental Protection Agency under Grant No. R804513. The
contents do not necessarily reflect the views and policies of
the Environmental Protection Agency.

INTRODUCTION

Pollutants are removed from the atmosphere by rain and snow and are transferred to soils, natural waters, and vegetation by wet and dry deposition processes. Through these processes, plants are exposed periodically to substances dissolved in atmospheric precipitation and to gaseous pollutants. The major soluble constituents in rain and snow in the eastern U.S. are hydrogen, sulfate, and nitrate ions and there is concern over the environmental influence of these substances, particularly acidity (Jacobson et al., 1976). Knowledge of plant nutrition and response to pollutants raises the question of whether it is necessary to consider the supply of nitrate and sulfate in rain and the concentration of ozone in the atmosphere when determinations are made of the effects of acidic precipitation on vegetation of the eastern U.S.

Nitrate and Sulfate

There are several reasons for suspecting that the supply of nitrate and sulfate in rain may be important for plant growth. Both elements are incorporated into metabolic processes and are essential for growth. They are the anions in highest concentration in rain of the eastern U.S. especially during the spring and summer seasons (Table I). Both substances are taken up through leaves as well as from the soil; in fact, nitrate and other nutrients have been applied, in solution, to foliage for use as fertilizers (Tukey,

Table I. Seasonal composition of individual rain and snow events (median values) in Yonkers, NY during the period of 1974 through 1977.

Variable measured	Spring	Summer	Fall	Winter	All Seasons
pH	3.97	3.93	4.22	4.22	4.10
Conductivity, μmhos	50	54	29	32	39
Sulfate, ppm	4.9	5.35	3.0	3.25	3.95
Nitrate, ppm	3.0	3.0	1.35	1.9	2.2
SO_4/NO_3 ratio[2]	1.79	1.96	2.0	1.74	1.88
Chloride, ppm	0.6	0.6	0.8	1.0	0.75

[1]Spring: 1 April - 30 June; Summer: 1 July - 30 Sept.; Fall: 1 Oct. - 31 Dec.; Winter: 1 Jan. - 31 Mar.

[2] Median ratios of sulfate to nitrate calculated from individual rain events decreased during the period of monitoring from 2.0 to 1.6.

1970). Atmospheric supplies of nitrate and possibly sulfate as well support plant growth in populations that do not receive soil applications of fertilizers such as in forests and raised bogs and with epiphytic vegetation (Tamm, 1958). An increase in supply of these nutrients from the atmosphere allows plant populations to flourish and exclusion of these nutrients can diminish growth (Cowling and Jones, 1970; Jones et al., 1972). Crop plants also respond in this way to deposition of sulfur compounds from the atmosphere when the soil is deficient in sulfur and atmospheric concentrations are not excessive (Thomas et al., 1943).

Although the supply of nitrate and sulfate in a single rain event is far less than provided in fertilizer applications, the repeated exposure of plants to rain and the opportunity for direct uptake through the leaves suggests that nutritional benefits may be significant even for some crop plants. The recent shift to high-analysis fertilizers with lower sulfur contents may increase the significance of sulfur from other sources for certain soils (Noggle et al., 1979). A final reason for giving consideration to the influence of nitrate and sulfate in rain is that the increasing utilization of coal and oil for energy will place greater quantities of these compounds into the atmosphere.

Ozone

In the last few decades, ozone has become the atmospheric pollutant of greatest concern to agriculture and forestry in North America. It occurs in highest concentrations during the growing season at considerable distances from the sources of precursor compounds. At many locations in both the eastern as well as western U.S., ozone concentrations periodically exceed both the old and new Federal ambient air quality standards of 0.08 and 0.12 ppm, respectively (Table II). Although there is considerable uncertainty over the impact of this pollutant on agriculture and forestry, experiments in many laboratories have established that a wide variety of commercially important plants are susceptible to injury by ozone (Jacobson, 1977). Reductions in growth at concentrations of ozone similar to those occurring in many portions of the U.S. have been demonstrated for alfalfa, beans, sweet and field corn, cucumber, onion, radish, soybeans, tobacco, tomato, and winter wheat (MacLean and Schneider, 1976; U.S. EPA, 1979). It is unlikely that ozone concentrations will decrease before the end of this century in rural areas of the U.S. so the combined response of plants to ozone and components of acidic precipitation is an important consideration.

Table II. Ozone concentrations (hourly averages) in Yonkers, N.Y.
           during the growing season (May to September) for the
           years 1970 through 1978[1]

Year	Maximum Conc. PPM	Mean Daily Maximum Conc. PPM	No. of Values Exceeding		No of Days Exceeding	
			0.08	0.12	0.08	0.12
			PPM		PPM	
1978	0.254	0.083	329	113	56	29
1977	0.26	0.096	493	114	80	36
1976	0.207	0.081	300	80	59	22
1975	0.15	0.054	128	8	35	5
1974	0.132	0.055	100	3	25	2
1973	0.23	0.077	316	62	59	20
1972	0.157	0.057	123	25	31	11
1971	0.178	0.047	58	9	17	4
1970	0.18	0.046	72	8	27	4

[1]Mast coulometric ozone meter used from 1970 through 1975. Monitor
Laboratories chemiluminescent analyzer (CL) used from 1976 through
1978. All instruments were calibrated by the neutral KI method.
Both instruments were used during 1976 and similar hourly average
values were obtained except that peak values by CL were slightly
higher than the Mast and minimum values by CL (nights and cloudy
days) were slightly lower than the Mast.

Acidity of Rain

     There are several descriptions of the response of plant foliage
to the application of acidified solutions which indicate that rain
events with pH values below 3 are likely to produce foliar lesions
on susceptible species (Jacobson, 1978). But there are other
important ways in which plants may be injured by direct contact
with acidic precipitation: essential elements such as potassium,
calcium, or magnesium may be leached from leaves; leaf cuticles may
be altered; stomatal function may be affected; or, pollination and
fertilization may be inhibited (Tamm and Cowling, 1976). These
possibilities can be identified but the likelihood of their occurrence in the ambient environment is not yet known. A continuing
series of experiments is being performed at the Boyce Thompson
Institute to determine the effects of acidic precipitation on
vegetation, and the results described here, dealing exclusively
with plant growth and yield, have been selected from large-scale
investigations, which will be reported in detail later on.

Methods

Separate experiments were performed to determine the effects
of nitrate and sulfate in rain and atmospheric ozone concentrations
on growth and yield of plants. In the first series, lettuce plants
(Lactuca sativa L. Oakleaf) were grown in the greenhouse in sand
culture using half-strength Hoagland's solution as the balanced
source of nutrients. Plants were exposed for 2-hour periods three
times at 5-day intervals to simulated rain containing the major
inorganic ions of ambient rain at a pH of 3.2. Three different
nitrate and sulfate concentrations were used giving nitrate to
sulfate mass ratios of 1:7.5, 2:1, and 20:1. Twenty-four hours
after the last treatment, the dry mass of leaves, stems, and roots
were measured.

In the second series of tests, field-grown soybeans (Glycine
max (L.) Merr. Beeson) were exposed to simulated rain at pH 2.8,
3.4, and 4.0 in paired chambers (Mandl et al., 1973) equipped with
covers to exclude ambient rainfall. One hour applications of 0.5
cm simulated rain were made 18 times on cloudy days from mid-July
to the end of September. Ambient air containing ozone, which
reached a maximum of 0.125 ppm during the experimental period,
continuously passed through one of each pair of chambers. Air
filtered through charcoal passed through the second chamber of
each pair. In these filtered-air chambers, hourly-average ozone
concentrations did not exceed 0.03 ppm.

The total amounts of nitrogen and sulfur applied in simulated
rain were approximately 1 and 3 kg/ha for the pH 2.8 treatment.
These figures compare to the application of approximately 56 kg/ha
of nitrogen and less than 1 kg/ha of sulfur in a single fertilizer
application early in the growing season. Leaves, stems, and pods
were harvested and dry weights were recorded at the end of September.

Results

The growth of lettuce plants, as measured by dry weight, was
affected by changes in the nitrate and sulfate concentrations of
simulated rain at pH 3.2 (Table III). Dry weights of apical leaves
and roots of plants were significantly increased after exposure
to simulated rain with the highest sulfate and lowest nitrate
concentrations compared to plants exposed to simulated rain containing lower ratios of sulfate to nitrate at pH 3.2 or 5.7.

The dry mass of beans (vertical scale, Figure 1) and leaves
plus stems (horizontal scale, Figure 1) are presented for soybean
plants exposed to simulated rain at three levels of acidity and
high and low concentrations of ozone. The weight of vegetative

Table III. Dry mass of lettuce plants exposed to simulated rain of
            different acidity, nitrate and sulfate concentrations.

		Dry weight (g) after exposure to:			
		pH 5.7		pH 3.2	
Tissue	NO_3:[1]	0.1	6.0	3.2	0.4
Sampled	SO_4:[1]	0.05	0.3	1.7	3.0
Plant		7.33	7.18	7.7	7.88
Root		1.35^{ab}	1.29^a	1.44^{bc}	1.49^c
Shoot		5.98	5.99	6.26	6.4
Stem		0.33	0.32	0.33	0.34
Leaves		5.65	5.67	5.92	6.06
Basal		1.65	1.64	1.61	1.61
Middle		2.3	2.32	2.2	2.26
Apical		1.7^a	1.7^a	2.11^b	2.19^b

[1] Nitrate and sulfate concentrations are in units of 10^{-4} $\underline{M}$
[2] Values across rows with the same superscript letter are not
    significantly different at : p = 0.05.

tissues was significantly reduced in the pH 2.8 and 3.4 rain treatment vs. pH 4.0 in the high ozone exposure. However, the differences in leaf and stem mass between simulated rain treatments in
the low ozone exposure were not statistically significant. Differences in bean yield were not significantly affected by acidity
of simulated rain either for the low or high ozone exposures.
However, exposure to elevated concentrations of ozone depressed
both growth and yield of soybeans with all three rain treatments
and the depression was greatest with the most acidic rain (Figure
1). Furthermore, it appears that a shift occurred in the partitioning of photosynthate from vegetative organs into beans in the
low ozone exposure but not in the high ozone exposure with increasing acidity of applied rain.

Discussion and Conclusions

     Although this report emphasizes the potential significance
of the nutrient supply of precipitation, it is important to recognize that the loss of nutrients from plants by leaching also has
been known for more than 70 years (Le Clerc and Breazeale, 1908).
Leaching of soluble nutrients from the interior of leaves and
removal by rain of materials deposited from the atmosphere or
secreted onto leaf surfaces are affected by the chemical composition of the cuticle, morphology of the leaf, age and stage of

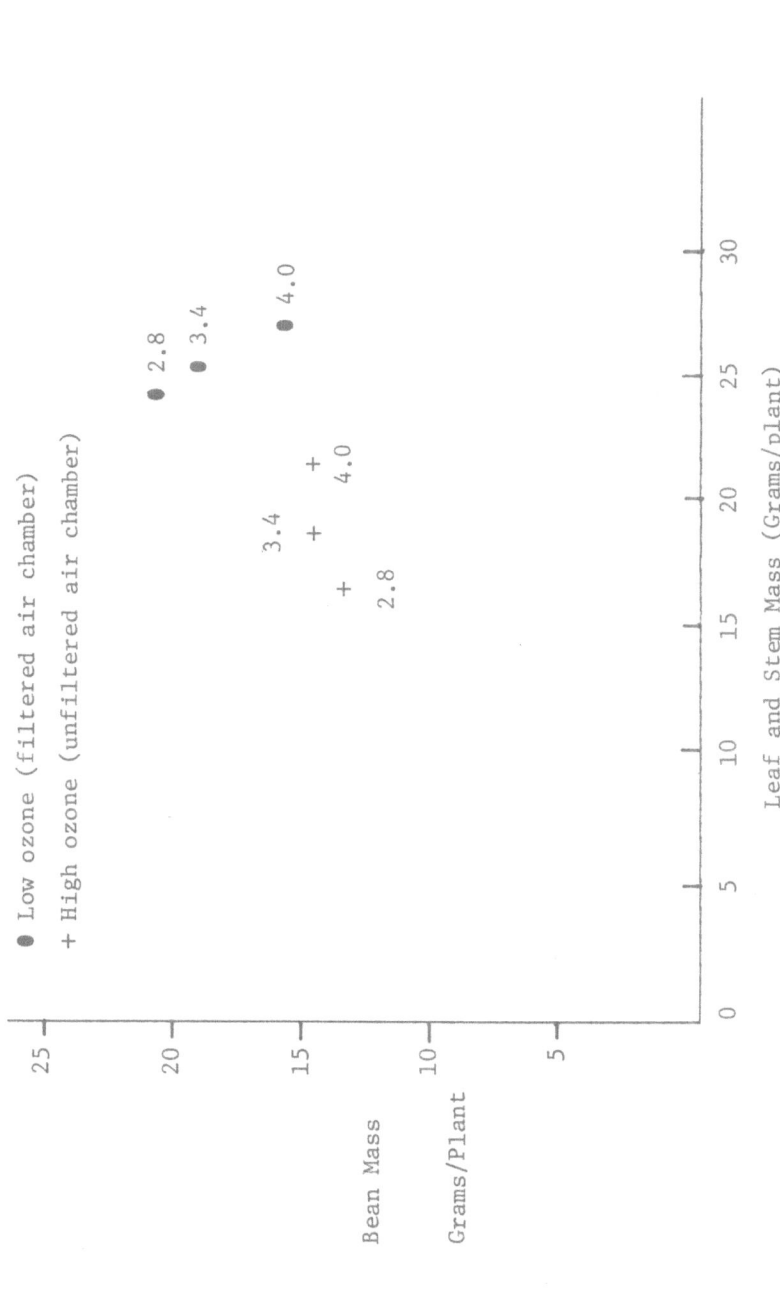

Figure 1. Effect of Simulated Acidic Precipitation and Level of Atmospheric Ozone on Growth and Yield of Soybeans.

plant development, and slope of the leaf surface (Tukey, 1970).
The balance between uptake and loss of nutrients in plants exposed
to rain depends on biological, environmental, and temporal factors
that vary with geographical location and the individual rain event.

It has been reported previously that simulated acidic rain
and multiple exposures to ozone decreased the foliar dry weight of
red kidney beans in laboratory tests (Shriner, 1978). The stimu-
latory effects of nitrate in simulated acidic rain on the growth of
trees also has been demonstrated (Tveite and Abrahamsen, 1978; Wood
and Bormann, 1977). Exposure to ozone shifted the partitioning of
photosynthate in parsley but, in this case, root growth was depressed
relative to the growth of aboveground parts (Oshima et al., 1978).

Our results indicate that both the exposure of plants to nitrate
and sulfate in rain and to ozone in the atmosphere are important
factors to consider when determinations are made of the effects of
acidic precipitation on vegetation. The qualitative character and
magnitude of effects of acidic precipitation may change with different conditions of plant nutrition and air pollution. Future
experiments should be performed with these factors controlled or
carefully measured before a full understanding of the effects of
acidic precipitation on vegetation can be obtained.

REFERENCES

Cowling, D.W., and Jones, L.H.P., 1970, Soil Sci., 110:346-354.
Jacobson, J.S., Heller, L.I., and van Leuken, P., 1976, Proc.
     First Internat. Symp. on Acid Precipitation and the Forest
     Ecosystem, USDA Forest Service Gen. Tech. Report NE-23.
Jacobson, J.S., 1977, VDI-Ber. Nr., 270:163-173.
Jacobson, J.S., 1978, Paper presented at meeting on: "Effects of
     acid precipitation on terrestrial ecosystems," NATO Advanced
     Research Institute Ecosciences Panel, Toronto, Canada, May
     22-26.
Jones, L.H.P., Cowling, D.W., and Lockyer, D.R., 1972, Soil Sci.,
     114:104-114.
Le Clerc, J.A., and Breazeale, J.F., 1908, in: "Yearbook of the
     USDA," pp.389-402, Washington, D.C.
MacLean, D.C., and Schneider, R.E., 1976, J. Environ. Qual., 5:75-
     78.
Mandl, R.H., Weinstein, L.H., McCune, D.C., and Keveny, M., 1973,
     J. Environ. Qual., 2:371-376.
Noggle, J.C., and Jones, H.C., 1979, Interagency Energy/Environment
     R+D Program Report, EPA-600/7-79-109.
Oshima, R.J., Bennett, J.P., and Braegelmann, P.K., 1978, J. Am.
     Soc. Hort. Sci., 103:348-350.
Shriner, D.W., 1978, Abstracts of 70th Annual Meeting of the Ameri-
     can Phytopathological Society, Phytopathol. News, 12:153.

Tamm, C.O., 1958, pp.232-242, in: The Encyclopedia of Plant
    Physiology, vol. IV, W. Ruhland, ed., Springer-Verlag, Berlin.
Tamm, C.O., and Cowling, E.B., 1976, pp.854-855, in: Proc. First
    Internat. Symp. on Acid Precipitation and the Forest Ecosystem,
    USDA Forest Service Gen. Tech. Report NE-23.
Thomas, M.D., Hendricks, R.H., Collier, T.R., and Hill, G.R., 1943,
    Plant Physiol., 18:345-371.
Tukey, H.B., Jr., 1970, Annu. Rev. Plant Physiol., 21:305-324.
Tveite, B., and Abrahamsen, G., 1978, Norwegian Forest Research
    Institute, SNSF-Contribution FA 29/78, Ås-NLH, Norway.
U.S. Environmental Protection Agency, 1979, OAQPS-78-8 IV-A-3,
    Office of Air Quality Planning and Standards.
Wood, T., and Bormann, F.H., 1977, Water, Air, Soil Pollut., 7:479-
    488.

DISCUSSION

EVANS: With your lettuce experiment, you used different ratios
of sulfate and nitrate, and you indicated that the values
obtained for the intermediate ratio of nitrate to sulfate were
significantly lower than those obtained for the two extremes,
on apical structures as well as all leaves. But if you
compared that value with the controlled pH level of 5.6, there
was no significant change. Is this interpretation true?

JACOBSON: That's correct in some cases, but not in all.

EVANS: Specifically I ask about the comparison between apical
leaves versus all leaves.

JACOBSON: In apical leaves and root mass.

EVANS: Why did the intermediate ratio give values that were
significantly lower than the pH control?

JACOBSON: I think the clearest way of describing the data in
Table III is that the applications of simulated rain with the
higher sulfate to nitrate ratios increased the dry mass of
apical leaves and roots of lettuce plants. Unfortunately, we
have no information on the sulfur content of the soil. Perhaps
addition of sulfate in rain can overcome sulfur deficiency in
the soil. Future experiments on the effects of acid rain
should always include measurement of soil sulfur.

EVANS: Nitrate is not very toxic to plants. Plants seem to
absorb large quantities of nitrate and just shunt it off into
the vacuoles. Any comment?

JACOBSON: The interpretation that the high sulfate to nitrate
ratio increased growth is indeed more reasonable than
suggesting that the high nitrate to sulfate ratio treatment
depressed growth.

McLEAN: Isn't it rather surprising that your growth is better
at the lower pH than at the control pH?

JACOBSON: No, it's not so surprising. There's precedent for
this. Dr. Abrahamson in Norway found when he applied sulfuric
acid to forest trees that in some treatments he got increased
growth (G. Abrahamson and G.J. Pollard (1978)) SNSF
Contribution FA 32/78). Wood and Borman (1977) in applying
simulated acid rain to tree seedlings in the greenhouse found
increased growth, and this is nitrogen fertilization. So it

isn't necessarily the acidity, <u>per</u> <u>se</u>, but of course you can't just put hydrogen ions in. You've got to put add anions as well.

<u>SHRINER</u>: Is it not possible that with high nitrogen treatment sulfur might be limited?

<u>JACOBSON</u>: It's possible but I doubt it because we were able to grow healthy lettuce plants with half-strength Hoagland's solution. It is used traditionally as a balanced source of nutrients, so that doesn't necessarily mean that it was, but I doubt it.

<u>MATTESON</u>: Where did you measure your pH?

<u>JACOBSON</u>: The lettuce plants in the greenhouse were exposed to rain from a nozzle which simulated both the intensity and the droplet diameter of ambient rainfall. We found in previous experiments that it's not only the acidity of rain that's important in determining the development of foliar lesions or growth effects; in addition, it's the duration and frequency of exposure, the intensity of rainfall, and, of course, the cultural conditions of the plant, which change their predisposition to injury.

        The chambers of the soybean experiment were in the field and had a cover to exclude ambient rain. The cover was placed on loosely so that air blown from the outside through filters and through the chamber can flow through the chamber. Inside the plexiglas chamber were rain nozzles pointed in the upward direction so that we obtained gravity fall of droplets.

        Now, notice how these soybean plants just about fill the bottom half of this chamber. These are control plants exposed to filtered air containing negligible quantities of ozone. This next picture, taken at the same time, shows plants exposed to ambient ozone; they had an increase in the number of senescent leaves as well as decrease in growth that was obvious to the eye.

        You may not be able to see but these are symptoms of ambient ozone injury, a kind of bronzing to the surface of the soybean leaves. And as I said, it also produced enhanced senescence in the soybean. There were very slight symptoms-lesions, as a result of exposure to the simulated acid rain at the most acid pH. They were so slight that you wouldn't be able to see them in a photograph. The symptoms are very small lesions on some leaves.

Droplets reside on a bean leaf and then produce necrotic areas of varying shape and size subsequent to the exposure to simulated rain. Spinach leaves also develop necrotic lesions. Droplets likewise form on a very waxy conifer, Eastern White Pine but note that in this case there was no "puddling" in the axils of the needles after periods of simulated rain exposure. The injury occurred in a very sporadic manner--red, brown necrotic lesions usually toward the tips of the needles, but there were also many needles that were entirely green.

MATTESON: I wanted to know when you measured the pH, before or after the ozone exposure.

JACOBSON: Before the ozone exposure. But, we have not measured to see if there is a change in pH as the droplets dry on foliage, which I think would be an interesting thing to do.

STENSLAND: With the soybeans, did you know the amount of sulfur in the soil? Production increased as you put on more sulfur.

JACOBSON: If you mean the sulfur going into the soil to be taken up by the roots and back up into the plants I doubt that we would have that effect. There was a large amount of soil and relatively small quantities of sulfate were added.

McFEE: It is an agronomic practice to add foliar fertilizers. And if you could add foliar nitrate and sulfate without damage to the leaf, you can frequently get an increase in yield, even if the soil is pretty well supplied with both.

JACOBSON: Right. Foliar fertilization is a well known practice in some areas.

STENSLAND: One solution you mentioned was a 5×10^{-4} molar nitric acid solution and it seems like that should be a pH of about 3.3 instead of 3. How you do you explain that?

JACOBSON: Some values in Table III were incorrect and will be corrected in the final publication.

STENSLAND: What did you add to bring the pH down to 3?

JACOBSON: We made up a standard rain solution then titrated it to the appropriate pH.

STENSLAND: If you added something like hydrochloric acid, you could get it down to pH 3 and then--

JACOBSON: But we wouldn't want very much chloride in there either. Chloride is not a very common constituent in high concentrations.

STENSLAND: Well, there is probably enough nitrate and sulfate in acid form to give you pH 3.

MILLER: Did that answer your question as to what he did use to titrate it down to pH 3.

STENSLAND: Well, my problem is I cannot see how he can have pH 3 for that particular solution. I'm wondering if a column is mislabeled or something like that. If it is 5×10^{-4} molar nitric acid, that would be a pH of between 3 and 4. And so I do not know how he got it down to pH 3.

JACOBSON: I think the only three anions, if I remember correctly, were chloride, sulphate, and nitrate. Because they are the only three that are really common; phosphate is really negligible.

WILSON: If you have a situation in which acid precipitation is striking a leaf surface, and then you have evaporation, presumably there is a salt crystalline film left. Are these typically hygroscopic?

JACOBSON: Leaves have a lot of materials on their surfaces from two other sources: (1) exuded from the interior, both organic acids, amino acids, as well as inorganic components, and (2) dry deposition, which provides particulate matter. When simulated rain is applied to a leaf surface, a mixture of all these materials may reside on the surface. Extended periods of exposure to simulated rain washes the surface of the leaf, removes a lot of these substances and results in not leaving very much except what may have been in the last part of the rain.

WILSON: It would seem, intuitively, that the presence of the salts may exacerbate transpiration and that in fact there may be moisture moving out from below the cuticular wax.

JACOBSON: There is one situation where that could occur, that is, a long dry period, where a lot of materials have accumulated on the leaf surface and then you have a dew or very light rain which does not wash the leaf but wets it. Then you might get a situation where there is water coming out of a leaf but with a lot of ions going into the leaf.

WILSON: This has particular implications for the Adirondacks where the higher peak areas involve fogs or clouds in addition to raw precipitation.

McFEE: Has the pH of the droplet on the leaf surface been measured as evaporation takes place?

JACOBSON: No. We have washed the leaf surface after a period of drying, subsequent to a period of exposure to simulated rain and found that the pH of the material washed off the surface afterwards is lower than that normally removed from the leaf, but is considerably higher than the pH of the rain.

SHRINER: Has anyone looked at the effects if the last bit of rain is deionized water? Does this type of exposure reduce the damage to leaf surfaces?

JACOBSON: I suspect that it does. I mentioned a list of factors that were important in determining plant growth in response to acid rain, and I think that one of them may be the fluctuations in concentration of components during the rain, which, incidentally would greatly complicate our experiments.

McFEE: Would you suspect the sulfur compounds, in whatever form they come off the leaf as they're exuded, would be a significant or a trivial impact on the actual accumulation of all the sources of sulfur to the surface of the leaf?

JACOBSON: I don't think the sulfur compounds would be important but there are many other materials that are exuded onto the surface of the leaf which may be alkaline or acidic, may act as buffers, and react with materials which are deposited on the leaves.

McFEE: Such as?

JACOBSON: Organic acids, amino acids, and some inorganic components which are exuded onto the leaf surface. It depends very much on the conditions. In the light rain or dew event these reactions would be much more important than in a heavy rain which is simply going to scrub everything off the leaf.

KADLECEK: Can you comment on the budget of sulfur and nitrogen that would be removed with the crust compared with what might be arriving from the atmosphere.

JACOBSON: Yes. We did that for the soybean experiment by
applying fertilizer according to information that we obtained
about what one should do for soybeans in this particular soil.
The amount of sulfur that one applies in fertilizer is far in
excess of sulfur deposited in an equivalent land area in a
single rain event. However, rain events occur repeatedly over
the course of a growing season and supply sulfur to the soil as
well as directly to the foliage in a very dilute form. By
comparing these two very different situations, one can develop
an opinion that the sulfur or nitrogen in rain is totally
irrelevant as far as plant nutrition is concerned to one where
it is possible that it has some importance because of the
different mode and pattern of application.

EFFECTS OF ATMOSPHERIC POLLUTANTS ON SOILS

William W. McFee

Agronomy Department
Purdue University
West Lafayette, IN 47907

ABSTRACT

The effects of two types of atmospheric pollutants on soils, acid precipitation and metals, are considered. Potential acid precipitation effects include soil acidification, increased loss of plant nutrients, accelerated weathering of mineral components, decreased rates of organic matter decay, changes in soil organism populations, mobilization of aluminum ions, and reduction in cation exchange capacity. Soils that are poorly buffered, i.e., have low cation exchange capacity due to low clay and organic matter contents, are most likely to undergo appreciable change due to acid inputs. Acid precipitation inputs experienced thus far are generally low compared to the effects of agricultural fertilization and liming practices on soil pH.

Results of field experimentation have not shown serious deleterious effects of acid precipitation on productivity or soil biota. The reaction of soils to acid inputs is complex and dependent on numerous soil parameters, such as type of clay present, base saturation, presence of easily weatherable minerals, and upon the ionic composition of the precipitation.

Soils are efficient collectors of metallic ions. Damaging levels of metal contamination of soils reported thus far are confined to urban areas and regions around point sources, such as smelters. Since metals are retained over long periods, recovery from metal contamination is slow.

The pollution of a soil is difficult to define because soils are natural receptors of wastes. Soils are sometimes referred to as "nature's garbage heap". Both the mineral and the organic portions of soils are products of weathering and decay. The weathering of rocks and minerals and the deposition of material by ice, water, and wind are the sources of the soil "parent material". To that, the remains of generations of plants and animals have been added, providing organic matter, or what is often called "humus". Add the present generation of living organisms: viruses, bacteria, fungi, tree roots, insects, burrowing rodents, etc., and you have the mix of living and inanimate materials that we call "soil". It is not a static system, but continually adjusting in response to natural and man-made forces.

When is a soil polluted or when is it being damaged? Since a soil is somewhat of a waste heap, what are the criteria for classifying some particular additions as pollutants? I would propose one based primarily on the edaphic approach. That is, if additions reduce the ability of the soil to support plants, reduce the utility of the plants produced, or seriously restrict the choice of plants that can be grown, then they are polluting. This must be applied both in the short term and long term. Secondly, I believe a hydrological approach has to be examined also. It may be that a pollutant that does not noticeably affect the productive potential of a soil may cause the passage of damaging materials from the soil to an aquatic system through leachate or run-off. For example, release of $Al+^3$ from an acid soil may not damage the soil, but reduce the receiving water's usefulness.

Soils are remarkable in their ability to withstand drastic treatments and to incorporate wastes and inputs of many kinds without serious change. Agriculturalists clear forests from virgin land, rip out the stumps or plow under the prairie sod, add fertilizer, lime, and sometimes herbicides and insecticides; yet the soil remains very much the same. Prairie and other dark soils lose some organic matter, but after 20 - 30 years a new equilibrium develops and a soil maintains most of its original character. The population of living organisms will have changed, more of some groups and fewer of others will be present, but if the original vegetation is allowed to return, and in many cases it will if simply left alone, the soil organism community and the physical and chemical properties begin to take on the original, uncultivated nature rather rapidly. This is the way we would like to see the soil, with its ability to respond and renew itself intact, and ready to furnish food and fiber for future generations or to grow forests or flowers for their enjoyment. That situation exists when the soil has been treated reasonably, not severely polluted or allowed to erode. The most widespread

serious damage to soil results from accelerated erosion. Soils
that have been severely eroded frequently are not capable of sustained productivity or return to their former use. Somewhere down
the scale in seriousness is the subject of today, soil pollution by
atmospheric deposition. I mention erosion because the worst cases
of atmospheric pollution may ultimately lead to accelerated erosion.

The annual addition of N fixed by lightning discharges is not
considered soil pollution nor is the elevated salt component of rain
near the coasts. Dust from the drier lands of the west, which over
geological time has been important in soil development is not a
pollutant, but the metal enriched dusts from smelters, roadways, and
industrial centers are potentially damaging. Rain enriched from
anthropogenic sources with oxides of nitrogen and sulfur that lower
its pH dramatically is considered polluted. I am not sure how much
acid from sulfur oxidation has been a regular addition to soils for
centuries, but certainly, there has been some. It is apparent that
the pollution of soils is, in most cases, a matter of degree or
amount of input not the absolute presence or absence of any specific
additions.

Obviously, the changes brought about by pollutants are not
equally reversible. Most added organics will be deactivated in
cultivated or uncultivated soils in a few years depending on amount
and nature of the material. On the other hand, metal loading or
radioactive fallout of long-lived isotopes is difficult to amelio-
rate and is corrected very slowly. Acidification is reversible
where the soil is managed, but is essentially permanent in uncultivated areas. I will not try to deal with all possible pollutants,
but will discuss acid precipitation effects primarily and add some
information on metal contamination.

Acidification of soils

In order to understand the effects of acid precipitation on
soils, it is necessary to consider first the chemical nature of
soils. I will narrow the discussion to the soils that are common
in the humid regions of the world. Acid precipitation may become
a problem elsewhere, but the present concern is concentrated in
eastern North America and Europe. Soils of these regions are
dominated by the residual silicate mineral and rock fragments that
have survived chemical and physical weathering processes, plus the
secondary clay minerals that have formed in the soil. Easily
weatherable minerals are absent or present in small quantities in
all except the recently deposited soils such as floodplains. All
have organic matter, but in various amounts. Commonly, one to five
percent of the uppermost layers are composed of organic matter and
the amount decreases rapidly with depth. A small percentage of the
landscape may contain "organic soils," those whose properties are

dominated by the organic constituents (20-30% or greater organic
matter). The chemically active constituents are the finely divided,
partially decayed, organic matter or "humus", and the colloidal
mineral material, the clay fraction. These materials, the clay and
humus, behave as cation exchangers (Figure 1). Cation exchange
capacity (CEC) is frequently reported in milliequivalents per 100
gram of soil (meq/100g). Many medium textured soils will contain
10-30 meq/100g in their surface layers, but sands with low organic
content may have only 1-2 meq/100g or less.

     The type and amount of clay and humus determine the CEC of a
soil which is a measure of its buffering capacity against changes
in pH, and thus, the effects of acid precipitation. The first
effects of acid additions are on the balance of cations on the cation
exchange capacity unless the soil contains excess base such as
calcium carbonate (only relatively new soils in humid temperate
regions). Soils with little clay and little humus have low CEC,
and thus, little resistance to pH changes. When the soil contains
moderate or high CEC, the effects of acid inputs will be slow or
even negligible. McFee and Kelly (1977) calculated the theoretical
effects of 100 years of pH 4.0 rain, 100 cm per year, on a soil with
20 meq/100g CEC and estimated a possible downward pH shift of only
0.6. This would result from a reduction in the proportion of basic
cations such as Ca^{+2}, Mg^{+2}, K+, and Na+, to the total exchange
capacity. This is usually referred to as the percent base saturation,
e.g., a soil with 18 meq of basic cations and 6 meq of H+ and Al^{+3}
would have a percent base saturation of 75%. The percent base
saturation controls the equilibrium concentration of ions in the
soil solution and thus controls the pH.

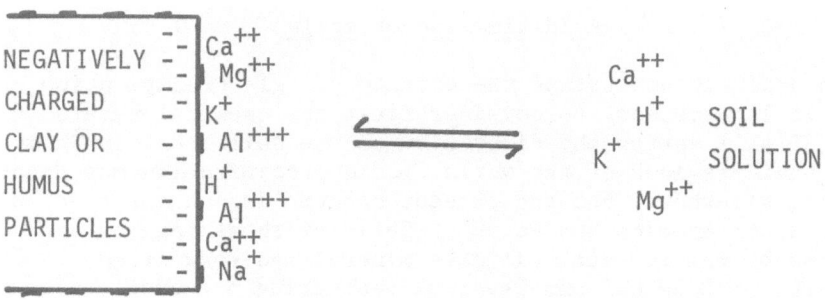

Figure 1. Illustration of soil cation exchange. Equilibrium
between cations adsorbed on soil colloid surfaces and those in
solution.

One cannot assume that all of the hydrogen ion added to the
soil via acid precipitation will exchange for basic cations on the
cation exchange capacity. Wiklander (1973-74) and Wiklander and
Andersson (1972) have pointed out that the efficiency of hydrogen
ions in replacing metal cations is strongly dependent on soil
properties, such as soil pH, and also dependent upon the accompanying ions in the acid precipitation. In soils with pH below about
4.5, one equivalent of hydrogen replaces less than one equivalent
of metal cations. This is predictable from several equations
designed to reflect mass action replacement of cations on soil
exchange complexes. There are several forms of these relationships
(Helfferich, 1962; Gapon, 1933), but all indicate that the extent
of replacement of one cation for another on the exchanger phase is
a function of the ratio of the ions in the solution inputs. In
soils with a pH above 5, the apparent replacing efficiency may be
one or greater. All of the basic ions removed from the exchange
capacity by hydrogen cannot be assumed to be lost. Some of them
would be absorbed by plants and recycled to the surface. Some of
them may be retained in the lower layers of the soil. Nutrient
loss through leaching is also somewhat offset by the weathering
processes. Acid precipitation may accelerate the weathering of
residual minerals, and thus, release basic cations that compete
for positions on the exchange sites. The gradual decomposition of
soil material also results in the consumption of hydrogen ions at
the same time it releases metallic or basic cations. There are so
many interacting factors that the precise effect of acid precipitation in acidifying soils is hard to predict. The general trend is
apparent and some rough limits can be put upon the rate of acidification.

There are at least four sources of hydrogen ions (hydronium
ions) that are responsible for soil acidity (Bache, 1979):

1) Carbonated water. $CO_2 + 2H_2O \rightleftharpoons H_3O^+ + HCO_3^-$. The carbon
    dioxide in the atmosphere is sufficient to bring water that
    is otherwise pure to a pH of 5.6. In surface soils,
    respiration creates a higher CO_2 level, that in equilibrium
    with water creates a pH < 5.0.

2) Nitrification of naturally occurring or added ammonium ions.
    $NH_4 + 2O_2 + H_2O \rightarrow 2H_3O^+ + NO_3^-$.

3) Organic acids from decomposition of plant residues. Many
    organic acids are formed by microbial action in soils.
    This is certainly an important process in the formation
    of acid podzol or spodozol soils.

4) Acid precipitation. Rainfall probably always contained
    small amounts of nitric and sulfuric acids from oxidation
    of nitrogen and sulfur compounds in photochemical and
    lightning induced processes, but increased levels are well

documented in recent years. A fifth source, oxidation of
reduced sulfur forms, such as pyrite, is important in
certain soils, but is not widespread.

To put these in perspective, it should be pointed out that soils
in humid regions naturally acidify, even without the addition of any
man-made pollution. The role of additional acidic precipitation is
simply to increase the rate of acidification. Our activities, such
as the addition of nitrogen fertilizers, accelerate acidification of
the soil much more rapidly than any recorded acid precipitation.
Reuss (1975) made it clear that in agricultural situations precipitation with a pH of 4 would play a minor role in determining the
soil pH. He pointed out that hydrogen ion release by nitrogen and
sulfur transformations commonly occuring in the soil are quite large
by comparison. The potential acidifying effects of 100 kg of nitrogen per hectare applied as ammonia exceed that expected in
annual precipitation by an order of magnitude. Furthermore, common
liming practices employed in agriculture are designed to apply basic
materials ($CaCO_3$) at levels far in excess of that needed to neutralize the cumulative acidity of several year's rain.

The effects of increased acidification of soils are generally
considered undesirable and include the following:

1) Loss of basic cations by leaching, many of which are plant
    nutrients (e.g. Ca^{+2}, Mg^{+2}, K^+). In soils containing free
    carbonates, this results from their solution by the acid;
    otherwise, it is the result of displacement of the basic
    ions from the exchange sites directly or by Al^{+3} freed by
    the action of the H^+ on the alumino-silicate minerals.

2) Reduced cation exchange capacity. Commonly, much of the
    cation exchange capacity in soils is pH dependent, i.e.,
    as the pH drops, some of the negative exchange sites are
    permanently occupied by H^+ and are no longer active. Thus,
    the ability to store nutrient ions and buffer against
    further change is reduced.

3) Increased aluminum availability. Since Al solubility is
    high around pH 4.0, as the soil acidifies soluble Al^{+3}
    occupies the exchange sites and dominates the soil system.
    It is, thus, more likely to create Al toxicity in sensitive
    plants.

Experimental results

Experimental results with artificial acidification of rainfall
have been carried out both in Europe and in North America. In Norway,
Abrahamsen et al. (1976) reported on lysimeter experiments which
demonstrated that dilute sulfuric acid applied to a podzol soil
resulted in a significant retention of hydrogen within the soil

column. These were sandy soils with relatively low cation exchange
capacity and would be expected to be susceptible to acidification. The amount of hydrogen ion passing through the lysimeter
amounted to only 12 percent or less. The retention increased with
the increasing concentration of hydrogen in the rain. This retention was undoubtedly due to the ion exchange processes and some
consumption of hydrogen ion by weathering processes. The acid
treatment did increase the output of aluminum ion from the lysimeters
which came from replacement on the exchange sites or release by
weathering.

Swedish studies by Tamm (1977) involving addition of sulfuric
acid to a podzol soil produced a reduction in the pH of the water
leaching from the soil from approximately 5 down to 4.5. At the
same time, the amount of basic cations in the upper portion of the
soil was reduced significantly. In general, the increased deposition of sulfate which is a part of the acidification of the precipitation has increased the leaching of sulfate from soils. This along
with the retention of hydrogen ions in the soil has produced an increased leaching of the nutrient cations, calcium, magnesium,
potassium, and manganese.

The release of metallic cations from weathering processes may
be greater than the input of hydrogen ions; therefore, there may
be situations where the increased output of metallic ions is greater
than would be expected from the increased acidity of the precipitation. Extended over very long periods of time, this could lead to
serious depletion of the soil nutrient supply.

Other effects of acid precipitation on the soil

It is a well-known fact that soil acidity tends to reduce nitrogen fixation by soil organisms living in symbiotic relationships with
leguminous plants. Shriner (1977) noted a reduction in nodule
formation on legumes that were exposed to simulated rain with a pH
of 3.2. Alexander (1979) pointed out some of the likely effects of
acidification on soil microorganisms. He indicated that nitrification processes are very sensitive both in culture and in nature to
increasing acidity, and that acidity is generally linked with decreasing rates of humus decomposition. He also indicated that de-
nitrification is likely to be reduced by increasing acidity, and
that microbial conversion of organic to inorganic forms of phosphorus in soil appeared to be lower in the more acid environments.
Experimental results available thus far, such as the work of Tamm
et al. (1977) and Abrahamsen et al. (1979), and Baath et al. (1978)
indicate that the decomposition of organic material in coniferous
forest soil is only slightly sensitive to acidification. The pH
of the rain had to drop below 3.0 before decomposition of fresh
litter was much affected. However, Tamm et al. (1977) did observe
reduced respiration in a laboratory study with soil samples acid-

fied with sulfuric acid or powdered sulfur. He observed increased
amounts of ammonia and decreased amounts of nitrate which were
apparently due to reduced immobilization and nitrification rather
than being due to increased ammonification.

Some of the Scandinavian studies (Hagvar, 1978; Baath et al.,
1979) have included investigations on the effects of acid precipitation on the abundance of soil animals. Since the soils they were
dealing with were already acid, the additional acid precipitation
seemed to have little ill effect on the population of collembola.
In fact, some species increased in response to acid inputs.

It is well known that some of the cation exchange capacity in
soils is pH dependent. That is, if the pH of the soil drops, there
is less available cation exchange capacity. Norton (1976) described
the potential effects of acid rain in destabilization and solution
of clay minerals leading to the loss of cation exchange capacity and
increased release of iron and aluminum. Overrein (1972) found only
slight increases in leaching at pH 4.0 or greater on several soil
types, but when the precipitation or input water was reduced to a
pH of 3.0 or lower, leaching losses increased rapidly. This is in
agreement with some of my own results where leaching columns treated
with dilute H_2SO_4 produce significantly more metallic cations.
Fisher et al. (1968) in a study of the effects of atmospheric inputs
at Hubbard-Brook, New Hampshire, concluded that the overall effect
of acid in precipitation is to leach base metal ions from watershed
minerals, and this is in agreement with most of the other work so
far. The results of the research can be summarized in these five
points:

1) Soils are not all equally susceptible to effects of acid
   precipitation.

2) Short term effects on biota due to soil changes brought
   about by acid inputs are difficult to measure.

3) Application of simulated acid precipitation increases the
   leaching of cations, but frequently in amounts much less
   than the added cations.

4) Acid rainfall treatments above the pH of 4.0 have not
   usually shown significant effects on soils in the limited
   time frame of experiments conducted thus far.

5) Decreased productivity has not been demonstrable thus
   far in natural conditions or under treatments with pH 4.0
   or higher.

It has been demonstrated quite dramatically that when acid rain
and air pollution were so severe that large areas such as that in
the Copper Basin of Tennessee and around Sudbury, Canada, lose most
of their vegetation, the soils may be severely degraded. Roughly,

a 160 square mile area around Sudbury is severely denuded and the
soils are apparently incapable of full recovery. Severe erosion has
probably been the most damaging, but there is evidence of acidification also. Hutchison and Whitely (1976) put it this way, "These
profound and damaging changes may be only indicative of extreme local
conditions, or they may be a harbinger of what can be anticipated in
other acid rain areas."

     In general, soils are very resistant to the effects of addition-
al acid inputs. The effects of liming and fertilization in agricultural situations will greatly overshadow any effects of acid rain on
the soil. Only those soils that are very low in their cation exchange capacity, such as light colored, sandy soils, are likely to
be significantly acidified by acid precipitation. The effects of
acid precipitation on soil organisms in most soils is likely to be
concentrated in the surface few centimeters of the soil. The effects
of additional acid inputs on soils that are already very acid is unclear, but is likely to lead to an increased release rate of aluminum
and other metallic ions from those soils. This may be more damaging
to the surrounding aquatic systems than it is to the acid soil. It
is apparent that a system of monitoring these effects in the field
is needed, and additional basic research should be performed on
soil-acid interaction to enable prediction of future effects more
accurately.

Metal additions to the soil

     Numerous elements are added to the soil by atmospheric deposi-
tion. Annual inputs of nitrogen, phosphorus, calcium, and magnesium,
as well as many metals that are considered toxic have been measured.
Leaching losses of the nutrient cations are frequently in the same
order of magnitude as the annual inputs in rural areas. The inputs
near localized sources may far exceed the normal losses. The amounts
of basic metallic cations in precipitation from several locales is
reported in Table I. Metals, such as Pb, Zn, Cu, Cd, Ni, and Hg,
in quantities high enough to be of concern have been reported in
some areas. Zinc and Cu are beneficial in small amounts, but like the
others listed may become toxic to plants and soil organisms at high
levels. Normal soil levels of metals are reported in Table II. In
contrast to this, Buchauer (1973) reported soils contaminated with
several metals to the point that vegetation was severely affected
near a smelter in Pennsylvania. Values as high as 50,000 ppm for
Zn, 900 ppm Cd, 600 ppm Cu, and 200 ppm Pb were found in the Al
horizon near the smelter, and some contamination extended beyond
15 km. Hutchison and Whitby (1976) also reported heavy metal contamination within 8 km of the Sudbury smelters, and Munshower (1972)
reported serious effects around a smelter in Montana. Other sources
of metal contamination are known. Lagerwerff and Specht (1970) reported metal contamination of soils along highways, and Purves (1972)
reported convincing evidence of metal contamination by Zn, Cu, and

Pb in urban and industrial areas of Great Britain. Parker, McFee, and Kelly (1978) reported 2,400 ppm Zn, 460 ppm Pb, 120 ppm Cu, and 10 ppm Cd in the surface soils of the urban area in northwestern Indiana with lesser, but detectable, contamination extending into the rural countryside (Pietz et al., 1978). All of the contamination is not restricted to urban or industrial areas. Reiners et al. (1975) reports that Pb, and perhaps, Zn may be accumulating in the remote areas of New Hampshire; and Lindberg's et al. (1977) data from east Tennessee shows metal inputs of Cd, Mn, Pb, and Zn that may cause some soil accumulation. In their report, they cite data from several areas both rural and urban that show even higher rainfall inputs. A report on the rural areas of Manitoba by Mills and Zwarich (1975) showed little evidence of metal contamination of the soils. However, Siccama and Smith (1978) found evidence of significant Pb deposition in remote areas of New England and Ruhling and Tyler (1968) found evidence of long distance transport of Pb and Cd to southern Norway and Sweden where it was accumulating in moss carpets.

TABLE I. Basic cations in precipitation from several locales

Location	Ca	K	Mg	Na	Source
	- - - kg/ha/yr - - -				
England	3.5	1.4	2.2	17.0	Carlisle et al. (1966)
Indiana	6.9	1.2	-	2.6	Carroll (1962)
Minnesota	3.5	1.1	0.7	1.1	Verry and Timmons (1965)
North Carolina	5.0	2.3	1.1	4.4	Henderson et al. (1978)
New Hampshire	2.7	0.7	0.6	1.6	Fisher et al. (1968)
Ontario	8.5	0.7	3.0	0.8	Schindler and Nighswander (1970)
Oregon	5.0	0.2	1.0	1.8	Henderson et al. (1978)
Tennessee	14.2	3.4	2.2	4.0	Henderson et al. (1978)

TABLE II. Normal soil levels of selected trace metals

Element		Average Concentration	Source
	- - - ppm - - -		
Copper	2-100	31	Reuther and Labanauskas (1965)
Cadmium	-	.06	Lisk (1972)
Lead	2-200	16	Brewer (1965)
Mercury	-	.03	Lisk (1972)
Nickel	5-500	100	Vanselow (1975)
Zinc	10-300	10	Chapman (1965)

Metals entering the soil from the atmosphere in soluble form
are likely to become exchangeable ions on the clay or organic exchange
sites or be complexed by organic compounds. Some of the input is
undoubtedly insoluble and may remain insoluble depending upon the
pH of the soil solution. In either case, there is a strong tendency
for the soil to hold the metals near the soil surface. Only a small
portion of that entering goes through the soil. Even on sandy
soils in northwestern Indiana, Zn has accumulated to greater than
1000 kg/ha in the surface layers and continues to be deposited at
the rate of about 1 kg/ha/yr, but only .15 kg/ha/yr is leaching
through the profile (Parker et al., 1978). The tendency for soils
to retain metals has both good and bad aspects. The tight retention
of metals causes the soil to be an effective filter, thus removing
the metallic cations from the water that passes through the soil.
This prevents high metal levels from accumulating in ground water
which furnishes much of our drinking water supply. On the other hand,
the ability of the soil to retain metals means that a soil highly
contaminated with a metal is likely to remain contaminated for a very
long time.

It appears that metal contamination of soil by aerial deposition has reached problem levels only in localized areas near urban
and industrialized sources. Hopefully, our efforts to reduce emissions in the U.S. and Europe will prevent any additional areas from
becoming seriously contaminated. Because of its rather permanent
effect, it is a type of contamination that bears careful monitoring.
We do not know the spatial trends over large areas nor the temporal
trends over long time periods on the order of decades. This information is necessary if we are to adequately assess the risk to our
resources.

It appears that high levels of metal deposition on soils are
most likely to occur in the same regions that are experiencing acid
precipitation. The accumulation of metals by soils is due to exchange phenomenon, involves organic-metal complexing, and sometimes
the presence of insoluble metal oxides. A change in soil pH would
influence all of these. Lower pH increases the solubility of most
metals, reduces the stability of organic metal complexes, and reduces the cation exchange capacity. Thus, an important interaction
of acid precipitation with metal contamination is likely to exist.
The mobility of metals in the soil is likely to be increased by
acid inputs, thus shortening the residence time of metals in the
soil and increasing the metal content of ground water.

Summary and conclusions

Atmospheric pollution of soils appears to be a problem where
susceptible soils coincide with zones of very acid precipitation
and anywhere toxic metals are accumulating. However, soils appear

to be much more tolerant to acid precipitation than many aquatic
and plant systems due to the chemical buffering capacity provided
by clay and organic matter. In unmanaged soils of forests and wild-
lands, soil acidification may be significantly accelerated by the
acid precipitation currently experienced if the dominant soils are
low in exchange capacity. It will probably require time spans on
the order of decades to measure the effects on soils even where they
are poorly buffered. The effects on site productivity are difficult
to predict due to the multiple effects of additional acid. Increased
weathering and the nitrate and sulfate ions in the precipitation may
provide nutrients that were formerly limiting, and thus, increased
productivity may be noted. Long term effects on the soil are generally considered to be negative since the loss of basic cations,
reduction in cation exchange capacity, and accelerated weathering
ultimately lead to soils of lower natural productivity. Long term
field studies are needed to determine if remedial action is needed
on soils that are presently unmanaged. Acid precipitation effects
on agricultural systems may be important in its direct effect on
plants, but the soil effects will be completely masked by normal
soil amendments already in use.

Metal contamination of soils needs to be monitored closely.
The potential effects are serious and long lasting. Reports available indicate low levels of metal accumulation are widespread in
industrial nations, but toxic levels are presently found only near
major point sources. Some of these point sources have influenced
significant areas. Other sources of metals added to the soil should
be considered also when evaluating tolerable atmospheric deposition.
There is likely to be an interaction between acid precipitation and
soil metal levels since acidity increases metal mobility in the soil.

Literature Cited

Alexander, M. 1979. Effects of acidity on microorganisms and micro-
    bial processes in soil. Proceedings of NATO Adv. Res. Inst.
    Effects of acid precipitation on terrestrial ecosystems.
    Toronto, 1978. (In Press)
Abrahamsen, G., K. Bjor, R. Horntvedt, and B. Tviete. 1976. Effects
    of acid precipitation on coniferous forest. In F. H. Brackke
    (ed.) Impact of acid precipitation on forest and freshwater
    ecosystems in Norway. F.R. 6/76, Oslo-As, Norway. pp. 33-63.
Abrahamsen, G., J. Hovland, and S. Hagvar. 1979. Effects of artifi-
    cial acid rain and liming on soil organisms and the decomposi-
    tion of organic matter. Paper presented at NATO Advanced
    Research Institute, Effects of Acid Precipitation on Terrestrial
    Ecosystems, Toronoto, May 22-26. (In Press)
Baath, E., B. Berg, U. Lohm, B. Lundgren, H. Lundkvist, T. Rosswall,
    B. Soderstrom, and A. Wiren. 1979. Soil organisms and litter
    decomposition in a Scots pine forest. Effects of experimental

    acidification. Paper presented at NATO Advanced Research
    Institute, Effect of Acid Precipitation on Terrestrial Eco-
    systems, Toronto, May 22-26. (In Press)
Bache, B. W. 1979. The acidification of soils. Proceedings of
    NATO Adv. Res. Inst. Effect of acid precipitation on terres-
    trial ecosystems. Toronto. (In Press)
Buchauer, M. J. 1973. Contamination of soil and vegetation near a
    zinc smelter by zinc, cadmium, copper and lead. Environ. Sci.
    Tech. 7:131-137.
Carlisle, A., A. H. F. Brown, and E. J. White. 1966. The organic
    matter and nutrient elements in the precipitation beneath a
    sessile oak canopy. J. Ecol. 54(1):87-98.
Carroll, D. 1962. Rainwater as a chemical agent of geologic proc-
    esses - a review. U.S. Geologic Survey Water - Supply Paper,
    1535 - G, 18 pages.
Fisher, D. W., R. W. Gambell, G. E. Likens, and F. H. Bormann. 1968.
    Atmospheric contributions to water quality of streams in the
    Hubbard Brook Experimental Forest, New Hampshire. Water
    Resources Res. 4(5):1115-1126.
Gapon, E. N. 1933. The theory of exchange adsorption in soils.
    J. Gen. Chem. (USSR) 3:144-152.
Hagvar, Sigmund. 1978. Effects of acidification and liming on
    collembola and acarina. SNSF Proj. IR 36/78, Oslo-As, Norway.
Helfferich, F. 1962. Ion Exchange. McGraw-Hill, NY. pp. 275.
Henderson, G. S., W. T. Swank, J. B. Waide, and C. C. Grier. 1978.
    Nutrient Budgets of Appalachian and Cascade Region Watersheds:
    A comparison. Forest Science 24:385-397.
Hutchinson, J. C. and L. M. Whitby. 1976. The effects of acid rain-
    fall and heavy metal particulates on a boreal forest ecosystem
    near the Sudbury smelting region of Canada. Proc. 1st Intern.
    Symp. on Acid Precipitation and the forest ecosystem. USDAFS
    Gen. Tech. Rep. NE-23.
Lagerwerff, J. V. and A. W. Specht. 1970. Contamination of road-
    side soil and vegetation with cadmium, nickel, lead, and zinc.
    Environ. Sci. Tech. 4:583-586.
Lindberg, S. E., R. R. Turner, N. M. Fergeson, and D. Malt. 1977.
    Walker Branch watershed element cycling studies: Collection
    and analysis of wetfall for trace elements and sulfate. Water-
    shed research in Eastern North America Workshop, Feb. 28 -
    March 3, 20 pp.
McFee, W. W., J. M. Kelly, and R. H. Beck. 1977. Acid precipitation
    effects on soil pH and base saturation of exchange sites. Water,
    Air and Soil Poll. 7:401-408
Mills, J. G. and M. A. Zwarich. 1975. Heavy metal content of agri-
    cultural soils in Manitoba. Can. J. Soil Sci. 55:295-300.
Munshower, F. F. 1972. Cadmium compartmentation and cycling in a
    grassland ecosystem in the Deer Lodge Valley, Montana. M.S.
    Thesis, Univ. of Montana, Missoula, MT.
Norton, S. A. 1976. Changes in Chemical Processes in Soils caused
    by acid precipitation. USDA Forest Serv. Gen. Tech. Rep. NE-23,
    pp. 711-724.

Overrein, L. N. 1972. Sulfur Pollution patterns observed: Leach-
    ing of calcium in forest soil determined. Ambio 1(4):145-147.
Parker, G. R., W. W. McFee, and J. M. Kelly. 1978. Metal distribu-
    tion in a forested ecosystem in urban northwestern Indiana.
    J. Environ. Qual.
Pietz, R. I., R. J. Vetter, D. Masarik, and W. W. McFee. 1978.
    Zinc and cadmium contents of agricultural soils and corn in
    northwestern Indiana. J. Environ. Qual. (In Press).
Purves, David. 1972. Consequences of trace element contamination
    of soils. Environ. Pollut. 3:17-24.
Reiners, W. A., R. H. Marks, and P. M. Vitousek. 1975. Heavy metals
    in subalpine and alpine soils of New Hampshire. OIKOS
    26:264-275.
Reuss, J. O. 1975. Chemical/Biological relationship relevant to
    ecological effects of acid rainfall. EPA - 660/3-75-032.
Ruhling, A. and G. Tyler. 1968. An ecological approach to the
    lead problem. Bot. Notiser 121: pp. 321-342.
Schindler, D. W., and J. E. Nighswander. 1970. Nutrient supply and
    primary production in Clear Lake, eastern Ontario. J. Fish.
    Res. Board, Canada. 27:2009-2036.
Shriner, D. S. 1977. Effects of simulated rain acidified with
    sulfuric acid on host parasite interactions. Water, Air, and
    Soil Pollution 8   : pp. 9-14.
Tamm, C. O. 1977. Skogmarkens forsurning, orsaker och motatgarder.
    Sveriges SkogsvFor. Tidskr. 75:189-200.
Tamm, C. O. and E. B. Cowling. 1977. Acidic precipitation and
    forest vegetation. Water, Air, and Soil Pollution 7 (4):
    pp. 503-511.
Verry, E. S. and D. R. Timmons. 1977. Precipitation nutrients in
    the open and under two forests in Minnesota. Can. J. Forest
    Res. 7:112-119.
Wiklander, Lambert. 1973/74. The acidification of soil by acid
    precipitation. Grundforbattring 26:155-164.
Wiklander, L. and A. Anderson. 1972. The replacing efficiency of
    hydrogen ion in relation to base saturation and pH. Goederma
    7 7: pp. 159-165.

DISCUSSION

JACOBSON: Individual rain events vary in acidity quite a bit, over 2 pH units. Is there any concern from the point of view of soils over this variability or is the yearly average acidity the important point to take into consideration.

McFEE: My first reaction is that, it is the total amount of acid received that affects the soil. However, to a soil microbiologist or one who is interested in soil insects that inhabit that surface layer, a very high percentage of them are right at the interface between the litter and the mineral soil, the episodal effects could be important.

VOLCHOK: What fraction of the total lead or zinc came in by atmospheric deposition, and how much was there?

McFEE: We don't know for sure except by going to the surrounding area and looking at soils that we consider to be the same soil, or very similar. When we did it appears that almost all of lead and zinc metal has been added. Of course, we don't know whether it came in as dust or how much of it came in with the rain pollution.

VOLCHOK: But you showed that the peak concentration was about $2\frac{1}{2}$ cm. below the surface which suggests the peak is moving down.

McFEE: The peak is in the very surface layer of mineral soil; the litter material did not hold as much on a weight basis.

VOLCHOK: We've been monitoring strontium 90, which is a pretty good analogue for calcium and maybe even for lead. We know the input source very well in terms of time. For a long time the strontium-90 peak occurred in the surface layer; it's now moved down. There are substantial amounts below 10 or 15 centimeters, and the peak may be 8 or 10 centimeters down, depending upon precipitation and soil type.

McFEE: We can expect this or might look for this to happen in this situation where the inputs are now reduced.

VOLCHOK: We know that happens with calcium; in places where you don't add calcium it eventually gets out of the root zone.

McLEAN: However, the heavy metals are more strongly complexed to the humic matter and therefore will not be transported through the soils, as rapidly as calcium, magnesium and strontium which are in solution.

VOLCHOK: Even the plutonium is moving down, very slowly.

McFEE: Once they move from that surface in this particular
soil, then they probably will move very rapidly because there's
very little to hold them. It wouldn't necessarily be true for
all soils.

HICKS: One of the figures we know very well from coring
studies for the area of interest here is probably the
deposition of lead into Lake Michigan. They compare very
well. The rural number you came up with is 221 grams per
hectare per year which would lead you to about 200 tons per
year over the Lake Michigan basin by my figures, and that is
about what is measured by the coring programs.

McFEE: That's encouraging.

JACOBSON: Are the soils you studied undisturbed for the whole
time?

McFEE: Yes, as far as we can tell. There's been some fire but
they've never been tilled.

TOMLINSON: You spoke about the addition of calcium carbonate
to agricultural soils. For the Cranberry Lake area we heard
from Dr. Bloomfield yesterday that the calcium composition of
the water in the rivers was essentially the same as in the
rain. But, some years ago, measurements were made on the
various minerals in forest soil in Northern Quebec. The amount
of calcium available was really not enough for another crop of
trees, but the area is being cropped. It was indicated in this
study that the inputs of calcium from the atmosphere would be
very helpful based on values that were averaged from England
and places where they'd measured dust fall. From more recent
measurements of the rain input into Northern Quebec, it's
evident there is very little calcium added in this way, far
less than there is in industrial areas closer to various
sources of calcium input to the atmosphere. In addition, there
is the leaching of calcium from the soil. I feel we may be in
for some major problems. Have you any specific data on that
type of area, where the calcium budget is low in the soil?

McFEE: No, I don't have any.

TOMLINSON: There's also the question of what the tree needs
the calcium for. It normally takes up more calcium than any
other mineral. But, for forest nutrition, when the tree

decays, the calcium in the soil presumably supplies part of the base for neutralization of the acids that are formed. I think that may be our big problem in the years ahead.

McFEE: There seem to be numerous cycling studies that have shown very, very tight budgets for some of these cations--calcium in some of those regions; potassium in the southeast and the southern pine region. The total potassium in the soil is very small compared to the potassium levels that we normally deal with. But yet a potassium response is not obtained when potassium is added. In other words, it would appear that the little bit of potassium cycling through the vegetation is being very efficiently reused and not much of it is actually escaping.

TOMLINSON: Of course the breakdown of organic materials is much slower in northern Quebec relative to the southern areas because of both the low pH and the cool temperatures.

McFEE: That makes a very good trap for any incoming metallic cations.

KRUPA: We conducted for two years measurements of transects from Bailey Station and also for Michigan City. The results from the transects do not parallel yours but go the other way from Indiana Dunes all the way to Chesterton and the other way. One of the interesting things we found was a lot of boron movement in the soils.

McFEE: Unfortunately we didn't look for boron at all.

SESSION IV:
ANTICIPATED PROBLEMS AS YET
NOT QUANTITATED

ON THE DRY DEPOSITION OF ACID PARTICLES TO NATURAL SURFACES

Bruce B. Hicks

Radiological and Environmental Research Division
Argonne National Laboratory
Argonne, IL 60439

ABSTRACT

Monitoring programs conducted over the northeastern continental USA during the past few years have indicated that sulfate particles present in air near the surface are often acidic. These particles, which are typically small and hygroscopic, might be expected to attach themselves to foliage, thus imparting a strong but very localized dose of acid. At this time, the efficiency with which small particles attach themselves to leaf surfaces and the conditions under which they might be re-emitted by abrasion, for example, are largely unknown, and so a considerable uncertainty must be associated with any evaluation of the net effect. Application of deposition velocities in the range presently advocated for sulfate particles suggests acid fluxes by dry deposition that average about two orders of magnitude less than those probably resulting from rainfall. This should not be interpreted as an indication that dry deposition effects can be neglected, since it is clear that acid particles might reside on surfaces for considerable times, perhaps until washed off by rain or sufficiently diluted by dewfall.

INTRODUCTION

In recent years, there has been an increasing recognition that the turbulent transfer of airborne pollutants to the surface of the earth, a mechanism known as dry deposition, is a significant sink for atmospheric contaminants on the one hand, and a contributing factor to ecological and structural damage on the other. Budget calculations show that the average removal rate of some contaminants by dry deposition is as great as (or

sometimes greater than) that resulting from precipitation. In
the case of sulfur, for example, a number of regional-scale
simulations agree that dry and wet removal mechanisms remove
about the same amount, within a factor of about two either way.
A study by Shannon (1) of the sulfur budget of the northeastern USA
indicates that the ratio wet/dry deposition for sulfur from
industrial sources is likely to average about one in summer, but
rise to about two during winter.

     While it is clear, therefore, that dry deposition cannot be
neglected as a potential impact upon ecological systems, it is
equally clear that the characteristics of the two processes are
fundamentally dissimilar. Dry deposition results from a continuing low rate of transfer, perhaps with a strong diurnal cycle (2)
but not varying too much from day to day. In comparison, wet
deposition is an intermittent and highly variable process, the
statistics of which are certainly not well understood.

     The present considerations must be confined to the deposition
of acid. In this case, we must refine the arguments derived from
sulfur budget considerations, firstly because not all of the
acidity in question is related to sulfur, and secondly because
most of the dry deposition is of sulfur dioxide.

     The present purpose is to set down some guidelines for assess-
ing the dry flux of acidic airborne contaminants to natural
surfaces, and to discuss their limitations. For now, this dry flux
will be taken to include all turbulent transfer of aerosol not
associated with rainfall; the reason for the subtlety will become
apparent later. The matter will be considered in the light of
three questions: (a) how much acidic aerosol is present in air,
(b) how is it transported to a surface, and (c) how can we estimate
the intensity and duration of typical impacts at the surface? A
fourth question on the effects associated with these fluxes will
be left entirely for others to address.

SOME CHARACTERISTICS OF ACID AEROSOL

     Over the last decade, improvements in chemical techniques
have led to a vast increase in the quantity and quality of
information regarding the constitution of the mixture of trace
gases and solid and liquid particles that make up the atmospheric
aerosol. The range of species that are acidic is quite wide,
including such benign compounds as carbon dioxide and such
strongly acidic species as bisulfates. While the scope of the
present considerations might well be interpreted to include
everything within these bounds, this is clearly counter to the
spirit of the overall endeavor and so we will concentrate on

those particularly bad "actors" that are likely to have some
environmental effects associated with their acidity. To a
considerable extent, this constrains us to considerations of
particles (both solid and liquid).

Technical questions associated with the possible production
of "artifact sulfate" on filter media used in early monitoring
networks have cast doubt on the usefulness of all except recent
observations of concentrations in air of sulfate, nitrate, etc.
The recent Sulfate Regional Experiment (SURE), promoted by the
Electric Power Research Institute, has documented particle air
chemistry at a large number of rural and semi-rural sites distributed across the continental USA, but with major emphasis on
the more polluted northeast. Mueller et. al. (13) have presented
frequency distributions of sulfate concentrations measured at
SURE sites, and have supported these data with nitrate and
ammonium concentrations. From these measurements, distributions
of acidic aerosol occurrences can be obtained but as yet are not
available.

Meanwhile, a considerably smaller sampling network has been
operated as part of the Multistate Atmospheric Power Production
Pollution Study, initiated by DOE and now under EPA sponsorship.
The MAP3S network has addressed directly the question of the
incidence of acid sulfate aerosol by sampling particles in a range
of sizes by subjecting the collected material to infrared spectro-
scopic analysis. The average acidity of submicron-size particles
is found to be 0.01 near Chicago, 0.04 at Rockport, Ind., 0.25 in
central Virginia, 0.33 in central Pennsylvania, and 0.27 in
central Long Island (4). These numbers, determined by analysis of
particles in the so-called accumulation size range (about 0.3
to 1.0 μm), represent the molar ratio of hydrogen to sulfate in
the collected sample. According to Kumar (5), the probability of
finding submicron particles to be acidic averages about 0.6 in
central Pennsylvania, and about 0.5 in both central Virginia and
on Long Island. Samples collected at Rockport are far less
frequently acidic (probability $\simeq$ 0.2). There does not appear to
be any consistent diurnal cycle in either the frequency of occurrence or in the degree of acidity, when it occurs.

The average acidities of the acid particles collected at the
four sampling sites are about 0.6 (Pennsylvania), 0.48 (Virginia
and Long Island), and 0.25 (Rockport). It appears, therefore, that in order to estimate the surface flux of acidic sulfate
particles over the more heavily impacted areas of the northeastern
USA, it is sufficient to base calculations on the sulfate frequency
distribution of Mueller et al. (3), on the assumption that this
sulfate is acidic on about 50% of occasions, and that the average

acidity is then about 0.5. A long-term average particulate hydrogen ion concentration of about 30 nM m^{-3} is then indicated.

DEPOSITION PROCESSES

In the case of those gases which are readily soluble (such as sulfur dioxide), our understanding of the mechanisms involved in dry deposition is well advanced (6,7,8,9). The total resistance to transfer (the reciprocal of the familiar "deposition velocity") is made up of a number of individual resistances, each of which can be associated with a specific part of the transfer process. Some of these resistances act in series, others in parallel. It is often not possible to say which resistance will dominate; in daytime biological factors might be most important, while at night the controlling considerations might be aerodynamic. In this case of soluble gas transfer, results obtained in recent field experiments have tended to confirm the predictions of models. Modelers and experimentalists also seem in substantial agreement when considering deposition of soluble or reactive gases. But there is a continuing disparity for the case of deposition of small particles.

The two principal factors that have concerned those who attempt to describe theoretically the flux of small particles to horizontal surfaces are their inertia and their low diffusivity. These limitations suggest that particles are incapable of responding to the high-frequency turbulent motions that transport material in the immediate vicinity of a surface. Moreover, transport of particles across any limiting laminar sublayer by molecular diffusion will be exceedingly low, should such a layer exist. As a direct result of considerations such as these it has been predicted that the effective deposition velocity of small particles to smooth surfaces would be small, perhaps as low as 10^{-4} cm s^{-1} (10).

An excellent review of these ideas appears in the introduction to a recent paper by Davidson and Friedlander (11). Friedlander and Johnstone (12), for example, assume that particles are carried with atmospheric turbulence until they are within one "stopping distance" of the surface, as determined by the particle inertia and its velocity. Sehmel (13), however, assumed an effective sink located at one particle radius above the surface. It is on the basis of studies such as these that very low deposition velocities have been predicted. Wind tunnel experiments have tended to support the prediction of comparatively small particle fluxes to smooth surfaces.

Many authors have considered vegetation to act as a filter which scavenges particles as air permeates through the canopy (e.g. 11,14). Again, there is wind tunnel work which supports these models, (e.g. 15,16), but in every case it has been necessary to

simplify the treatment in order to make the problem tractable.
The results demonstrate the importance of considering such matters
as the morphology of the vegetation, the density of the canopy,
and the water vapor distribution, but they give little information
on the roles of such factors as plant physiology, electrostatic
effects, and atmospheric stability.

Atmospheric stability deserves special attention in considering the "filtration efficiency" of tall vegetation. When the
surface is heated by insolation, pockets of warm air develop
within the canopy and eventually rise through it. These buoyant
bubbles carry heat and moisture with them, and give rise to thermal
plumes in the surface boundary layer above the vegetation. Air
removed from beneath the canopy by the rising bubbles is replaced
by surrounding air, carrying with it the aerosol particles of
interest here, and providing a mechanism for airborne pollutants
to be entrained deeply within the canopy. Here they remain until
they are deposited on one of the many available surfaces or are
"scavenged" and recirculated upward by another rising bubble of
warm air. This process clearly bears little resemblance to the
picture of high-frequency turbulent transfer painted by some
modellers.

It is of passing interest to note that the advective-convective
process described above continues above the canopy and as such is
fairly typical of events in the surface boundary layer over open
ocean, where the causative factor can also include the buoyancy
of water vapor as well as the surface heating and direct thermal
effects.

The transfer of particles to open water surfaces is also
worthy of special attention, since in this case very little field
data is available. In the lack of alternative information, most
workers have accepted wind tunnel evidence of very low deposition
velocities; at times values as low as 10^{-5} cm s^{-1} have been
suggested. The few direct measurements that have recently been
made (17) support the wind tunnel predictions in light winds. It
might be expected, however, that breaking waves will provide an
efficient mechanism for transferring particles through the surface
diffusive layer that otherwise limits the magnitude of the fluxes.

Detailed study by use of optical and electron microscopy
has shown that sulfate particles are often present in liquid or
semi-liquid form. This feature has formed the basis for techniques
by which the chemical "speciation" of atmospheric aerosol can be
investigated (e.g. 18). It is now well accepted that acidic sulfate
particles (which are typically composed of sulfuric acid or ammonium
bisulfate) are hygroscopic and tend to leave impaction patterns
indicative of a liquid shell, perhaps surrounding a solid core.

It is possible that these small, moist particles might have considerably different impaction (and hence deposition) properties
from completely solid particles. Although details of the processes
involved in the surface deposition of small particles are as yet
far from clear, several field experiments have indicated that the
deposition of sulfate particles is not necessarily at the slow rates
suggested by studies of other particles.

    As mentioned previously, early studies of the deposition of
small particles indicated low deposition velocities, on the basis
of which values of the order of 0.1 cm s^{-1} have been adopted for
use in typical calculations. However direct measurements of the
flux of natural aerosol particles to natural surfaces have indicated
deposition velocities that might be considerably higher (2). In
the case of sulfur, Hicks and Wesely (19) find that the deposition
velocity for total sulfur (both gaseous and particulate) often
exceeded 1.0 cm s^{-1} over a pine canopy, even when no significant
amounts of SO_2 could be detected. A study of sulfate concentration gradients reported by Everett et al. (20) also tends to confirm
the hypothesis that natural sulfate aerosol is deposited more
efficiently than is indicated by the familiar 0.1 cm s^{-1} deposition
velocity assumption.

ESTIMATED SURFACE FLUXES

    While it is true that more confirmation is needed before
these few experimental indications can be extended to other circumstances with confidence, it is also apparent that the presently
available data point to an hypothesis suggesting larger dry
deposition rates of acidic aerosol than might otherwise be anticipated. The lack of certainty in this regard clearly detracts
from the significance of any conclusions made regarding the relative
importance of dry deposition. Accordingly, three alternative
assumptions will be made regarding the appropriate deposition
velocity v_d:

    1. $v_d = 0.1$ cm s^{-1}, thus representing the conventional
        viewpoint;

    2. $v_d = 0.4$ cm s^{-1}, representing the most likely long-term
        average value for the northeastern continental U.S.A.,
        based on the recent experimental work and extrapolated
        by Sheih et al. (21); and

    3. $v_d = 1.0$ cm $^{-1}$, representing the highest value likely to
        be appropriate as a long-term average over any natural
        surface in the continental U.S.A.

If the average hydrogen ion concentration content of air is assumed to be about 30 nM m^{-3}, then application of the three alternative deposition velocities listed above indicates long-term average surface fluxes in the range 1-10 mM m^{-2} yr^{-1}, but probably averaging about 4 mM m^{-2} yr^{-1}. In comparison, rainfall delivers a much greater quantity of acid. Measurements of hydrogen ion concentration in rainfall collected at the sites of the MAP3S precipitation chemistry network indicate great variability across the northeastern USA, but suggest that an average concentration of about 100 μM m^{-2} might be appropriate in more heavily impacted areas (22). If an average annual rainfall of 1 m is assumed, then an annual hydrogen ion flux of about 100 mM m^{-2} yr^{-1} results.

The long-term average acidic "dose rate" from dry deposition is clearly one or two orders of magnitude less than that due to rainfall. However it is not appropriate to conclude that effects related to dry deposition can be neglected in comparison to the larger dose delivered by rain, since (a) particles are scavenged by foliage whereas rainfall might primarily affect the underlying surface; and (b) acid delivered to foliage in particulate form results in a small area of high dosage that might continue for an extended period, perhaps until cleansed by rain or distributed by dew. The inter-relation between wet and dry deposition has been emphasized by some recent field experiments conducted at the Walker Branch Watershed in Oak Ridge, Tennessee. Lindberg et al. (23) comment that a major effect of falling rain is to remove trace metals from foliage, thus enriching precipitation arriving at the ground b factors typically of about 3. In the special case of sulfate the effect is not only to wash accumulated particles from leaves but also to leach sulfate from them, at a rate which is itself dependent upon the rainfall acidity. A combination of cation-exchange and leaching appears to take place, associated with a net uptake of hydrogen ions by the irrigated foliage. It is clear that much more work needs to be done before this simple result (for a deciduous forest) can be extended to other canopies.

CONCLUSIONS

It is not possible to estimate, with assurity, what doses will be imparted to foliage as a result of dry deposition of acid particles. Certainly, the fluxes of material will vary according to the availability of appropriate constituents in the air (whether they be nitrates, chlorides, or sulfates as assumed here) and to many atmospheric and surface characteristics. The physical picture is one in which acidic particles, which might often be liquid or at least covered with a liquid film, impact on exposed surfaces and may reside there for a considerable time, perhaps until washed off by rain. The "residence time" for particles on leaves might then be estimated as about half the average time lapse between inundating rain events, perhaps of the order of a few days.

The average exposure to acidity resulting from dry deposition
is much less than that due to acid rain, probably by about two
orders of magnitude. The substantial differences in the nature of
the exposure, however, suggest that it might be unwise to neglect
one effect at the expense of the other. A possible exception
is the case of transfer to water surfaces, in which case there is
no exposed vegetation potentially susceptible to highly localized
but intense doses and where, as a consequence, the influence of
wet deposition seems likely to be overwhelming.

ACKNOWLEDGMENTS

        The dry deposition field studies which formed the basis for
this work were supported jointly by the U. S. Environmental
Protection Agency and the U. S. Department of Energy. Thanks are
due to Dr. Romesh Kumar for supplying the sulfate speciation results
derived from the MAP3S surface air chemistry network.

REFERENCES

1.   Shannon, J. D. (1979). The advanced statistical trajectory
         regional air pollution model, Proceedings, Fourth
         Symposium on Turbulence, Diffusion, and Air Pollution,
         American Meteorological Society, Boston, 376-380.
2.   Wesely, M. L. and B. B. Hicks (1979). Dry deposition and
         emission of small particles at the surface of the earth
         Proceedings, Fourth Symposium on Turbulence, Diffusion
         and Air Pollution, American Meteorological Society,
         Boston, 510-513.
3.   Mueller, P. K., G. M. Hidy, T. F. Lavery, K. Warren, and
         R. L. Baskett (1979). Some early results from the
         Sulfate Regional Experiment (SURE), Proceedings, Fourth
         Symposium on Turbulence, Diffusion, and Air Pollution,
         American Meteorological Society, Boston, 322-329.
4.   MacCracken, M. C. (1979). MAP3S update: progress report for
         FY-1977 and FY-1978, in preparation.
5.   Kumar, R. (1976). Private communication.
6.   Garland, J. A. (1977). The dry deposition of sulphur dioxide
         to land and water surfaces. Proc. R. Soc. Lond. A., 354,
         245-268.
7.   Garland, J. A. (1978). Dry and wet removal of sulphur from
         the atmosphere. Atmospheric Environment, 12, 349-362.
8.   Wesely, M. L. and B. B. Hicks (1977). Some factors that
         affect the deposition rates of sulfur dioxide and similar-
         gases on vegetation. J. Air Pollut. Cont. Assoc., 27,
         1110-1116.
9.   Wesely, M. L., J. A. Eastman, D. R. Cook, and B. B. Hicks
         (1978). Daytime variations of ozone fluxes to maize,
         Boundary-Layer Meteorol., 15, 361-373.

10. Sehmel, G. A. and W. H. Hodgson (1976). Predicted dry depo-
       sition velocities, Proceedings, Symposium on Atmosphere-
       Surface Exchange of Particulate and Gaseous Pollutants,
       NTIS CONF-740921, ERDA Symposium Series 38, 399-419.

11. Davidson, C. I. and S. K. Friedlander (1978). A filtration
       model for aerosol dry deposition: application to trace
       metal deposition from the atmosphere, J. Geophys. Res.,
       83, 2343-2352.

12. Friedlander, S. K. and H. F. Johnstone (1957). Deposition of
       suspended particles from turbulent gas streams, Ind.
       Eng. Chem., 83, 1151-1156.

13. Sehmel, G. A. (1970). Particle deposition from turbulent air
       flow, J. Geophys. Res., 75, 1766-1781.

14. Slinn, W. G. N. (1977). Some approximations for the wet and
       dry removal of particles and gases from the atmosphere.
       Water, Air, and Soil Polln., 7, 513-543.

15. Chamberlain, A. C. (1974). Mass transfer to bean leaves,
       Boundary-Layer Meteorol., 6, 477-486.

16. Wedding, J. B., R. W. Carlson, J. J. Stukel, and F. A. Bazzaz
       (1977). Aerosol deposition on plant leaves, Water, Air,
       and Soil Polln., 7, 545-550.

17. Williams, R. M., M. L. Wesely, and B. B. Hicks (1978).
       Preliminary eddy correlation measurements of momentum,
       heat and particle fluxes to Lake Michigan, Argonne
       National Laboratory, Radiological and Environmental
       Research Division Annual Report (Part III), in press.

18. Butcher, S. S. and R. J. Charlson (1972). An introduction to
       air chemistry, Academic Press, New York, 241 pp.

19. Hicks, B. B. and M. L. Wesely (1978). Recent results for
       particles deposition obtained by the eddy-correlation
       method, presented at AIChE. National Meeting, June 4-8,
       1978, Philadelphia, Penn.

20. Everett, R. G., B. B. Hicks, W. W. Berg, and J. W. Winchester
       (1979). An analysis of particulate sulfur and lead
       gradient data collected at Argonne National Laboratory,
       Atmospheric Environment, in press.

21. Sheih, C. M., M. L. Wesely, and B. B. Hicks (1979). Estimated
       dry deposition velocities of sulfur over the eastern
       United States and surrounding regions, Atmospheric
       Environment, in press.

22. Dana, M. T. (1979). The MAP3S precipitation chemistry network:
       Second periodic summary report (July 1977-June 1978),
       Pacific Northwest Laboratory Report PNL-2829, 121 pp.

23. Lindberg, S. E., R. C. Harriss, R. R. Turner, D. S. Shriner,
       and D. D. Huff (1979). Mechanisms and rates of
       atmospheric deposition of selected trace elements and
       sulfate to a deciduous forest watershed. Oak Ridge
       National Laboratory Report ORNL/TM-6674, 514 pp.

DISCUSSION

KRUPA: What does the ozone profile look like near the surface?
The reason I ask is that I am preparing a paper where we measured
ozone at 10 feet and 35 feet at nine locations. When the wind
speed went to below five miles an hour, and this seems to occur,
at least in our area, in the middle of the night, the ozone flux
between 10 and 35 feet may be as different as 100%.

HICKS: At night, flow might be highly stable. We have measured
the ozone deposition as a function of time of day; the flux goes
down to zero or very close to zero at night. Even though the
flux is low, you can still get very large concentration
differences because the atmospheric diffusivity is also very
low. The basic question is: how do you interpret these
concentration differences? The problem is one of stability and
of how you estimate the diffusivity in very stable air. It's
always been a problem.

McLEAN: I didn't catch your reasoning for why you're leaving
nitrate out of this consideration.

HICKS: Partly because the concentration of nitrate is much lower
than sulfate, but also because we are just starting to determine
what's going on with sulfates; nitrates are way down the line,
I'm afraid.

EVANS: Bruce, I think you said the sulfate aerosols were sort of
"sloppy". What do you mean by that?

HICKS: The vision I have is one in which the sulfate, if it is
acidic, is probably quite hygroscopic. I'm not sure to what
extent it can be classified as either liquid or solid phase.
That's what I meant by "kind of sloppy". And it was
intentionally sloppy terminology. I'm at a loss to know how to
classify these aerosols. I am inclined to guess that the
sulfates have a solid core with some sort of liquid exterior.

KRUPA: Culvert, Larsen and Charlson did work on the St. Louis
plume and they do find that sulfuric acid is by light scattering
hygroscopic and probably liquid. In fact the way they measured
it is by using a nephelometer and looking at the amount of light
scattered.

HICKS: You think it's liquid?

KRUPA: Maybe with a solid core.

EVANS: That's only in the optical size range, though. How about the suboptical particles in the plume?

KRUPA: The sulfuric acid aerosols are very easily in the submicron range.

HICKS: That's right. And eventually they might become largely ammonium bisulfate.

KRUPA: That's also hygroscopic.

MORROW: Since particles are so small in the atmosphere, how do you get such high deposition velocities? Are we talking about motion of air?

HICKS: Yes, it's definitely a dynamic transfer process. I think the picture we've had in the back of our minds for a number of years is one of transfer to pipes, flat plates, and smooth surfaces. I don't think that's really appropriate when we're talking about a surface like a forest. One might ask whether we should wash the inside of the collecting buckets and the outside, too? It's a difficult situation and I don't really know how to solve it.

McLEAN: You talked about the problems in visualizing particulate deposition into lake water. Have you considered the problems that might be involved in deposition on snow pack?

HICKS: Yes, we've several sets of measurements of particle deposition onto snow, and we're torn between believing them and not believing them. The results are difficult to explain. Firstly, the particle fluxes are small. Secondly, they're quite often of a sign we would not expect. And we do not know whether that is simply a case of there being a very low flux and we're looking at a random statistical variation about it, or whether in fact there is something real in the measurements.

VOLCHOK: Regarding dry deposition on the ocean, we know there is dry deposition and we know it is the right amount as shown by mass balance studies. The only explanation I've heard is that something like 5-10% of the total ocean at any one time is white caps. These are breaking waves which emit droplets of water at high velocity, and also giant salt particles which scavenge particles near the surface.

HICKS: The model we have for the transfer to water, actually calls for zero deposition velocity until you reach the whitecapping point, which is typically about 5 meters per second. At that point you might take the whitecap area to have

zero "holdup" resistance. Now, all of a sudden you're coming up
with numbers that will account for what's going on; for example
it might account for the fact that apparently the lead deposition
over land in the vicinity of Lake Michigan is the same as what
goes into the lake itself, per unit surface area. So we're
getting a little bit of confidence there. Unfortunately, it's
going to mess up a lot of our thinking about global budgets
because all of a sudden we have to consider a major source and
sink of particles in the ocean in the "Roaring Forties", with not
much going on in the tropics. That's going to have some
repercussions that are rather interesting to think about.

KADLECEK: Have you taken vertical flux measurements over forest
canopies?

HICKS: Yes, the sulfur data that I mention was obtained over a
pine plantation.

KADLECEK: As you approach the canopy from above, how does the
vertical component change?

HICKS: As far as we're concerned not too much, because the
effective zero plane for the material transfer, where you might
expect the vertical component to die away, is located, in our
case, about some 4-5 meters down into the canopy. In experiments
of this kind, you certainly want instrumentation to be above the
level at which treetops are likely to interfere, and so data are
obtained well above the region in which difficulties might be
expected.

BERG: Your velocity measurements are made well above ground, and
in your system, unlike Paul Morrow's system, you don't have the
benefit of earth inhaling the air. So you can introduce the
concept of a zero plane. Now, how do you calculate the velocity
of the particle hitting the ground through that zero plane where
air is reflected.

HICKS: Our approach is to measure the total flux at a point in
space. The need then, and what really you're getting down to, is
how can we be assured that that net flux at that point in space
is the same as what the surface is receiving through its various
processes.

BERG: Or, what's the transfer function?

HICKS: It is necessary to address that by considering equations
of continuity, and the laws of conservation. You must consider a
volume around the location of measurement, and then show that you

do not have any terms that are going to cause you to lose flux or gain extra flux.

There is no net transfer of air or else we'd be running out of air to breathe. The whole system is moving with the wind speed and migrating past a single fixed point in space.

ALUMINUM TOXICITY TO BROOK TROUT (<u>SALVELINUS</u> <u>FONTINALIS</u>)

IN ACIDIFIED WATERS

Carl L. Schofield and John R. Trojnar

Department of Natural Resources
Cornell University
Ithaca, N. Y. 14853

Abstract

Aluminum was identified as a primary toxicant present in acidic, snow-melt runoff and lakes in the Adirondack Mountain region of New York. Covariance analysis of water quality data from fifty three Adirondack lakes suggested that stocked brook trout survival was determined primarily by aluminum concentrations, rather than pH or calcium levels. Lakes not supporting brook trout had a mean aluminum concentration of 0.29 mg/liter, as compared to 0.11 mg/liter of aluminum in lakes where stocked brook trout survived. Comparisons of mortality and gill pathology of brook trout exposed to acidic synthetic solutions and natural Adirondack waters with aluminum levels above 0.2 mg/liter, indicated a specific toxic response to aluminum at pH levels down to 4.4. An apparent increase in toxicity of fixed total aluminum levels with increasing pH suggested that changes in speciation involving hydroxy complexing enhances the toxicity of the aluminum cation.

Introduction

Acidified waters in regions impacted by acid precipitation characteristically exhibit elevated aluminum concentrations (Wright and Gjessing 1976). Aluminum concentrations found in acidified lakes of southern Norway (Wright and Henriksen 1977), Sweden (Dickson 1975), and the northeastern United States (Schofield 1976a) were five to ten times higher than aluminum levels in circumneutral waters in these same areas. In those regions with acidic, base deficient soils, aluminum appears to be the primary element mobilized by strong acids of meteoric origin (Cronan 1978).

Increased aluminum loading to aquatic ecosystems in acidified regions has significant chemical and biological implications. Loss of bicarbonate alkalinity, due to strong acid inputs, with associated increases in aluminum concentrations represents a major shift in the pH range of maximum buffering intensity ($pK_{H_2CO_3}$ = 6.3, pK_{Al} ≃ 4.9). The primary biological implication of this process may be the development of aluminum toxicity to fish at pH levels which are otherwise not directly harmful. Adverse effects of acidification on fish populations, which have been reported recently in Scandinavia and eastern North America (Schofield 1976b), have been generally attributed to the interference of hydrogen ion, at low calcium concentrations, with the osmoregulatory mechanism controlling sodium balance at the level of the gill (Leivestad et al. 1976). The potential for aluminum intoxication as a cause of fish mortality in these acidified waters has not previously been considered.

The chemical speciation of aluminum in natural waters is complex and there is a lack of information concerning the relative toxicity of free aluminum, cationic hydroxy complexes, aluminosilicates, aluminum fluoride species, and organically bound aluminum, all of which may be present at varying levels in the acidic waters of concern. Burrows (1977), in reviewing the toxicology of aluminum, emphasized that the effects of decreasing pH, below the isoelectric point, on aluminum toxicity were essentially unknown. Jones (1964) demonstrated toxicity of aluminum in acidic solutions to the stickleback (Gasterosteus aculeatus) at concentrations as low as 0.1 mg/liter, but pH was not controlled in his experiments. Freeman and Everhart (1971) investigated the toxicity of aluminum to rainbow trout (Salmo gairdneri) at alkaline pH levels and found both acute and chronic responses to the aluminate anion and neutral hydroxy precipitates. They suggested that a "safe" level of dissolved or suspended aluminum, in the alkaline pH range, was well below 0.5 mg/liter.

Our study concerned the potential for aluminum toxicity to brook trout (Salvelinus fontinalis) in acidified waters of the Adirondack Mountains of New York State. In this paper we present an analysis of field observations on brook trout survival in acidified Adirondack waters and comparative laboratory bioassay data for evaluation of aluminum effects on brook trout survival and growth.

Methods

Field Studies

During the period 24-27 June, 1975 a water chemistry survey was conducted on 219 Adirondack lakes above 610 meters elevation.

Water samples collected at 0.5 meter depths from each lake were
analyzed for pH, conductivity, alkalinity, Ca, Mg, Na, K, Fe, Mn,
Zn, Al, SO_4, NO_3, and Cl. Analytical methods and discussion of
these data were reported by Schofield, 1976a. During the summers
of 1975-76, fifty-three of the high elevation lakes, which had been
stocked annually for the previous five years with fall fingerling
brook trout, were surveyed to determine the status of the stocked
trout populations. Standardized, overnight gill net sets were made
in each lake using graded mesh nylon nets. The amount of net utilized was proportional to the lake surface area. Water samples were
collected for determinations of pH, alkalinity, and conductivity.
Occurrence or absence of brook trout in the netting surveys was
used as a basis for comparison and evaluation of water quality
parameters potentially significant to trout survival.

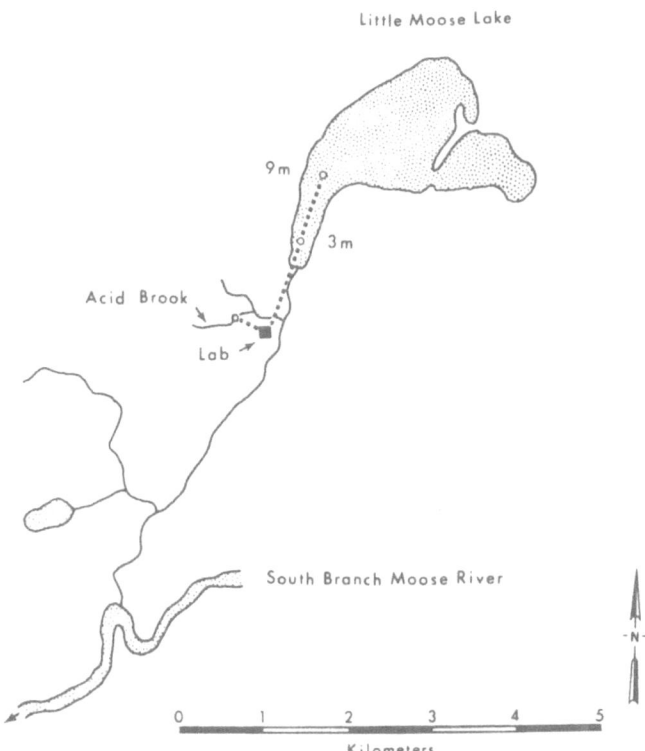

Figure 1. Little Moose Lake study area near Old Forge, N. Y.
           Dashed lines represent pipelines to the laboratory
           from the lake (intakes at 3 meters and 9 meters depth)
           and Acid Brook.

Episodic, snow-melt effects on stream water quality and toxicity to brook trout were investigated at Little Moose (figure 1). Experimental fish populations were held in water supplied to the Little Moose field station by gravity flow through PVC and steel pipe from the 3 meter depth intake in Little Moose Lake (figure 1). Stocks of brook trout eggs, yearlings, and adults were transferred to the laboratory in October-November 1976 and maintained in Little Moose Lake water throughout the winter. Water samples were collected on a weekly basis and more frequently during thaws from Little Moose Lake outlet, the laboratory intakes from Acid Brook, and Little Moose Lake. Routine analyses included pH, alkalinity, conductivity, and reactive aluminum (Smith 1971). Additional analyses for major ions (Ca, Mg, Na, K, SO_4, NO_3, Cl) and trace metals (Fe, Zn, Cu) were obtained during thaw periods (APHA, 1971).

Bulk precipitation was collected at weekly intervals in an open PVC container (62 cm. diameter) mounted 1.5 meters above ground level in an open area at the Little Moose lab site. The collector was protected by an alter shield. The collector was removed at weekly intervals, covered, and allowed to thaw in the lab at room temperature. A clean collector was then placed back at the collection site. Total volumes of weekly precipitation samples were determined and aliquots removed for analyses of pH, SO_4, NO_3, alkalinity, and conductivity.

Six vertical cores of the snowpack were obtained each week utilizing a 1.5 meter long plexiglass tube 6.5 cm in diameter. Snow depth was recorded, the cores melted down completely at room temperature, individual volumes recorded, and the meltwater samples were then composited and an aliquot removed for chemical analysis. Meltwater leaving the snowpack during thaws was sampled from a lysimeter, comparable in design to that described by Johannessen, et al. 1976. The lysimeter was installed in level ground near the precipitation collection site just prior to the first snowfall. Meltwater entered the lysimeter through a perforated plate, which supported the overlying snowpack, and had no contact with the soil. Water samples were removed during thaws with a peristaltic pump, through tygon tubing which extended to the bottom of the lysimeter.

Laboratory Experiments

The toxicity of aluminum, at controlled pH levels, to brook trout fry was determined in the laboratory utilizing synthesized aluminum solutions and Adirondack stream water. The experiments were conducted from January through March 1977 at the Cornell Fishery Laboratory, Ithaca, N.Y. The synthetic water used was a mixture of tap and deionized water and had a mean base conductivity of 27.50 µmhos/cm (range 25.35-29.46) in order to approximate the dilute waters commonly found in the Adirondack mountains.

Aluminum stock solution of 2.5 g/l was prepared by dissolving ACS
grade aluminum wire in concentrated hydrochloric acid. Aliquots of
the stock solution were added to 100 l volumes of water to obtain
the desired aluminum concentrations. The water was further acidified with 6N sulfuric acid to obtain the necessary pH levels; aerated
for at least two days before being placed in 76 l plastic containers
which delivered it through capillary and tygon tubing to 7.6 l polyethylene test units. The turnover time of water in the units was
approximately 24 hr. The test units were placed in a temperature
controlled recirculating water bath which maintained temperature
between 9.5 and 11°C. Daily samples of treatment water without
added aluminum were composited for analysis of major cation and
anions. The pH of two replicates from each treatment was recorded
daily.

Trials A, B, and C had respective nominal pH levels of 4.0,
4.4, and 4.9 with added aluminum concentrations of 0, .1, .5, and
1.0 mg/l. Based on the results of these treatments trial D was
conducted at pH 4.9 with .25 and .50 mg/l aluminum added. Trial E
had nominal pH 5.2 and added aluminum of 0, .1, .25, and .5 mg/l.
Trial F compared water mixed in the laboratory to that of Adirondack
water. Water was obtained from Acid Brook (Figure 1) on March 9,
1977 at the beginning of peak spring runoff. This water had a pH
of 4.8 and total aluminum concentration of .63 mg/l. Brook trout
are occasionally seen near the mouth of the brook during summer,
but it does not harbor a permanent population. Trial F treatments
consisted of laboratory water with added aluminum concentrations of
0, .32, and .65 mg/l and brook water at a nominal pH of 4.9 and
aluminum concentration of .63 mg/l.

Brook trout fry were obtained from the U.S. Fish and Wildlife
Service Tunison Laboratory of Fish Nutrition, Cortland, N.Y. Fry
obtained for trial A on January 12, 1977 had just started to feed.
Fry in subsequent trials were progressively larger. Trial F ended
April 4. New fry were obtained from Tunison Laboratory for each
trial, except trial F which used extra fish not used in the previous
trials and which had been held at the Cornell Fishery Laboratory.
Tunison Laboratory has water of pH 8.3-8.4 and a conductivity of
approximately 470 μmhos/cm. The fry were placed in dilute water of
pH 7.1 and a conductivity of approximately 27 μmhos/cm for 24 h prior
to the start of the experimental trials. Twenty fish were placed
in each of 3 replicates of each treatment. Fish were fed a dry
trout diet four to five times a day on weekdays and once a day on
weekends. Fecal material and waste food were siphoned out twice
daily on weekdays and once a day on weekends. Trials lasted until
50% mortality occurred or were terminated in 14 days if this point
was not reached. Only the .25 mg/l treatment of trial D was allowed
to go for 23 days. Fish were judged to be dead when opercular movement had ceased and no swimming response could be elicited through
stimulation of the caudal peduncle. If the time of 50% mortality

was reached during the night, it was estimated by assuming a constant rate of mortality during that period. Surviving fish were
placed in 10% formalin, following anesthesia in MS-222, for later
histological examination. In trials C, E, and F, fish in treatments
which did not experience 50% mortality were measured to the nearest
mm and weighed on a Mettler balance to the nearest .0001 g before
being placed in formalin. A sample of 20 fish were also weighed and
measured at the start of trials E and F.

Results

Lake water chemistry and trout survival

Physical and chemical characteristics of the Adirondack lakes
surveyed to evaluate trout survival are given in table 1. The lakes
were segregated into two categories based on the occurrence or
absence of brook trout in the netting surveys. The two groups of
lakes did not differ significantly in mean size, elevation, or
stocking rate. Lakes in which stocked brook trout were not recovered
exhibited significantly lower mean pH, calcium, and magnesium concentrations and higher mean aluminum concentration in June 1975.
Mean pH levels measured during the 1975-76 summer surveys did not
differ significantly (P .05) from the June, 1975 means. Calcium,
magnesium, and aluminum concentrations are correlated with pH levels
in Adirondack waters (Schofield 1976a), hence the relative significance of these parameters to trout survival was evaluated by covariance analysis. Adjusted mean calcium and magnesium levels, for
lakes with or without trout survival, were not significantly different (F = 0.207, P .05) when analyzed as dependent variables of pH.
In contrast, adjusted mean aluminum concentrations, for lakes with
and without trout survival, were significantly different (table 2)
when considered in relation to pH. Also, substitution of pH for
aluminum as the dependent variable in the analysis indicated that
adjusted mean pH levels were not significantly different (F = 0.992,
P <.05).

Toxicity of stream runoff

During the winter of 1976-77, there was no significant snow-
melt from mid-December through February at the Little Moose Lake
station. During this period snow-pack storage of strong acid increased to a maximum of 95 H^+- equivalents/hectare in late February.
The first major thaw in early March (Figure 2) resulted in the release of 80% of this stored acid in snow-melt within a one-week
period. The pH levels of water leaving the snow-pack during thaw
periods are shown in figure 3. Cores taken from the snow-pack just
prior to the thaw had an average pH of 4.4, when melted down completely. Lower pH levels obtained from the first melt water samples,
followed by progressively higher pH levels illustrate the phenomenon

Table 1. Comparative data for Adirondack lakes stocked annually
with brook trout for five years and then surveyed to
evaluate survival. Significance noted by ** for p.01 and
* for p.05.

	No Survival	Survival	
N	25	28	
	$\overline{x}_1$	$\overline{x}_2$	$t_{\overline{x}_1-\overline{x}_2}$
Surface Area (hectares)	12.7	21.9	−1.34
Max. Depth (meters)	8.3	9.5	−0.93
Elevation (meters)	726	689	2.01
Stocking rate (No./hectare)	110.9	102.3	0.51
pH (June 1975)	4.79	5.25	−5.44**
mg/l			
Ca	1.622	1.870	2.30*
Mg	0.299	0.382	3.78**
Na	0.394	0.450	−1.99
K	0.222	0.255	−1.48
SO_4	6.57	5.94	1.97
NO_3	1.130	1.320	−0.80
Cl	0.368	0.342	0.84
Fe	0.148	0.041	1.61
Mn	0.045	0.049	−0.72
Zn	0.023	0.020	1.66
Al	0.286	0.111	6.27**
pH (July-Sept. 1975-76)	4.94	5.34	3.50**

Table 2. Analysis of covariance comparing pH (X) and log Al (Y)
concentrations in stocked brook trout lakes exhibiting
survival (ST+) and no survival (ST-).

Treatment	f	ΣX^2	ΣXY	ΣY^2	Reg. Coeff.	f	Σd^2	MS
1 +ST	27	3.60964	2.08428	2.27123	0.57742	26	1.06772	0.04106
2 -ST	24	1.23840	0.64028	0.88865	0.51702	23	0.55761	0.02424
3 Within						49	1.62534	0.03317
4 Reg.Coeff.						1	0.00336	0.00336
5 Common	51	4.84804	2.72456	3.15988	0.56199	50	1.62870	0.03257
6 Adj.Means						1	0.25080	0.25080
7 Total	52	7.66189	5.40548	5.71417	0.70550	51	1.26019	

$$F_{6,5} = 8.346** \text{ (Adjusted Means)}$$

$$F_{4,3} = 0.101^{NS} \text{ (Regress. Coeff.)}$$

$$F_{1,2} = 1.694^{NS} \text{ (Homog. Variance)}$$

of ion separation which occurs as a result of leaching by melt water
percolating down through the snow pack (Johannessen, et al., op
cit.).

Discharge of this acid melt water resulted in sharp drops in
pH of stream and surficial lake water (figure 4). Although the pH
drop in the lake was not as marked as the stream water change, alkalinity was reduced by 80% (from 100 µeq/liter prior to thaw to 20
µeq/liter) to a depth of 3 meters. The corresponding change in
lake aluminum concentration was from 20 to 320 µg/liter. Increases
in brook aluminum levels were even greater (figure 5) and the high
correlation between brook pH and aluminum concentration (figure 6)
suggests that the process of exchange of melt water hydrogen ion
for soil aluminum exerts a significant control over stream runoff
pH during thaws.

Captive populations of adult, yearling, and larval brook trout,
which had been maintained over winter in Little Moose Lake water

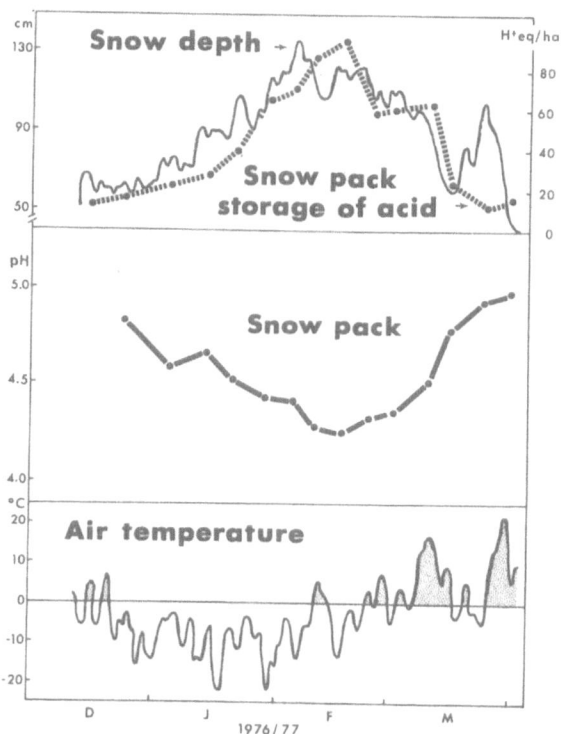

Figure 2. Snow pack storage levels of strong acid, pH, depth, and
maximum daily air temperature in the vicinity of Old
Forge, N.Y. during the winter of 1976/77. (Daily snow
depth and air temperature readings obtained from United
States Weather Bureau data.)

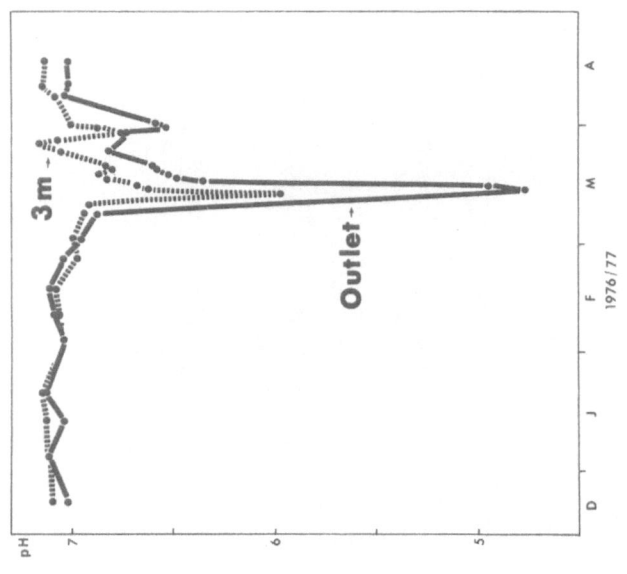

Figure 4. Levels of pH measured in Little Moose Lake Outlet and in water entering the laboratory from the 3 meter intake depth pipeline from Little Moose Lake (refer to figure 1 for locations).

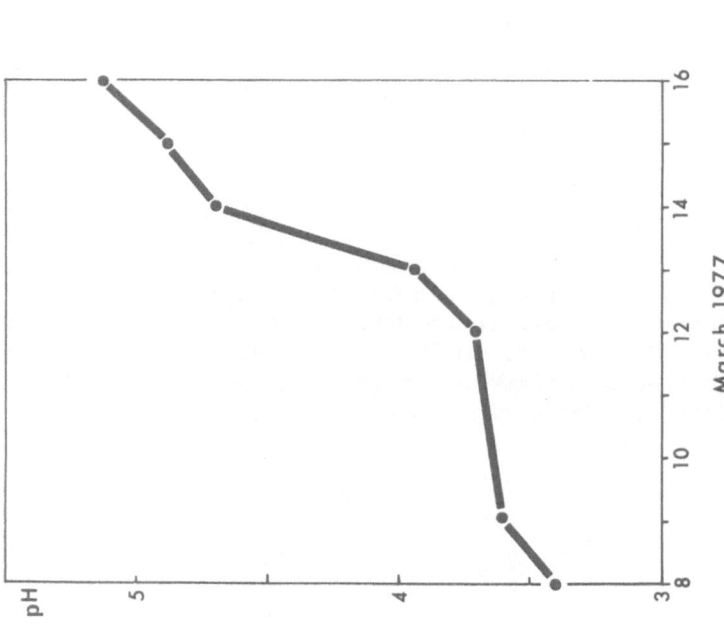

Figure 3. Lysimeter snow-melt pH levels during thaw period of March 8-16, 1977 in the watershed of Little Moose Lake, Old Forge, N.Y.

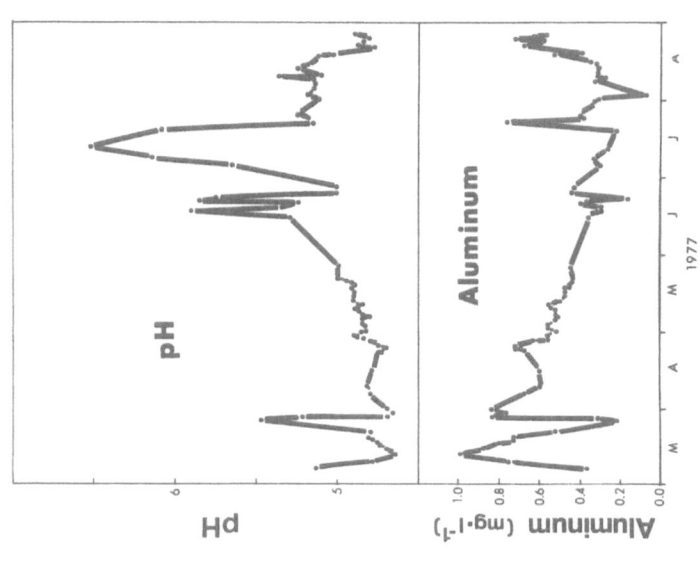

Figure 5. Aluminum and pH levels measured in Acid Brook during the period March-August, 1977.

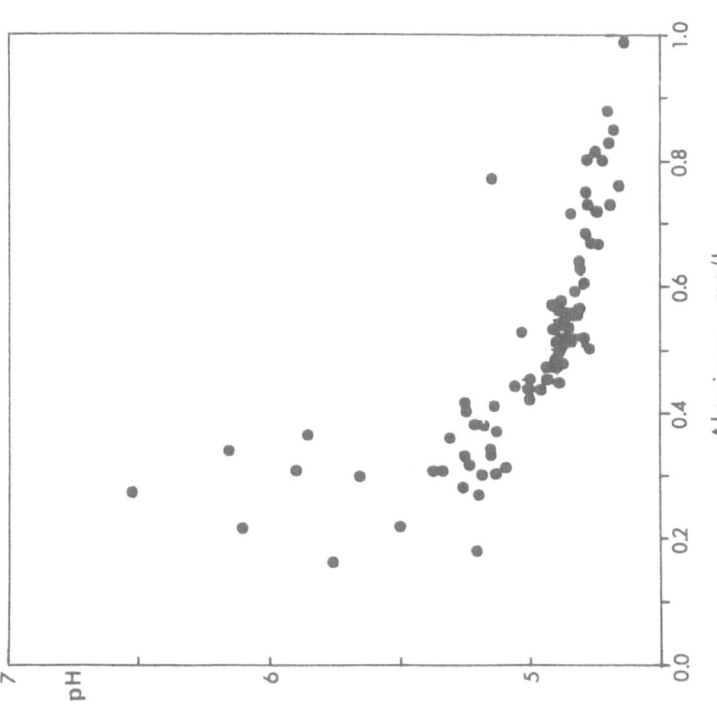

Figure 6. The relationship of pH and aluminum levels measured in Acid Brook (see figure 5). The regression of pH vs. log aluminum is $Y = 4.57 - 1.15 \log X$, $r = 0.940$ ($P < .001$).

without incident, experienced distress and mortality during the early
March episode of water quality change (figure 4). Mortalities included 3 adult brook trout (25-30 cm), 25 yearlings (13-17 cm), and
an undetermined number (~50,000) of recently hatched fry, also a
round whitefish (Prosopium cylindraceum) and a smallmouth bass
(Micropterus dolomieui) (Personal communication, Professor Dwight
A. Webster, Dept. Natural Resources, Cornell Univ.). Eyed brook
trout eggs exposed to the same water did not experience significant
mortality. Yearlings and adults exhibited irregular opercular move-
ments, lassitude, and cessation of feeding for the five day period
of March 13-17, during which the mortalities were incurred. Moribund
yearling trout exhibited gill damage similar to that observed in the
laboratory experiments described below. The minimum pH measured
during this period was 5.9 on March 13 and corresponding lake water
aluminum concentration was 320 µg/liter. No stress or mortality
was observed in surviving trout during the second major thaw at the
end of March. Minimum alkalinity observed during this period was
43 µeq/liter (March 29), which was sufficient to prevent any marked
drop in pH. The maximum aluminum concentration was 170 µg/liter.

Laboratory studies of aluminum toxicity

The means and ranges of daily pH measurements in the treatments
are given in Table 3 and the major ion concentration in Table 4.
Synthetic treatments acidified with H_2SO_4 had higher SO_4/NO_3 ratios
than Acid Brook water, which exhibited a very high NO_3 concentration.

When fish were first placed in the test units they swam actively
for several hours in all treatments. Symptoms of stress were a
darkening of skin coloration and cessation of feeding. Fish exhibiting these symptoms became inactive and rested on the bottom. These
symptoms were characteristic of all fish at pH 4.0 and 4.4; however,
they took longer to develop at pH 4.4 with 0 and .1 mg/l aluminum
and did not develop at pH 4.9 and 5.2, with 0 and .1 mg/l aluminum.
Darkening of the skin and cessation of feeding occurred at .25 mg/l
and higher concentrations of aluminum at all pH levels. Heavy
accumulations of mucous and cellular debris on the gills were present
on dead and surviving fish at .25 mg/l aluminum and higher at pH 4.4
and above, but not at pH 4.0. Fish in this condition would occasionally flare their opercles and shake their head. At .25 and .32
mg/l aluminum fish exhibited these stress symptoms, but after one
week lightened in color and resumed feeding to a limited extent.

Gills removed from fifteen preserved survivors of each treatment were scored on a scale of 0-3 for relative levels of gill
damage. The distribution of these scores in the various treatments
is given in Figure 7 and Table 5. Scoring was based on the condition
of the gill filaments and lamellae, as observed at magnification of
10-70X under a dissecting microscope. A score of zero indicated no
visible damage to the gills. Fish exhibiting cell proliferation

Table 3. Means and ranges of daily pH measurements in treatments.

Nominal pH	Trial	Al mg/l	Mean pH	Range
4.0	A	0.0	4.002	3.97 - 4.03
		0.1	3.986	3.95 - 4.05
		0.5	4.010	3.97 - 4.04
		1.0	3.994	3.97 - 4.04
4.4	B	0.0	4.424	4.36 - 4.53
		0.1	4.412	4.35 - 4.49
		0.5	4.447	4.43 - 4.47
		1.0	4.460	4.40 - 4.53
4.9	C	0.0	4.897	4.68 - 5.16
		0.1	4.884	4.76 - 5.12
		0.5	4.904	4.87 - 4.94
		1.0	4.989	4.97 - 5.00
	D	0.5	4.926	4.90 - 4.96
		0.25	4.921	4.77 - 5.09
5.2	E	0.0	5.295	5.13 - 5.64
		0.1	5.257	5.10 - 5.51
		0.25	5.211	5.12 - 5.39
		0.5	5.222	5.20 - 5.25
4.9	F	0.0	4.887	4.73 - 5.11
		0.32	4.944	4.76 - 5.16
		0.65	4.923	4.78 - 5.00
		0.63 (Brook)	4.958	4.87 - 5.08

Table 4. Ionic composition of treatment waters without added aluminum (analyses of daily composites from treatments).

								mg/l				
Trial	pH	Ca	Mg	Na	K	Fe	Al	Cu	Zn	SO_4	NO_3	Cl
A	4.0	3.5	<.5	1.0	0.6	.031	.020	.022	.075	12.8	0.35	1.51
B	4.4	3.7	<.5	0.8	0.4	.028	.020	.015	.030	10.5	0.38	1.48
C,D,F	4.9	3.7	<.5	0.8	0.2	.012	.010	.015	.075	10.3	0.35	1.48
E	5.2	3.7	<.5	1.2	0.1	.015	.020	.021	.020	9.2	0.40	1.18
F	4.9	3.7	<.5	1.6	0.2	.028	.630	.021	.075	4.5	6.60	0.60
Acid Brook 3/9/77												

Table 5. Frequency of levels of macroscopic gill damage observed in surviving fish from trial F. (See text for explanation. Fifteen fish from each treatment examined.)

Treatment	Gill damage score				Days exposed
	0	1	2	3	
Controls	100%				0
Brook water		47%	37%	16%	12
pH 4.8, Al 0.0	100%				14
pH 4.8, Al 0.32	20%	67%	13%		14
pH 4.8, Al 0.65				100%	6

and clubbing at the distal ends of the gill filaments were assigned
a score of one, indicating "slight" damage. Extensive clubbing and
lamellar fusion and edema over the entire length of the filaments
was considered as "moderate" damage and assigned a score of two. A
"severe" damage score of three was assigned to fish exhibiting
epithelial desquamation, filament collapse, and general loss of gill
structure. At pH 4.0 there was no evidence of gill damage in fish
exposed to aluminum concentrations of 0-1.0 mg/liter. At pH 4.4 and
higher, moderate to severe gill damage was predominant at aluminum
levels of 0.5 and 1.0 mg/liter and the severity of damage appeared
to increase with increasing pH. Levels of gill damage observed in
survivors of the Acid Brook water treatment were intermediate to
those found in the 0.32 and 0.65 mg/l aluminum synthetic treatments
(Table 5).

Histopathological changes observed in sections of gills from
fish exposed to aluminum levels $\geq$ 0.5 mg/l included cell proliferation at the distal ends of gill filaments, lamellar edema and
fusion, epithelial desquamation, filament collapse, and general loss
of gill structure. Fish exhibiting these pathological changes in
gill structure also showed evidence of degenerative changes in kidney structure, such as tubular nephrosis and destruction of intra-
renal tissue (Personal communication, Drs. Louis Leibovitz and
James Carlisle, College of Veterinary Medicine, Cornell University,
Ithaca, N.Y.).

Mean lengths and weights of fish in treatments lasting 14 days
are given in Table 6. Mean weights were different between fish in
0 and 0.1 mg/l aluminum at pH 4.9 in trial C (t = 4.93, p .001),
but weight differences were not as significant at pH 5.2 in trial E
(t = 1.87, p .10). Fish in .25 mg/l aluminum at pH 5.2 experienced
a loss in weight over 14 days (t = 3.14, p .01). A decrease in
weight was not as significant for fish exposed to 0.32 mg/l aluminum at pH 4.9 in trial F (t = 1.76, p .10) but there was certainly
no significant growth in comparison to fish in the same water with
no aluminum added.

Times to 50% mortality in the treatments are given in Table 7
and cumulative mortality rates are illustrated in Figure 8. In
trial A at pH 4.0 there were no discernible negative effects of
aluminum on fish survival. In fact, initial mortality rates were
lowest at the 1.0 mg/l aluminum level (Figure 8). Although the 50%
mortality times were shorter at 0.5 mg/l aluminum, the times for 0,
.1, and 1.0 mg/l were similar. In Trial B at pH 4.4 toxicity specific to aluminum is apparent. The 50% mortality times at 0.5 and
1.0 mg/l aluminum are much lower than those at 0 and 0.1 mg/liter.
In trials C, D, and E at pH 4.9 and 5.2, aluminum levels of 0.5
mg/l or greater yielded 50% mortality times of 1.6-3.8 days, while
mortality at 0 and 0.1 mg/liter was minimal (0-10%) over 14 days.

Table 6. Length and weight data for surviving fish from trials C,
          E, and F.

Trial	Nominal pH	Aluminum mg/1	Length Average mm	Length Range	Weight Average mg	Weight Range
C	4.9	0.0	30.2	25–34	219.5	101–310
		0.1	28.8	21–32	173.7	56–272
E	5.2	0.0*	27.5	23–31	147.2	68–233
		0.0	31.1	24–32	212.3	78–444
		0.1	30.3	25–32	191.6	97–316
		0.25	27.3	21–31	111.1	44–162
F	4.9	0.0*	34.9	31–38	320.0	217–423
		0.0	37.3	29.42	401.5	152–595
		0.32	35.3	27–41	286-3	112–458

Table 7. Times in days to 50% mortality for fish in trials A-F.
Percent survival is given in those trials where the 50%
mortality level was not reached after 14 to 23 days
exposure.

Trial	Nominal pH	Aluminum mg/1	Replicate			
			A	B	C	$\bar{x}$
A	4.0	0.0	4.1	4.4	5.1	4.5
		0.1	5.4	4.8	5.4	5.2
		0.5	1.9	2.6	4.0	2.8
		1.0	4.8	5.4	4.9	5.0
B	4.4	0.0	9.7	10.0	11.4	10.4
		0.1	11.2	11.2	12.0	11.5
		0.5	1.5	1.8	2.9	2.1
		1.0	3.7	3.3	3.9	3.6
C	4.9	0.0	100%	100%	100%	100%
		0.1	95%	95%	95%	95%
		0.5	2.1	2.1	2.8	2.3
		1.0	1.7	1.6	2.1	1.8
D	4.9	0.25	90%	90%	90%	90%
			(80%)*	(70%)*	(75%)*	(75%)*
		0.5	3.1	3.1	3.8	3.3
E	5.2	0.0	90%	90%	100%	93%
		0.1	100%	95%	90%	95%
		0.25	65%	70%	55%	63%
		0.5	1.6	1.7	1.5	1.6
F	4.9	0.0	100%	90%	100%	97%
		0.32	95%	75%	100%	90%
		0.65	4.7	4.1	5.3	4.7
		0.63**	11.9	5.4	4.5	7.3

*23 day survival   ** Brook water

CARL L. SCHOFIELD AND JOHN R. TROJNAR

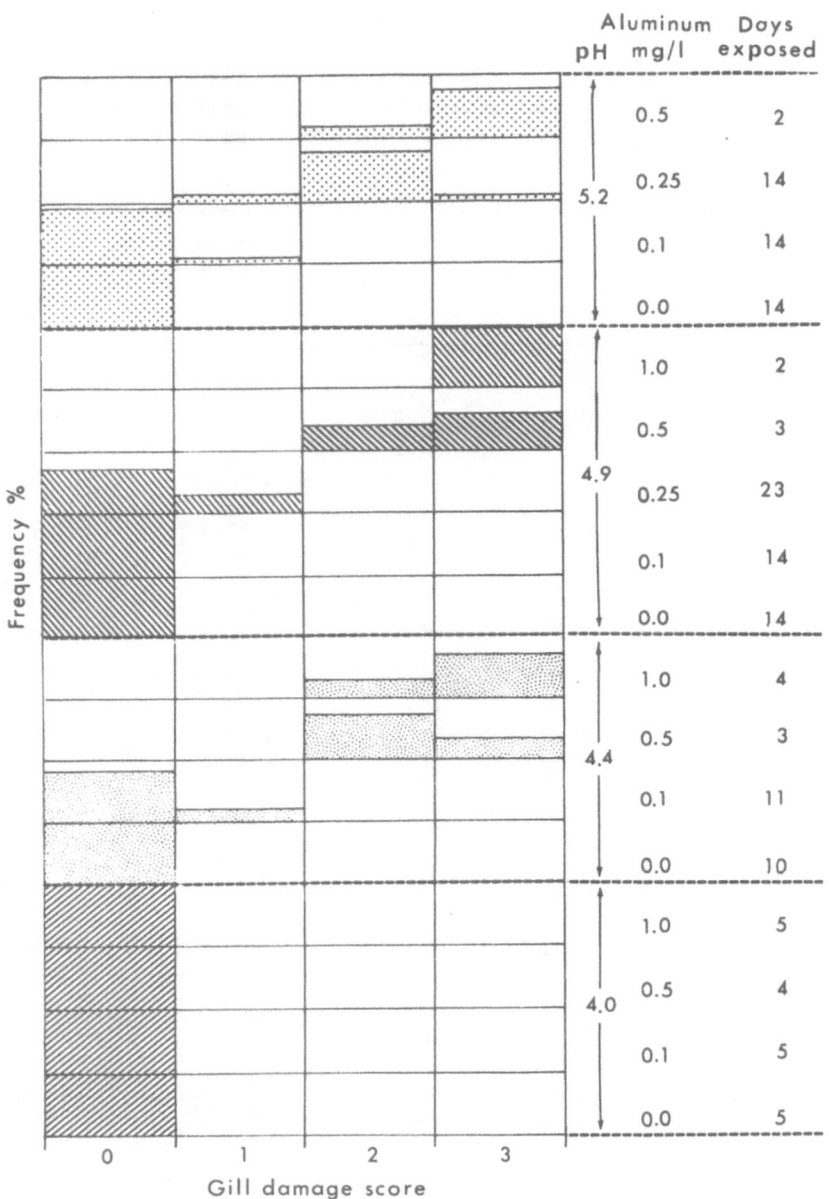

Figure 7. Relative levels of gill damage observed in survivors from
           synthetic pH-aluminum treatments. Fifteen fish from each
           treatment examined.

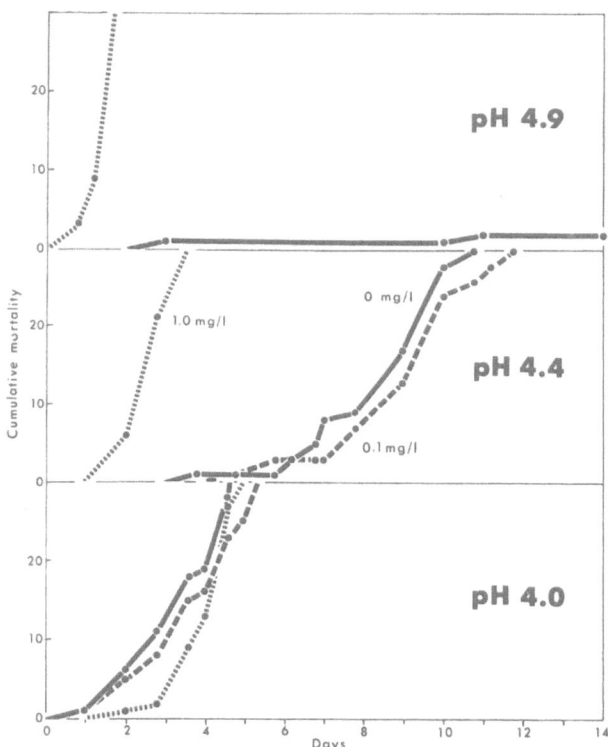

Figure 8. Cumulative mortality to the level of 50%
mortality in combined replicates at three
pH and three aluminum levels. Mortality
at pH 4.9, Al – 0 mg/l and pH 4.9, Al –
01 mg/l did not reach the 50% level after
14 days exposure.

Aluminum concentrations of 0.25 and 0.32 mg/1 were not sufficient to induce 50% mortality within 14 days; however, there were indications that delayed mortality would occur at these concentrations. Fish in trials E and F were in poor condition at these concentrations after 14 days exposure and the additional mortality after 23 days in trial D of the 0.24 mg/1 treatment supports this conclusion. In trial F 50% mortality times were similar between the 0.65 mg/1 treatment and Acid Brook water, except for replicate A of the brook water.

Discussion

The covariance analysis of water quality data obtained for the high elevation Adirondack lakes suggests that aluminum intoxication may have been a primary factor involved in determining the success or failure of brook trout stocking. Conversely, the relatively high mean pH and calcium levels and lack of significance of these adjusted variables in covariance analysis, indicates that low pH - calcium levels alone were probably not a significant source of trout mortality. A comparable analysis of water quality data in Norwegian lakes suggested that calcium and pH, but not aluminum, were the parameters significant for brown trout (Salmo trutta) survival (Wright and Snekvik 1978). Calcium and aluminum levels in the Norwegian lakes are considerably lower than levels observed in Adirondack lakes, probably because of the higher proportion of exposed bedrock and shallow soils in the Norwegian watersheds (Wright and Gjessing, op. cit.). Norwegian lakes not supporting trout populations had calcium concentrations of 0.3-1.0 mg/1. Temporal and spatial changes in water quality, which trout populations would experience in both areas, have not been considered in the survey analyses discussed, but may be important in determining survival.

Comparisons of mortality and gill pathology of brook trout exposed to acidic synthetic solutions and natural Adirondack water with aluminum levels above 0.2 mg/liter, indicated a specific toxic response to aluminum at pH levels down to 4.4. Significant, sublethal effects of aluminum on brook trout growth in the laboratory were found at aluminum concentrations of 0.1-0.3 mg/liter. Although both hydrogen-ion and aluminum toxicity were evident in laboratory experiments at pH 4.4, none of the mortality at pH 4.0 could be attributed to aluminum at concentrations up to 1.0 mg/liter. The lag in initial mortality rate in the latter treatment (figure 8) suggested an antagonistic effect of Al^{+3} [1] to hydrogen ion toxicity, comparable to that exerted by calcium. This apparent antagonism

---

(1) At pH 4.0 approximately 95% of the aluminum would be present as Al^{+3}, yielding an equivalent Al^{+3}/H^+ ratio of 1.1 for the 1.0 mg/liter treatment.

may not be specific to either ion, but rather may reflect the role of polyvalent cations in general on limiting membrane permeability and ion transport (Burrows, op. cit.).

Gill damage and mortality attributable to aluminum became evident in the pH range of 4.4-5.2. Increased aluminum toxicity with increasing pH suggests that changes in speciation involving hydroxy complexing in some way enhances the toxicity of the aluminum cation. If the mechanism of gill damage involves binding and transformation (polymerization and/or precipitation) of monomeric aluminum species at the gill surface, then increasing pH should enhance this process. The absence of gill damage at pH 4.0 (apparent mutual antagonism of H^+ and Al toxicity) and increasing severity of gill damage from pH 4.4-5.2 at aluminum concentrations above 0.2 mg/liter, suggests that this hypothesis may be applicable in solutions oversaturated with aluminum. The synthetic solutions used in the experimental trials and early spring stream run-off in the Adirondack region, but represent non-equilibrium conditions with respect to aluminum solubility and species distribution. Additional studies are in progress to test this hypothesis of aluminum intoxication, together with further investigation of aluminum speciation in acidified waters.

Acknowledgements

Financial support for this study was provided by grants from the USDI-OWRT (grant #14-34-001-9034) and from Federal Aid in Fish Restoration Project F-28-R, New York. The authors wish to thank the Adirondack League Club (Old Forge, New York) for the provision of laboratory facilities and appreciated the interest and cooperation extended by the members during the course of this study. Special thanks to Leo Demong and Art Stemp for assistance with the field studies. The cooperation of the New York State Department of Environmental Conservation in obtaining field survey data was also appreciated. Finally, we wish to thank Professor Dwight A. Webster for his review of this paper and for his advice and assistance during the study.

Literature Cited

Burrows, W.D. 1977. Aquatic aluminum: chemistry, toxicology, and environmental prevalence. CRC Critical Reviews in Env. Control. V. 7(2):167-216.

Cronan, C.S. 1978. Solution chemistry of a New Hampshire subalpine ecosystem: biogeochemical patterns and processes. Ph.D. Thesis, Dartmouth College, Hanover, N.H., 248 pp.

Dickson, W. 1975. The acidification of Swedish lakes. Inst. Freshwater Res. Drottningholm Rep. 54, p. 8-20.

Freeman, R.A. and W.W. Everhart. 1971. Toxicity of aluminum hy-
    droxide complexes to rainbow trout. Trans. Amer. Fish Soc.
    100(4):644-658.
Johannessen, M., Dale, T., Gjessing, E., Henriksen, A., and R.
    Wright. 1975. Acid precipitation in Norway: The regional
    distribution of contaminants in snow and the chemical concen-
    tration processes during snow-melt. Proc. Int. Symp. on
    Isotopes and Impurities in Snow and Ice, Int. Assoc. of Hydrol.
    Sci., Grenoble, France. August, 1975. (In press)
Jones, J.R.E. 1964. Fish and river pollution. Butterworth Sci.
    Publ., Washington, D.C., 203 pp.
Leivestad, H., Hendry, G., Muniz, I., and E. Snekvik. 1976.
    Effects of acid precipitation on freshwater organisms. In:
    SNSF Research Report 6/76, Ed. F. H. Braekke. pp. 87-111.
Schofield, C.L. 1976a. Acidification of Adirondack lakes by
    atmospheric precipitation: extent and magnitude of the
    problem. Final Rep. D.J. Proj. F-28-R, NYS Dept. Env. Cons.,
    11 pp.
Schofield, C.L. 1976b. Acid Precipitation: Effects on fish.
    Ambio. 5(5-6):228-230.
Smith, R.W. 1971. Relations among equilibrium and nonequilibrium
    aqueous species of aluminum hydroxy complexes. pp. 250-279.
    In: Nonequilibrium systems in natural water chemistry. ADV.
    Chem. Ser. 106, ACS, Washington, D.C.
Wright, R.F. and E.T. Gjessing. 1976. Acid precipitation: changes
    in the chemical composition of lakes. Ambio 5(5-6):219-223.
Wright, R.F. and A. Henriksen. 1978. Chemistry of small Norwegian
    lakes, with special reference to acid precipitation. Limnol.
    Oceanogr., 23(3):487-498
Wright, R.F. and E. Snekvik. 1978. Acid Precipitation: Chemistry
    and fish population in 700 lakes in southernmost Norway.
    Vert. Internat. Ver. Limnol., V. 20, (in press)

DISCUSSION

McLEAN: Carl, could I ask you if the aluminum toxicity problem
is as serious with the older fish as it is with the young
fish? Have adequate fish population studies been done?

SCHOFIELD: Yes, we've been working with several life history
stages with both brook trout and white sucker. And we see
similar responses that seem to be relatively independent of age
and size, for brook trout at least. Now, that's contrary to
some of the data for hydrogen ion stress which suggests that
size is an important factor.

McLEAN: This is an interestig point in relation to Jay
Bloomfield's observation that they were unable to find any
small-mouth bass in Stillwater Lake in 1979 after successful
catches in earlier years. Is it possible that elevated
aluminum levels rather than elevated hydrogen ion levels are
causing fish mortality in such situations?

SCHOFIELD: Well, we haven't made that kind of evaluation in
terms of what the significance is regionally in terms of loss
of fish population. We know that currently it's acting as a
stressing agent in many of the lakes and streams in this region
where we try to re-introduce fish where there are currently
none. That seems to be one of the primary factors of limiting
re-establishment of fish population.

McLEAN: From some of your data on the seasonal variation of
the input of aluminum to the waterways, it appeared that there
was actually a bigger input of aluminum to the waterway in the
early autumn than in the spring. Is this the case?

SCHOFIELD: No, actually the larger inputs occurred during the
spring, during the snow melt.

McLEAN: We've been carrying out similar studies on the
variation in the mercury flux in waterways throughout the
year. We have noticed in our study area that there is a higher
flux of mercury in the water at the times of the heavy rains in
the late autumn than over the period of spring run off. Your
data seemed to indicate a similar situation with aluminum
levels in waterways.

SCHOFIELD: There are temporal increases during rain events in
tributary systems of aluminum, but the magnitude is generally
less than what we see during snow melt.

MILLS: Do you find, Carl, that this toxicity affects other fish in some of these lakes, like yellow perch or bass?

SCHOFIELD: Well, certainly there are other species that have been impacted in the region, but we haven't done any experiments with species other than the brook trout and white sucker, the two I mentioned at the present time. Those two species seem to represent the relative extremes in tolerance of the indigenous species in the region.

MUHLBAIER: Is this unique to aluminum or do other metals act the same way?

SCHOFIELD: I believe that other metals, particularly some of the heavy metals, can exert a similar response, in terms of causing gill tissue damage, for example. Particularly when they tend to form hydroxy complexes, or react with the mucous secretions on the surface of the gill. That's not an unusual occurrence. I don't know whether other metals react in the way we're seeing at lower levels, in terms of causing a stress on the osmoregulatory system.

TORIBARA: Is there any uptake of the aluminum by the flesh?

SCHOFIELD: We've done analyses of the various organs for aluminum uptake and there doesn't seem to be any significant internal accumulation, at least, in terms of these short-term exposures we're working with. Most of the aluminum concentrates in the gill, and particularly at the surface of the gill.

McLEAN: It would be rather surprising to find the same type of accumulation of aluminum in fish flesh as one finds with some heavy metals. Those which concentrate in fish flesh form relatively more stable complexes with sulfur donors (i.e. soft bases), whereas aluminum certainly doesn't. If forms relatively very stable complexes with oxygen donors (or "hard" bases).

McFEE: Do you understand the physiology of the aluminum being retained in the gills? Why doesn't it just go on through with the water? Is there a pH change at the gill surface?

SCHOFIELD: There is some pH difference between the ambient water and the gill surface. At the present time, we don't have any means of measuring this. We haven't attempted it. There are some people in England looking at transepithelial potentials in relation to stresses from hydrogen ion which suggest that, yes, there could be significant changes in surface pH which would tend to influence the behavior of aluminum and other comparable metal ions at the surface of the gill. But we haven't got down to that level in terms of experimental investigation.

TOMLINSON: Aluminum has a property of coagulating pulp fibers at low pHs, and this is always related to the charge of the aluminum positive ion on a negatively-charged fiber. To my mind it looks like that type of coagulation is in your very interesting photographs.

SCHOFIELD: Yes, certainly that occurs. But we don't know whether it's occurring in the solution the fish is exposed to, or right at the gill surface.

McLEAN: Do I understand you to suggest that the fish toxicity problem arises from the formation of polymeric particulate aluminates which interfere with the proper function of the gills.

SCHOFIELD: Yes, we certainly find that in the water itself where we get oversaturation. We certainly believe that these particulates are coagulating or adhering to the gill surfaces.

McLEAN: This would be another distinction between aluminum and the heavy metals, because few of the latter form stable polymeric complexes in aqueous solution.

GUARI: Is the source of aluminum here from the deterioration of the bedrock or it is being derived by the further decay of the clay minerals?

SCHOFIELD: I can't answer that specifically. It would certainly appear, from the kinds of seasonal patterns we're seeing in aluminum release and in the specification of aluminum, that the reaction has to be rather fast. So we're not dealing with dissolution of silicate minerals for these very rapid changes in concentration, but rather a reaction with amorphous forms of aluminum or exchangeable aluminum in the soil.

McFEE: I would suspect that it would be primarily exchangeable aluminum that comes out; during the time there's not much water moving through the soil, there's probably a replenishment of exchangeable aluminum through the weathering and decomposition of the silicate minerals. So it's ultimately coming from the weathering rock.

SCHOFIELD: That would be its ultimate source, but not for immediate reactions.

GUARI: If this is now from the clay minerals itself, not from the parent minerals like feldspar, then of course there is considerable leeching of the soil that's taking place which would seem somewhat different from what we heard this morning.

EVANS: Would that be easy to measure in a watershed? Whether it's coming from clay or from parent material?

McFEE: I wonder how you'd separate them?

McLEAN: Could I ask you how long the liming projects have been going on and what kind of results you had?

SCHOFIELD: The state has been practicing liming lakes for the purpose of maintaining fish populations as far back as the 1950's. It's only been I'd say in the last five years that they organized programs developed specifically to address the problem of acid precipitation.

TOMLINSON: Have the results of the study been published, Carl, in any sort of a report?

SCHOFIELD: It's strictly a management effort; it's not an experimental study.

OXIDANTS IN PRECIPITATION

Edward L. Mills
Research Associate
Department of Natural Resources
Cornell University Biological Field Station
R. D. #1
Bridgeport, New York 13030

Cornelius B. Murphy, Jr.
Vice President
O'Brien & Gere Engineers, Inc.
Syracuse, New York 13201

Jay A. Bloomfield
Research Scientist
NYS Dept. of Environmental Conservation
Albany, New York 12233

ABSTRACT

Large quantities of chlorine are produced in the
United States each year. A by-product of industrial use
and manufacture is that some chlorine gas escapes into
the atmosphere. However, a general lack of information
exists on precipitation washout of these oxidant residuals
despite recognized emissions into the atmosphere. It
is hypothesized that acid precipitation is only an indicator of other less obvious phenomena, and the pH of
precipitation may well serve as an index of a variety
of atmospheric pollutant problems.

Significant inputs of chlorine gas are emitted into
the atmosphere in New York State; the major source of
chlorine emissions occurs in Niagara County, New York.
Oxidant residuals have been measured in rainfall from
Central New York. Maximum residuals in rainfall coincided with depressed pH values and maximum rainwater

temperature. The observed chlorine residuals and dominance of NH3 ions in rainfall suggest the residual may be a combined form of the type NH_xCl_y. One hypothesis is that industrial emissions of this gas combine with NH3 in the atmosphere to form NH_xCl_y compounds or chloramines which are later removed by atmospheric washout.

Laboratory and field studies reported in the literature suggest that oxidant residuals are toxic to biological organisms and the impact is highly dependent on concentration and duration of exposure. In dilute aquatic systems of low buffering capacity, measured oxidant residuals can potentially affect biological organisms and may pose a new element of concern.

INTRODUCTION

In recent years, the significance of interactions between the atmosphere and natural ecosystems has received widespread attention. Efforts to understand the atmospheric environment have focused on the meterological and hydrologic aspects of air-land interactions (Neiburger, Edinger, and Bonner, 1973), and only recently have the chemical interactions received attention. Acidification of extensive freshwater areas by acid precipitation in Europe and the United States pioneered exploration into the chemistry of rainfall, and extensive literature supports its effects on aquatic and terrestial ecosystems (Giddings and Galloway, 1976).

Chemicals introduced into the atmosphere creating acid precipitation are usually anthropogenic in nature, with coincident depressed pH concentrations of rainfall. Acidification of lakes in the Adirondacks of New York State has resulted in not only obvious economic and social impact, but also depletion of certain fish stocks creating altered problems of ecosystem structure and function (Schofield, 1976). Acid-induced mortality of fish populations is of major concern in these and other freshwater lakes; however, the cause of death of these fish warrants further attention.

Variations in pH in experimental studies have often altered the pattern of response of other environmental parameters. For example, a decrease in pH changes the anion-cation balance and the level of toxicity; the availability of such microcontaminants as heavy metals is often pH-dependent (Giddings and Galloway, 1976). It is possible that synergistic effects enhanced by

acid conditions have led to the decline of particular
fish species in freshwater lakes.

     This logic has led to the hypothesis that acid
precipitation is only an indicator of other less obvious
phenomena, and the pH of precipitation may well serve as
an index of a variety of atmospheric pollutant problems.
This paper represents a nucleus of information of one
group of microcontaminants, namely oxidant residuals,
in the environment including their character, sources,
and potential washout into aquatic ecosystems.

Occurrence in the Environment

     Large quantities of chlorine are produced each year
for both industrial use as well as the treatment of wastewater and water supplies. (Anonymous, 1976). In 1972,
sixty-eight industries were known to produce chlorine
in the United States and production increased 6.8 percent
over a ten-year period, 1962-1972. Nearly 70 percent
of the total consumption of this chemical is utilized
by the chemical and pulp and paper industries. A by-
product of industrial use and manufacturing is that some
chlorine gas escapes into the atmosphere.

     The burning of large quantities of chloride and
chlorine containing fuels (coals and fuel oil) (Stahl,
1969) and combustion of hydrocarbon fuels in transportation (Anonymous, 1976) are major sources of hydrogen
chloride gas to the environment. The average chlorine
content of coal is 0.13%,and nearly 95% of this component is emitted into the atmosphere. The emission of
chlorine and its derivatives to the atmosphere from
coal burning activities may take on future significance
in the United States because of its demand as an energy
source. Waste disposal by incineration of chlorine
containing plastics is also of concern since they generate hydrogen chloride gas on burning. In urban areas,
chlorine containing compounds exist in gaseous form or
combined with aerosols (Anonymous, 1976). The dispersion of natural and anthropogenic chlorine containing
compounds in the atmosphere is considered to follow
patterns similar to other trace contaminants (Strom,
1968; Wanta, 1968). However, a general lack of information exists on precipitation washout of these oxidant
residuals despite recognized emissions into the atmosphere.
Recent evidence by Lunde (et.al. 1977)in Norway showed that
a wide variety of organic micropollutants were found
in precipitation and many of these chemicals were of

industrial origin. One chemical group, PCB'S, were 5-
10 times higher in precipitation than observed in sea
water.

Chlorination of municipal water supplies and wastewater effluent in the United States is the most common
method of disinfection (Moore, 1951; Morris, 1971) and
represents a second major source to the environment.
Chlorine is a strong oxidizing agent and is widely used
as an antifouling agent in numerous industrial and water
treatment processes. The use of chlorine for these purposes is common and, consequently, water quality criteria
standards have been established. Recommended concentrations of residual chlorine entering stream waters are
0.5 to 1.0 mg/l after disinfection and 0.05 - 0.5 mg/l
after antifouling of intake structures and cooling systems (Brungs, 1973).

Natural sources of chlorine and its derivatives to
the atmosphere have been reviewed by Duce (1969). Sea
salt particles and volcanism have been suspected as
natural sources to the atmosphere. However, the most
commonly accepted view is that gaseous chlorine compounds
found in uncontaminated air result from sea-salt particles (Zafiriou, 1974; Erikson, 1960) and that photochemical reactions play an important role in the chemistry
of atmospheric chlorine compounds (Petriconi and Papee,
1972). Other natural sources of gaseous chloride include
emissions from forest and grass fires (NAS, 1976).
Atmospheric gaseous chloride has been measured in coastal
regions off Florida and Hawaii (Junge, 1956, 1957; Duce
et al, 1965) and, as might be expected, gaseous forms
have longer retention time in the atmosphere than residuals adsorbed to particulates.

Chemical Oxidants in an Aqueous Environment

A wide range of chemical oxidants are emitted into
the atmosphere and these microcontaminants could potentially be scrubbed out as a result of precipitation
washout. Potential oxidants include oxides of sulfur
and nitrogen, ozone, peroxides, and the various free
and combined forms of the halogens (F_2, Br_2, Cl_2, and
I_2). The strength and stability of each potential
oxidant depends on the pH of precipitation, the presence
of reducing agents, and the total ionic strength.

Chlorine exists in waters as Cl_2 and $HOCl^-$, with the
exact distribution of these species dependent upon the

characteristics of the aqueous system. Chlorine rapidly
hydrolyzes in water to form hydrochloric acid and hypochlorous acid.

$$Cl_2 + H_2O \rightarrow HCl + HOCl$$

The hypochlorous acid then disassociates to give
hydrogen ions and hypochlorite

$$HOCl \rightarrow H^+ + OCl^-$$

The three forms of free available chlorine involved
in these reactions, molecular chlorine (Cl_2), unionized
hypochlorous acid (HOCl) and the hypochlorite ion (OCl^-),
exist together in equilibrium. Their relative proportions are determined by pH, temperature, and dissolved
solids.

The available forms of free chlorine are both
extremely reactive and unstable. In the presence of
ammonia, chlorine reacts to form a series of chloramines
which are also oxidizing agents but considerably more
stable than the free forms of chlorine. Under neutral
and alkaline conditions and at chlorine to ammonia nitrogen ratios below 5, chlorine forms monochloramines by
the following reaction:

$$NH_3 + Cl_2 \rightarrow NH_2Cl + HCl$$

Under conditions of higher chlorine to ammonia
nitrogen ratios the dichloramines and trichloramines
are formed via the following mechanisms.

$$NH_2Cl + Cl_2 \rightarrow NHCl_2 + HCl$$

$$and$$

$$NHCl_2 + Cl_2 \rightarrow NCl_3 + HCl$$

Chemically, chloramines can form in the atmosphere
and may be washed out in rainfall; chlorine contributed
by anthropogenic sources of industrial origin can combine with ammonia in precipitation resulting in the
formation of a fairly stable oxidizing agent.

Analytical Techniques

The analytical techniques available for the measurement of free and combined forms of chlorine are adequately

described in Standard Methods (American Public Health Association, 1971). The available techniques include:

Iodometric Titration

Amperometric Titration

Orthotolidine Methods (Simple Acid Test, Modified Acid Test, Orthotolidine Arsenite Test, Drop Dilution Method, and the Neutral Orthotolidine Titration Method)

Leuco Crystal Violet Method

Methyl Orange Method

Syringe Aldazine Procedure

DPD Methods

Of all the conventional analytical techniques, the DPD method is unique in providing a range of differential procedures, not only for the various forms of residual chlorine but for their mixtures with other types of residuals including chlorine dioxide, chlorite, bromine and ozone. The Analytical Reference Service of the U. S. Environmental Protection Agency (EPA, 1969) in their first report published in 1969, concluded that "the best overall accuracy and precision was shown by the DPD titrametric method."

For the data compiled and discussed in this paper, the Orthotolidine and DPD Methods were primarily utilized. The modified procedures were used for the separate determination of free chlorine and the combined forms (monochloramine, dichloramine and nitrogen trichloride). U.V. spectroscopy was applied for the qualitative determination of the presence of both free and combined forms of chlorine. The spectra were compared with data presented by Zimmerman and Strong (1957) for absorption in the range of 200-400 mu.

To date, most analytical techniques reveal the presence of a class of oxidant residuals but are not sensitive to delineate such specific constituents as Cl_2, ClO^-, ClO_3^-, Cl_3, NH_2Cl, $NHCl_2$, NCl_3, NH_3, and amines. Future techniques may necessitate the use of reflectance ultraviolet and infrared applied to liquid and frozen samples as well as GC/MS as applied to gas

purged samples. These approaches have the advantage of
scanning a wide range of precursors and combined forms
of chlorine necessary for the development of a mechanism
of formation, transport, and impact on the aquatic environment.

Sources and Emissions of Chlorine into the Atmosphere in New York State

The production and use of chlorine in New York
State is significant since seven of the largest manu-
facturers in the United States have plants in New York
State which produce or use chlorine. Major producers
and users within New York are listed in Table 1. Most
industries involved in either the manufacture or use
of chlorine are concentrated in Western New York State;
furthermore, all of these plants are localized within
Niagara County and, therefore, would be suspected as
a significant source of chlorine gas to the atmosphere.

Chlorine emissions in tons per year are summarized
by county in Figure 1; these estimates may be conservative but do reflect potential source emissions. As
suspected, the major source of chlorine emissions occurs
in Niagara County, emitting approximately 185 tons per
year; Hooker Chemical Company of Niagara Falls, alone,
emits 125 tons per year and is considered the largest
chlorine emitter in the State. Other major centers of
chlorine emissions include portions of Central and
Northern New York. These findings indicate that significant quantities of chlorine gas are emitted into
the atmosphere within New York State; the possible
transport and washout in rainfall of these emissions
warrants further attention.

Residual Chlorine in Precipitation in Central New York State

Oxidant residuals in the form of total residual
chlorine were first reported by the New York State
Department of Environmental Conservation in rainfall
and lake waters of the Western Adirondacks of New York
State in 1976-77. Chlorine residuals analyzed by the
orthotoline colorimetric procedure developed by Palin
(1957) were detected in rainfall at Boonville and
Stillwater, New York and in Echo Lake, Hinckley Reservoir and Stillwater Reservoir (Figure 2). Total residual chlorine in rainfall near Boonville, New York

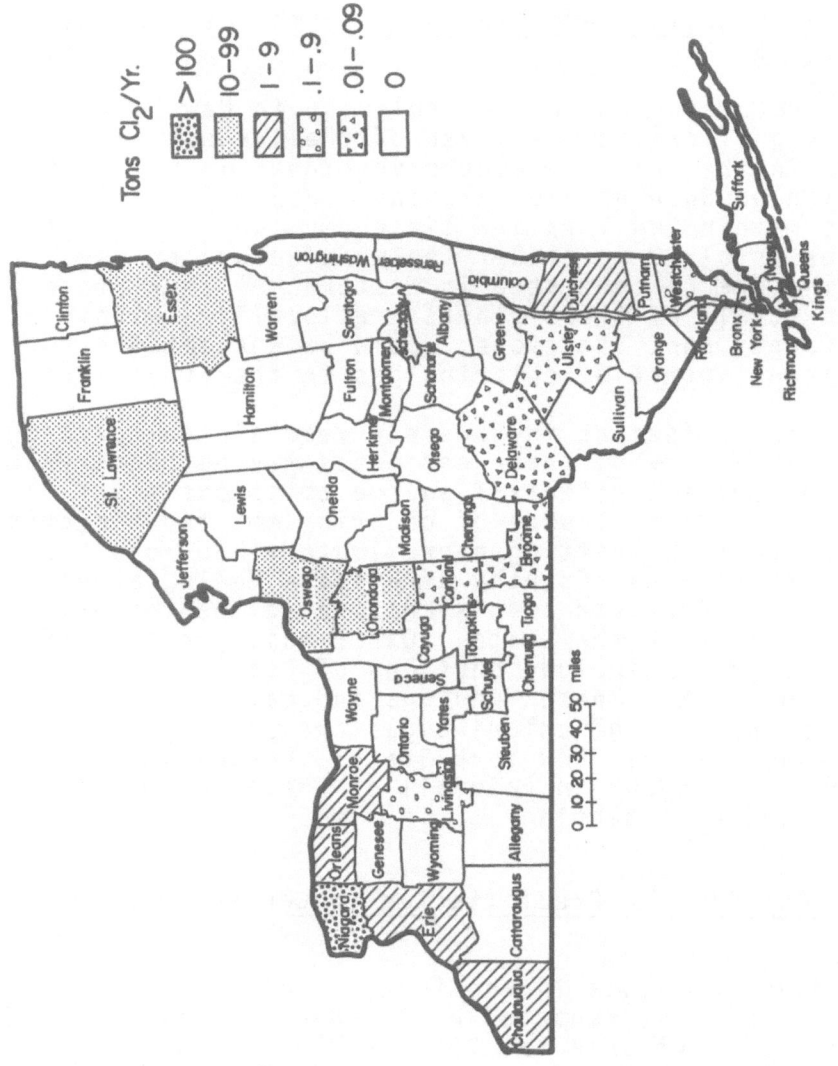

Figure 1. Chlorine emissions for each county within New York State. Estimates are in tons of Cl_2 per year. (Information provided by New York State Dept. of Environmental Conservation

Figure 2. Map of New York State outlining area in which chlorine residuals have been detected in rainfall.

Table 1. Major users or producers of chlorine in New
          York State.*

	Source	County	
1.	Hooker Chemicals and Plastics	Niagara	producer and user
2.	Olin Chemical	Niagara	producer and user
3.	E.I. DuPont de Namours Co.	Niagara	producer and user
4.	International Minerals and Chemicals	Niagara	producer and user
5.	Airco-Speer Corp.	Niagara	user
6.	Van De Mark Chemicals	Niagara	user
7.	Vanchlor Co.	Niagara	user
8.	Stauffer Chemicals	Niagara	user
9.	Allied Chemical Corp.	Niagara	producer and user
10.	Newton Falls Paper	St. Lawrence	user
11.	Alcoa	St. Lawrence	user
12.	International Paper Co.	Essex	user

* Information provided by New York State Department of
  Environmental Conservation.

ranged between 0.01 to 4.5 mg/l. Detectable levels of
oxidant residuals were also observed in nearby Echo Lake
waters following rain storm events. Echo Lake is a 39
hectare deep water lake (10-20 meters) of moderate buffering capacity (88mg/l total alkalinity) which had a
thriving brook trout population in 1900; the present
species complex of fish consists of largemouth bass,
yellow perch, suckers, sunfish, and bullheads. Declining fish populations of other poorly buffered surface

Table 2. Concentration of various chemicals in rainwater collected at Cornell Field Station on
Oneida Lake, New York during August-September,
1978*

Conductivity (Microhms/in @ 25°C)	Soluble Reactive Phosphorus μg/l	Total Phosphorus μg/l	Nitrate Nitrogen μg/l	Nitrite Nitrogen μg/l
93.2	10.9	25.0	931.4	3.1

Ca mg/l	Mg mg/l	K mg/l	Na mg/l	MN mg/l	NH_3N μg/l	SO_4 mg/l
1.2	0.39	0.11	0.18	0.01	580.0	5.3

* Values represent means for the period studied.

waters of the Adirondack Mountains of New York State have
been attributed to increased acid precipitation in water
of low inherent buffering capacity (Schofield, 1976).
It is likely, in the case of Echo Lake, that the loss
of certain fish populations were acid-induced; however,
the apparent atmospheric washout and the intermittent
presence of residual oxidants in this natural water
body raise new questions on the potential impact of
synergistic effects of microcontaminants on biological
organisms at depressed pH's. Experimental evidence has
shown that oxidant residuals are toxic to freshwater
organisms (Brungs, 1973). Further research is needed
to determine the exact nature of these microcontaminants
and examine the significance of these residual contaminants as an additional factor affecting food web yield
and population structure.

Subsequent Central New York rain samples were
monitored August through November of 1978 at the Cornell
University Biological Field Station located approximately 11 miles northeast of Syracuse, New York on the
south shore of Oneida Lake (Figure 2.). Oxidant residuals and other chemicals were measured during major
storm events as well as rainfall amounts which ranged
between 0.42 and 2.3 inches. DPD measured (Titrametric
method) oxidant resuduals were highest in early August
(Figure 3) with subsequent decline in late August through
September to near non-detectable levels in October
through November. maximum oxidant residuals coincided

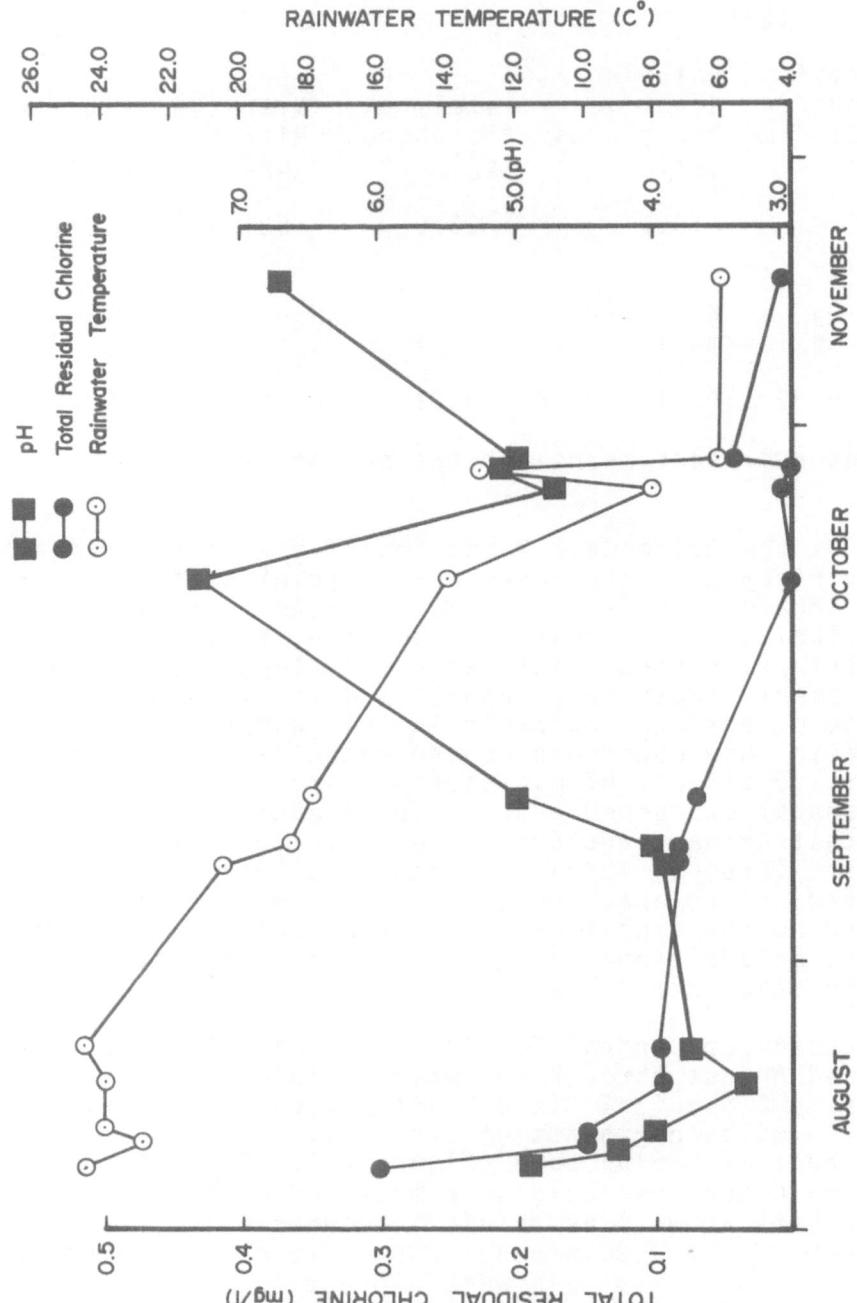

Figure 3.   Total residual chlorine, pH, and temperature of rainfall
collected near Oneida Lake, August through November, 1978.

with depressed pH values (3.4-5.1) and maximum rainwater temperatures (24°C). As cold air masses began to dominate during fall, total residual chlorine declined, the range of pH increased to 5-7, and rainwater temperatures declined to near 6°C. Further characterization of rainfall at this site indicated that nitrate and ammonia ions were dominant in Central New York rainfall (Table 2). Nitrate nitrogen concentrations mirrored the pattern of oxident residuals; nearly 1 mg/l of nitrate-nitrogen was measured in rainfall, August through early September, declining to nearly 0.3 mg/l in late fall. Nitrate is one of the dominant ligands of Adirondack rainwater as observed at Hinckley, New York (U.S. Department of Interior, 1972-1976).

Atmospheric washout was monitored during one rainfall event in mid-September in which 2.3 inches of rain fell over a 20 hour period (Figure $\underline{4}$). Maximum chlorine residuals and redox potential occurred during the early stages of the rainfall event. As the storm intensified, oxidant levels declined but complete washout was not observed; in fact, a slight increase was observed as the storm system weakened and moved beyond the region.

U. V. spectral analysis exhibited peak absorbances between 210 and 215 nanometers suggesting the oxidant residual may be a combined form of the type NH_xCl_y. These findings would lend support to observed chlorine residuals and the dominance of NH_3 ions in rainfall. Furthermore, there are several major emitters of chlorine gas in western New York State, and long-range transport of chlorine and its reaction products cannot be ruled out. One hypothesis is that industrial emissions of this gas combine with NH_3 in the atmosphere to form NH_xCl_y compounds or chloramines which are subsequently removed by atmospheric washout. However, further work is needed to characterize the microcontaminants and assess their impact on the aquatic environment.

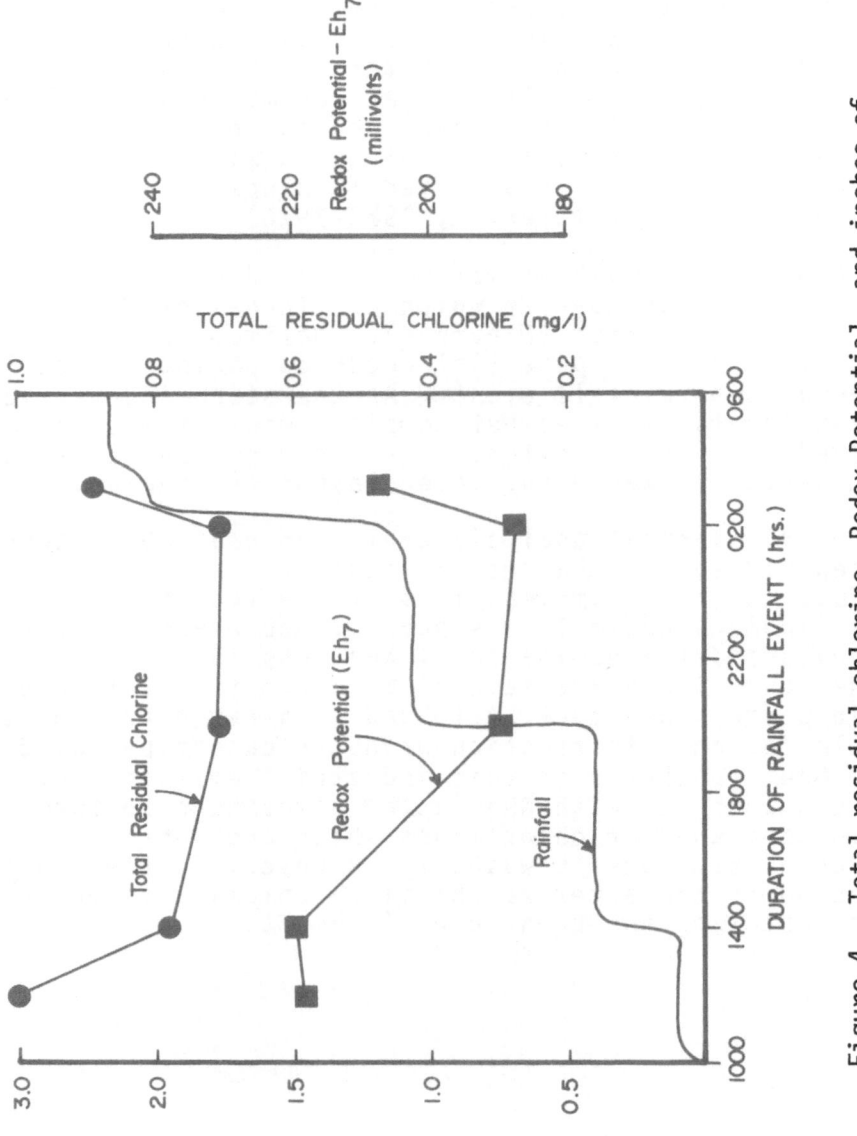

Figure 4. Total residual chlorine, Redox-Potential, and inches of
rainfall measured during a storm event, Sept. 11-12, 1978.

The Response of Food Web Organisms to Oxidants

The impact of toxicants on aquatic organisms is of immediate concern since the loss of a particular species will likely alter productivity as well as the energy flow of the system. Intermittent exposures of oxidant residuals derived from atmospheric washout and concurrent effects on biological organisms in the natural environment have not been evaluated. However, the potential effects of these contaminants on aquatic systems can be gleaned from the literature due to the importance of these chemicals as disinfectants, antifouling agents, and biocides. Consequently, production and use of these chemicals has stimulated research in areas concerning exposure, lethal levels, and the toxicity of chloro-derivatives.

The action of chlorine compounds is not specific to target pest organisms, and the response of a biological system depends on dose, duration of exposure, and the mode of action. In living tissues, chlorine converts to hypochlorous acid (Zillich, 1972) which, in turn, penetrates the cell wall reacting with cytoplasmic proteins to form N-chloro derivatives that destroy cell structure (Lawrence and Block, 1968). Considerable evidence has indicated that chlorine and associated reaction products are toxic, in low concentrations, to many types of aquatic organisms (Brungs, 1973). Merkens (1958) found that toxicities on aquatic biota were a function of levels of the residual component and the relative composition of free chlorine and chloramines; residuals of chloramines were dependent on concentrations of ammonia and chloramines as well as the pH and temperature of receiving waters. The toxicity of residual chlorine was higher at depressed pH levels and the effects to fish of free chlorine and chloramines were similar; these findings were consistent with those of Doudoroff and Katz (1950). However, the relative toxicity of free and combined chlorine is subject to debate (Brooks and Seegert, 1976). Most workers (Merkins, 1958; Eren and Langer, 1973) consider free chlorine to be more toxic while others (Holland et al 1960) consider the impact of chloramines to be greater. Recent evidence by Heath (1977) suggested, depending on fish species, that free chlorine was 3-14 times more toxic than mono-chloramine. Experiments by Brooks and Seegert (1978) concluded that free chlorine was more toxic than mono-chloramine but not necessarily more toxic than dichloramine.

A number of investigators have provided experimental evidence suggesting that brief exposures of oxidant residuals are toxic to fish. As shown in Table 3, these studies indicate that cold water trout species are more sensitive to oxidant residuals than such warm-water forms as yellow perch, bass and fathead minnows. Bosch and Truchan (1976) found a variety of warmwater fish species (mainly centrarchids and bullheads) that were able to survive repeated 30 minute exposures of chlorine levels up to 0.5 mg/l. Yellow perch are extremely resistant to residual chlorine at low levels (Arthur et al 1975; Brooks and Seegert, 1977), while avoidance by rainbow trout has been observed at levels of 0.001 mg/l (Sprague and Drury, 1969).

Fish appear to be most sensitive to chlorine and reaction products at higher temperatures (Brooks and Seegert, 1977). Furthermore, alewives and yellow perch exhibited a ten-fold increase in sensitivity as test temperatures were elevated from 10 to 30°C. Stober and Hanson (1974) showed that a synergistic toxic phenomenon occurred between temperature and oxidant residuals; these findings are relevant to observed oxidant residuals in Central New York rainfall since concentrations of these compounds are highest during periods when lake water temperatures in the region are maximum.

Brooks and Seegert (1977) suggested that indirect effects of exposure of fish to residual chlorine might increase mortality due to predation; they observed that fish became lethargic when exposed to elevated levels of chlorine and thereby increased their vulnerability to piscivorous fish.

Physiologically, the formation of methemoglobin is a result of chloramine-induced oxidation of hemo-globin causing an inability of blood to carry an ade-quate supply of oxygen. Buckley (et al. 1976) observed anemia in coho salmon (Oncorhynchus Kasutch) resulting from reductions of hemoglobin following exposure to chlorinated sewage effluent. Abnormalities of the blood were attributed to the oxidative nature of the chlorine residual.

The literature is relatively devoid of information regarding exposures of residual chlorine to invertebrates. Arthur and Eaton (1971) found that the 96 hour TL-50 for the amphipod Gammarus pseudolimnaeus was 0.220 mg/l and that reproduction was reduced at 0.0034 mg/l; these authors also reported that Daphnia magna was extremely

Table 3. Response of fish to brief exposures of residual chlorine (modified after Brungs, 1973, Table 2, page 283).

Fish	Response	Duration of Experiment	Residual Chlorine Concentration (mg/l)	Reference
Fathead Minnow	TL 50	1 hr.	70.79	Arthur, 1972
Fathead Minnow	TL 50	12 hr.	0.26	Arthur, 1972
Largemouth Bass	TL 50	1 hr.	70.74	Arthur, 1972
Largemouth Bass	TL 50	12 hr.	0.37	Arthur, 1972
Yellow Perch	TL 50	1 hr.	70.88	Arthur, 1972
Yellow Perch	TL 50	12 hr.	0.50	Arthur, 1972
Trout fry	Lethal	Lethality immediate	0.30	Coventry, 1935
Rainbow Trout	TL 50	7-day	0.08	Perkins, 1958
Rainbow Trout	Slight avoidance	10 min.	0.001	Sprague & Drury, 1969
Brook Trout	Median survival time	18 hr.	0.080	Dandy, J.W.T. 1972
Brook Trout	Median survival time	48 hr.	0.040	Dandy, J.W.T. 1972

sensitive to low concentrations of chloramines, and exposure of 0.001 mg/l killed all individuals over a 3-5
day period. Brooks and Seegert (1977) found that the
copepod, Cyclops bicuspidatus, was able to withstand a
wider range of residual chlorine than Limnocalanus
macruus.

Lamb (1978) introduced chlorine into a trout stream
in Central New York to evaluate the toxicity to various
stream organisms. Periphyton density was reduced only
at high chlorine residuals; fish were sensitive to low
concentrations; and total residual chlorine concentrations of less than 0.01 mg/l caused catastrophic drift
of sensitive stream insects.

Recent studies have shown that oxidant residuals
influence phytoplankton populations by reducing chlorophyll concentrations and rates of photosynthesis (Fox
and Meyer, 1975; Eppley et. al., 1976). Furthermore,
Brooks and Seegart (1977) investigated the effects of
chlorine on natural populations of Lake Michigan phytoplankton; intermittent 30 minute exposures reduced
chlorophyll a, increased the degradation product,
phaeophytin, and reduced primary productivity as measured
by Cl4. Chlorine concentrations exceeding 0.5 mg/l had
a pronounced impact on algal populations while a slight
decrease in chlorophyll a was noted at concentrations
less than 0.1 mg/l. In fact, estimates of primary
productivity showed full recovery following 30 min. exposures to chlorine levels less than 0.1 mg/l.

Inference from laboratory and field studies described above suggests oxidant residuals are toxic to
biological organisms, and the impact is highly dependent
on concentration and duration of exposure. In dilute
aquatic systems of high surface area/volume ratios and
low oxidative buffering capacity, measured oxidant
residuals can potentially affect biological organisms
and may pose a new element of concern.

References

American Public Health Association, 1971. Standard
        Methods for the Examination of Water and Wastewater.
        13th ed. Amer. Public Health Assoc., Washington,
        D.C. 874p.

Anonymous. 1976. Chlorine and hydrogen chloride.
    Medical and biologic effects of environmental pol-
    lutants. National Academy of Sciences. Washington,
    D.C. 1976.
Arthur, J. W. 1972. Progress Report. National Water
    Quality Lab., EPA, Duluth, Minn. (1971-1972)
Arthur, J. W. and J. G. Eaton, 1971. Chloramine toxicity
    to the amphipod (Gammarus pseudolimnaeus) and the
    fathead minnow (Pinephales promelas). J. Fish.
    Res. Bd. Canada. 28:1841-1845.
Arthur, J. W., R. Andrew, V. Mattson, D. Olson, G. Glass,
    B. Halligan, and C. Walbridge. 1975. Comparative
    toxicity of sewage-effluent disinfection to fresh-
    water aquatic life. Environmental Protection Agency,
    Ecological Research Series, Duluth, Minn. 62 pp.
Bash, R. E. and J. G. Truchan, 1976. Toxicity of chlor-
    inated power plant cooling waters to fish. Environ-
    mental Protection Agency, Ecological Research Series
    R-800700. Duluth, Minn. 75 pp.
Brooks, A. S. and G. L. Seegert, 1976. Toxicity of
    chlorine to freshwater organisms under varying
    environmental conditions. Proc. Confr. on environ-
    mental impact of water chlorination, Oak Ridge
    Nat'l Lab., Oak Ridge, Tenn.
Brooks, A. S. and G. L. Seegert, 1978. The effects of
    intermittent chlorination on selected warmwater
    fishes. Center for Great Lakes Studies, Univ. of
    Wisconsin-Milwaukee. Special Report No. 35. 45 pp.
Brungs, W. A. 1973. Effects of residual chlorine on
    aquatic life. Jour. Water Pollution Control Fed.
    45 (10): 2180-2193.
Buckley, J.A., L. M. Whitmore, and R. I. Matsuda. 1976.
    Changes in blood chemistry and blood cell physiology
    in Coho salmon (Oncorhynchus kisutch) following
    exposure to sublethal levels of total residual
    chlorine in municipal wastewater. J. Fish. Res.
    Board Can. 33: 776-782.
Coventry, F. L. 1935. The conditioning of a chloramine
    treated water supply for biological purposes.
    Ecology. 16:60.
Dandy, J. W. T. 1972. Activity response to chlorine in
    the brook trout. Salvelinius fontinalis (Mitchell).
    Can. Jour. Zool. 50:405.
Dondoroff, P. and M. Katz, 1950. Critical review of
    literature on the toxicity of industrial wastes and
    their components to fish. Sew. and Ind. Wastes
    22:1432.
Duce, R. A., J. W. Winchester, and T. W. Van Nohl, 1965.
    Iodine, bromine, and chlorine in the Hawaiian marine
    atmosphere. J. Geophys. Res. 70:1775-1799.

Duce, R. A. 1969. On the source of gaseous chlorine in
    the marine atmosphere. J. Geophys. Res. 74:4597-
    4599.
Environmental Protection Agency. 1969. Water Chlorine
    (Residual) Study No. 35. Cincinnati, Ohio.
Eppley, R. W., E. H. Renger, and P.M. Williams. 1976.
    Chlorine reactions with seawater constituents and
    the inhibition of photosynthesis of natural marine
    phytoplankton. Est. and Coastal Mar. Sci. 4(2)
    147-161.
Eren, Y. and Y. Langer, 1973. The effect of chlorina-
    tion on Tilapia fish. Bamidgeh. 25:56-60.
Erikson, E. 1960. The yearly circulation of chloride
    and sulfur in nature; meteorological, geochemical,
    and pedological implications. Part II. Tellus
    12:63-109.
Giddings, J. and J. N. Galloway, 1976. The effects of
    acid precipitation on aquatic and terrestrial
    ecosystems. p. 1-32. In Literature reviews on
    acid precipitation. Center for environmental
    quality management, and the water resources and
    Marine Sciences Center. Cornell University,
    Ithaca, N.Y.
Heath, A. G. 1977. Toxicity of intermittent chlorina-
    tion to freshwater fish; influence of temperature
    and chlorine form. Hydrobiologia 56:39-47.
Holland, G. A., J. E. Lasater, E. D. Newmann, and W. E.
    Eldridge, 1960. Toxic effects of organic and
    inorganic pollutants on young salmon and trout.
    State of Wash., Dept. Fish; Res. Bull. (5) 198-214.
Junge, C. E. 1956. Recent investigations in air chemistry.
    Tellus 8:127-139.
Junge, C. E. 1957. Chemical analysis of aerosol particles
    and of gas traces on the island of Hawaii, Tellus.
    9:528-537.
Lamb, N. J. 1978. The effects of experimental chlorina-
    tion on freshwater organisms in McCorn Creek, New York.
    PhD. Thesis. Cornell University. Ithaca, New York.
    198 pp.
Lawrence, C. A. and S. S. Block, 1968. Disinfection,
    sterilization, and preservation. Lea and Febiger.
    808 pp.
Lunde, G., J. Gether, N. Gjøs, and M. B. S. Lande. 1977.
    Organic micro pollutants in precipitation in Norway.
    Atmos. Environ. 11 (11) 1007-1014.
Merkens, J. C. 1958. Studies on the toxicity of chlorine
    and chloramines to the rainbow trout. Water and
    Waste Trt. Jour. 7:150.
Moore, E. N., 1951. Fundamentals of chlorination of
    sewage and waste. Water and Sewage Works. 98:130.

Morris, J. C. 1971. Chlorination and disinfection -
    State-of-the-Art. Jour. Amer. Water Works Assn.
    63:769.
Neiburger, M., Edinger, J. G., W. D. Bonner, 1973. Under-
    standing Our Atmospheric Environment. W. H. Freeman
    and Co., San Francisco.
Palen, A. T. 1957. The determination of free and combined
    chlorine in water by the use of diethyl-p-phenylene
    diamine. J. Amer. Water Works Assoc. 49(873).
Petriconi, G. L. and H. M. Papee, 1972. On the photolytic
    separation of halogens from sea water concentrates.
    Water, Air, Soil Pollut. 1:117-131.
Schofield, C. L. 1976. Acid precipitation. Effects on
    Fish. Ambio. 5 (5-6): 228-230.
Sprague, J. B. and D. E. Drury, 1969. Avoidance reactions
    of salmonid fish to representative pollutants. In
    Advances in Water Pollution Research. Proc. 14th
    Intl. Conf. Water Poll. Res. Pergamon Press, London,
    England, 169 pp.
Stahl, Q. R. 1969. Preliminary air pollution survey of
    chlorine gas. A literature review. National Air
    Pollution Control Administration Publication No.
    APTD69-33. Springfield, Virginia: Clearinghouse for
    Federal Scientific and Technical Information. 79 pp.
Stoler, Q. J. and C. H. Hanson. 1974. Toxicity of
    chlorine and heat to pink (Oncorhynchus gorbuscha)
    and chinook (O. tshawytscha) salmon. Trans. Amer.
    Fish. Soc. 103:569-576.
Strom, G. H. 1968. Atmospheric dispersion of stack
    effluents p. 227-274. In A. C. Stern, ed. Air
    Pollution. Vol. I. Air Pollution and its effects
    (2d Ed.) Academic Press, New York.
Wanta, R. C. 1968. Meteorology and air pollution. p. 187-
    226. In A. C. Stern. ed. Air pollution. Vol. I.
    Air Pollution and its effects (2d ed.) Academic
    Press, New York.
Zafiriow, O. C. 1974. Photochemistry of halogens in the
    marine atmosphere. J. Geophys. Res. 79:1730-2732.
Zillich, J. A. 1972. Toxicity of combined chlorine
    residuals to freshwater fish. J. Water Poll. Control
    Fed. 44:212-220.
Zimmerman, G. and F. C. Strong. 1957. Equilibria and
    Spectra of Aqueous Chlorine Solutions. J. Phys.
    Chem. 79:2063.

DISCUSSION

McLEAN: Do you have any evidence for the formation of monochloramine or dichloramine?

MILLS: The only evidence that we have is in the UV spectral analysis. The next step is to try to analyze this material chemically to try to determine what the species are.

McLEAN: Are mono - and dichloramine stable under these conditions and do they not hydrolyze fairly rapidly?

MILLS: It could be. Dr. Murphy might want to comment.

MURPHY: The chloramines are actually quite a stable supply of oxidants. In fact, the chloramines in stability terms, are probably 10 to 100 times more stable than chlorine.

McLEAN: Are organic chloramines possible?

MILLS: They could be. At this point we really don't know the actual structures.

McLEAN: With nontoxic chloramines generally what one is usually talking about is a phenyl-nitrogen ring. And these are the kind of materials we're examining. These are relatively stable, but what about the simple chloramines themselves?

MILLS: No, we're not suggesting that it's a very simple chloramine. The fact is the next necessary thing is to try to determine what the components are and whether they are toxic.

TOMLINSON: What analytical method are you using for establishing this concentration of up to 1 mg/liter?

MILLS: This is primarily the DPD method. It's an orthotolidine method.

MURPHY: It's a method that was advanced by Palen and is a colorometric technique using a dye which is specific to the determination of chlorine and chlorine types of compounds. That's the other thread of evidence that may indicate that they are in fact chloramines. The DPD test upon certain sets of test conditions will test specifically for combined forms of chlorine.

McLEAN: However, that method distinguishes between chlorine and hypochlorite.

MURPHY: Yes, it does.

BLOOMFIELD: Have you tested for such common oxidants as nitric acid, hydrogen peroxide, ozone, etc?

MILLS: We haven't actually measured them in rainfall, but John Kadlecek did suggest that they have been able to measure peroxides.

KADLECEK: In mountain water there are moderately high concentrations of hydrogen peroxide, and in rain water over clean areas they maintain moderate levels, and over urban areas, the hydrogen peroxide levels drop to only a few parts per billion.

MILLS: Actually, we would be very interested in some of your analyses, at least in various parts of the country, if you do look at this component to see if in fact you do see it.

McFEE: How strong an oxidant is the chloramine? Is it as strong as hydrogen peroxide? Where does it fit?

MURPHY: It's a weaker oxidant than halogens, peroxide and ozone.

McFEE: Weaker than hypochlorous?

MURPHY: Yes. It would come right underneath hypochlorous. There is very strong evidence that it is a combined form of chlorine which is the likely measured species. As to whether it's organochloramine or the inorganic chloramine we really don't know.

SCHOFIELD: Are the high nitrates or nitric acid components a problem in terms of being absorbed in the UV range?

MILLS: Nitrates do absorb down in that range as well.

McLEAN: I think in the UV range you quoted, you find a lot of species which absorb.

MILLS: Absolutely true.

KRUPA: Do you propose to do mass spectrometry?

MILLS: Yes, I think this should be the direction.

McLEAN: I should like to inform you of a new tool that has recently been developed in Toronto. It is based on a GC mass spectrometer system with which you can do direct real time measurements at atmospheric pressure and it has a very low detection limit for oxidants and other trace species in the atmosphere.

WILSON: I notice you had begun to work out some sense of the potential sources in terms of chlorine users. I would suggest that a spinoff might be to run some estimates of loading. Also not only look at the loading in terms of the larger industries, but consider the various modes of combustion, ie., types of fuel, coal or oil, incineration, and also the types of sources, whether they are stationary or mobile, such as vehicles.

MILLS: Yes I think this would be very useful because of the trends in terms of incineration and burning of certain materials.

THE TOXICOLOGICAL IMPLICATIONS OF BIOGEOCHEMICAL STUDIES

OF ATMOSPHERIC LEAD

Robert W. Elias* and Clair C. Patterson**

*Biology Department
Virginia Polytechnic Institute and State University
Blacksburg, Virginia 24061

**Division of Geological and Planetary Sciences
California Institute of Technology
Pasadena, California 91125

ABSTRACT

More than ninety-nine percent of the 400,000 metric tons of lead entering the atmosphere each year are from anthropogenic sources. Mass balance estimates of removal by precipitation and dry deposition are roughly equivalent to these inputs, with removal by dry deposition in remote terrestrial ecosystems being about twice that of precipitation. The impact of this anthropogenic lead, described in related studies, has served to elevate the amounts of lead in the critical storage reservoirs of plants, animals and humans by factors of 2-2000. From the biogeochemical reconstruction of Pb/Ca ratios, using the principle of biopurification, the levels of natural lead can be determined. These reconstructions suggest that lead levels in humans, both North Americans and Europeans, have been elevated about 2000-fold above natural prehistoric levels.

Since laboratory animals are contaminated by factors similar to humans, no research on the biochemical effects of lead has yet been conducted using controls free of anthropogenic lead. To determine the initial biochemical response to lead above natural levels, it is essential to raise control organisms in protected environments using special techniques to remove anthropogenic lead from the air, food, and water consumed by the organisms.

The History of Lead in the Atmosphere

As metallurgical technology for the smelting of lead from sulfide ores, and for the cupellation of silver from lead developed approximately 4500 years ago, the anthropogenic input of lead into the atmosphere increased in proportion to the rate of smelting activity. Spurred by the desire for silver, these Old World metallurgists produced 400 parts of lead for each part of silver, which resulted in the innovative industrial use of this surplus lead and a marked increase in atmospheric lead. Lead production rose from 200 tons/year 2600 years ago to 80,000 tons/year during the Roman era, 2100 to 1800 years ago, then declined slightly during medieval times (14,15). As the industrial requirement for lead outstripped the quantities produced as a byproduct of silver production, the mining and smelting of lead became an established industry in its own right, increasing from 100,000 tons/year at the beginning of the industrial revolution 300 years ago to 1 million tons/year 50 years ago. With the use of lead alkyls in gasoline, the current world production is about 4 million tons/year.

As production rates grew, smelter efficiencies increased correspondingly. Losses to the atmosphere during refinement decreased from 5% in the Roman era to about 0.5% with present technology. This decrease, however, was offset sharply by the introduction of lead alkyls as antiknock additives in gasoline, resulting in a 3 to 4-fold increase in the amount of lead introduced to the atmosphere during the last 30 years. The increases in atmospheric lead concentrations due to the industrial revolution and to leaded gasoline have been documented by the chronological records of lead in ice strata (12) and pond sediments (17). Geochemical evidence based on differences between lead ore bodies and crustal rock, and evidence from mass balance inventories confirm the conclusion that atmospheric concentrations of lead have risen in direct correlation to the anthropogenic production of lead. Today, more than half the lead produced enters the human environment in some form. Nearly 10% of the mined lead (300×10^3 t/yr) enters the atmosphere during smelting processes or from auto exhausts, while other anthropogenic sources not directly connected to the lead industry account for an additional 104×10^3 tons of lead each year (Table I).

The data of Table I show that the emissions to the atmosphere from anthropogenic sources exceed those from natural sources by a factor of 400. Air concentrations in the Northern Hemisphere vary from 0.5 ng Pb/m^3 in Northern Greenland (12) to 10 ng Pb/m^3 in remote terrestrial ecosystems of North America (6,7), to 4500 ng Pb/m^3 in densely populated urban areas. The prehistoric atmosphere probably contained less than 0.1 ng Pb/m^3 (17).

TABLE I.   Atmospheric lead from natural and anthropogenic emissions.
Data modified from Nriagu (13).

SOURCE	TOTAL PRODUCTION 10^6 t/yr	LEAD EMISSION FACTOR* g Pb/kg emission	LEAD TRANSFER TO ATMOSPHERE 10^3 t/yr
NATURAL EMISSIONS			
Volcanic and windblown dust	100	1×10^{-2}	1
Sea spray	1000	$<1 \times 10^{-7}$	<0.1
Forest foliage	100	$<1 \times 10^{-5}$	<1
Volcanic sulfur	6	2×10^{-4}	0.001
Total natural emissions			1
ANTHROPOGENIC EMISSIONS			
Lead alkyls	0.4	(70%)	280
Iron smelting	780	0.6	47
Lead smelting	4	6.0	24
Zinc and copper smelting	15	2.8	42
Coal burning	3300	4.5×10^{-3}	15
Total anthropogenic emissions			408

* See Nriagu for explanation of emission factors.

Atmospheric lead can be removed by two natural processes:
precipitation and dry deposition. Removal rates by precipitation are
a function of such factors as air concentration and the intensity of
rainfall. In the remote terrestrial ecosystem discussed above, the
rate of transfer from the atmosphere was 5 g Pb/ha-yr as winter snow
and 4 g Pb/ha-yr as summer rain. These values appear to be typical of
remote areas and are much lower than values reviewed by Nriagu (13) for
urban areas, where precipitation inputs reach 100-300 g Pb/ha-yr.

Very little reliable data are available for transfer by dry
deposition. In the remote terrestrial ecosystem, total dry deposition,
on all surfaces, amounted to 20 g Pb/ha-yr. Although the mechanisms
of deposition are not well established, this research has shown that,
for most of the year, deposition rates on smooth, flat artificial
surfaces were consistently between 0.08 and 0.12 ng Pb/cm^2-day (6).
This range is typical for particles less than one micron in diameter
and is characteristic of aerosol particles transported over great
distances (4). Although these values represent the lower limit for

rates of dry deposition in terrestrial ecosystems of the Northern Hemisphere, a total input of 20 g Pb/ha-yr into 120×10^6 km^2 of terrestrial area accounts for most of the 400×10^3 t Pb/yr emitted to the atmosphere. For urban and rural systems, where lead studies have been conducted in greater detail, the total input of lead from the atmosphere is not precisely known because of some confusion between precipitation and dry deposition. Where the collecting device has been exposed to the atmosphere between episodes, precipitation measurements may actually contain a fraction of dry deposition, but the size of this fraction is unknown because the devices are not efficient collectors of dry deposition.

From this evidence, it is clear that natural ecosystems have been exposed to anthropogenic lead for more than four millenia, and that this exposure has exceeded natural inputs by severalfold since the beginning of the industrial revolution. Since there are no organisms living today which have not experienced this exposure, no research on the biochemical effects of lead has been performed using control animals with natural levels of lead.

Natural Levels of Lead in Animals and Humans

Since human tissue and environmental samples taken from rural ecosystems generally contain only 10-50% of the lead of comparable samples in urban environments, these rural samples have been taken to represent the 'natural background' levels of lead for animals and humans. It would seem reasonable, therefore, to conclude that humans in urban environments are contaminated by a factor of 2 to 10. But such is not the case. The levels of natural lead in contemporary animals and humans can only be estimated from theoretical reconstructions based on biogeochemical principles. *The key principle is that of biopurification, which states that, at successively higher trophic levels, reservoirs of nutrient metals such as calcium become progressively purified of non-nutrient metals with similar chemical properties (8).*

The principle of biopurification has been used to demonstrate the degree of contamination of plants and animals in remote ecosystems by the analogy of lead to the non-pollutant metals, strontium and barium, expressed as an atomic ratio with respect to calcium (11). This analogy predicts the maximum value for the atomic ratio of natural Pb/Ca for certain plant and animal tissues assuming no contamination on the surfaces of plants and animals and in soil moisture, the nutrient medium of plants. To obtain the natural value, the Pb/Ca ratio is systematically corrected for known contaminant lead, such as lead in soil moisture (originating from litter and humus) and lead on plant surfaces and animal fur (7). Since these measurements do not include the effects of inhalation or grooming, the actual degree of contamination is probably greater than estimated. The graphical summary of this data in Figure 1 illustrates the principle of biopurification of strontium, barium and lead with respect to calcium, showing the predicted and actual atomic Pb/Ca ratios for plants, animals and humans. The contamination factor for herbivores and carnivores in remote ecosystems represents the minimum degree of contamination for lab animals. Three factors would increase this estimate of the degree of contamination for experimental laboratory animals:

1) Measurements of lead inhaled by breathing or ingested by grooming, not included in the above study, would further decrease the corrected Pb/Ca ratio for animals in natural ecosystems;

2) The laboratory animals breathe air 10 to 100 times more contaminated than in this remote ecosystem;

3) The food and water consumed by laboratory animals is generally contaminated (10), perhaps by a factor of ten more than the animal food of this remote ecosystem;

The natural lead in humans is even more difficult to determine. Although the atomic Sr/Ca and Ba/Ca ratios observed in Americans are nearly identical to those of carnivores in the remote ecosystem, the observed Pb/Ca ratio of humans is two orders of magnitude greater than that of the carnivores (Figure 1). Since these carnivores are contaminated at least 50-fold, it would seem that the appropriate correction for humans is 5000. Data from another source (9) show that ancient Peruvians, exposed to very little anthropogenic lead 1600 years ago, have a skeletal Pb/Ca of 3×10^{-8}, compared to that of the average American today, $2100-3500 \times 10^{-8}$. This Peruvian bone value sets an upper limit to the prehistoric human Pb/Ca ratio, since a small but undetermined amount of anthropogenic lead was present. The adjusted carnivore bone value, if fully corrected for inhalation and grooming, would represent the lower limit for humans, by analogy to the Sr/Ca and Ba/Ca ratios. It is likely that the actual prehistoric human value is greater than the carnivore value, due to an age accumulation factor. In this study and others, younger animals of the same population generally show a lower Pb/Ca ratio. Projecting this age factor to the average age of Americans gives a predicted prehistoric value of 1.5×10^{-8} for the Pb/Ca ratio, or 2000 times less than the present observed values.

Biochemical Effects of Lead

These biogeochemical reconstructions of natural lead in animals and humans raise the question: what is the initial biochemical target

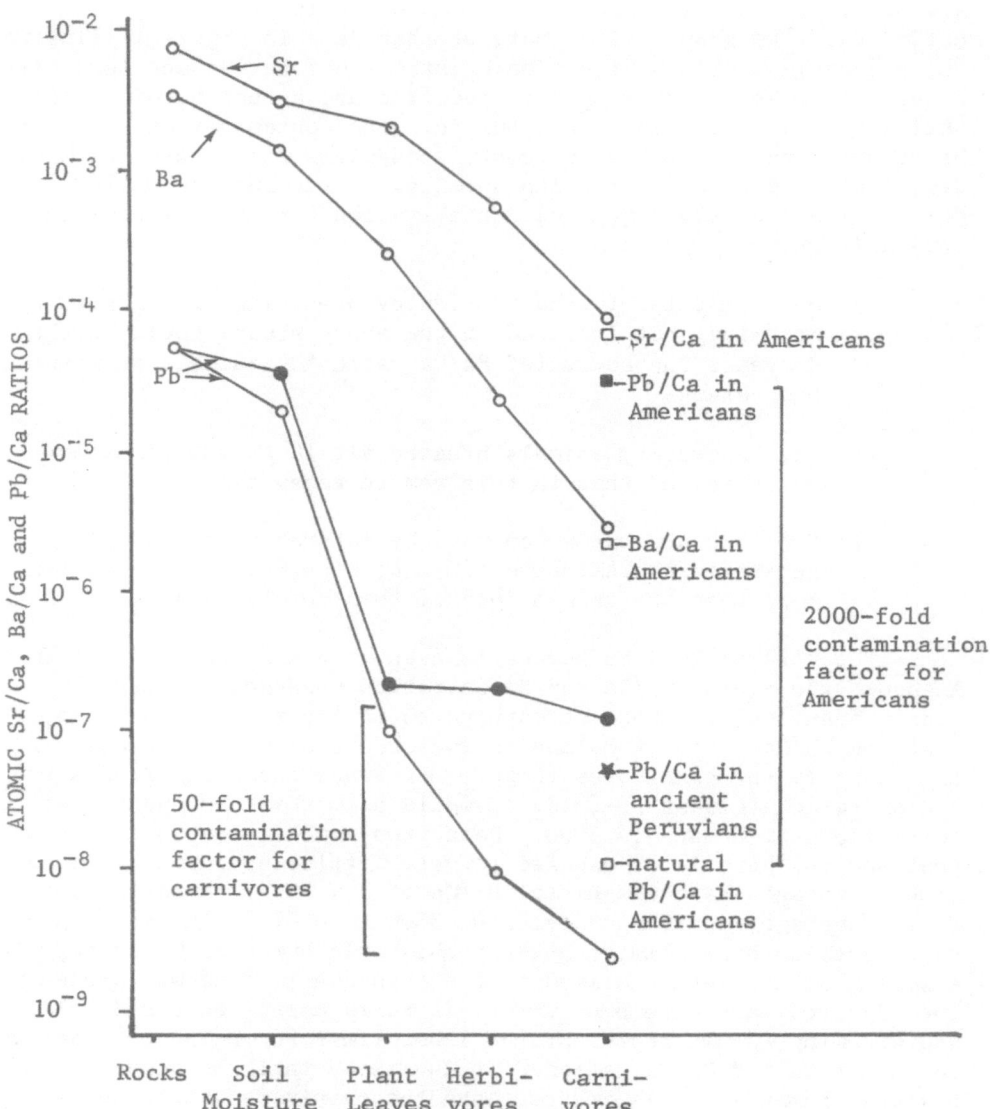

Figure 1.   BIOPURIFICATION IN A REMOTE TERRESTRIAL ECOSYSTEM AND IN
HUMANS.   The observed Pb/Ca ratios in this ecosystem (●) were
corrected for known contaminant lead (o), showing that these carnivores are contaminated 50-fold.   The atomic ratios for Americans (□)
have been normalized to crustal abundances and show that, whereas
the Sr and Ba ratios for Americans are similar to those of carnivores, the Pb ratio in Americans is 5000 times higher.   Since the
ancient Peruvian Pb/Ca ratio sets the upper limit for humans, the
natural Pb/Ca ratio for Americans is probably between Peruvians
and carnivores.   Data are from (9,11).

of lead at any concentration above natural background levels? From
the discussion above, it is possible to conclude that:

1) no biochemical studies on the effects of lead have been
   conducted using control animals free of anthropogenic lead;

2) laboratory animals have undergone many generations of
   breeding during continuous exposure to excessive levels of
   anthropogenic lead.

Although classical lead poisoning has been studied throughout
much of medical history, the effects of low level continuous exposure
are not known. Whereas high doses of lead are known to affect the
renal, hemopoietic and central nervous systems, lower levels of lead
are believed to affect mostly the central nervous system. The hyper-
kinesis reported by David (3) was not fully substantiated, according
to Posner et al. (16) and the brain damage during fetal development
and early childhood, suspected by many authors, defies agreement due
to inconsistent results or non-parallel experiments (16). Models
for the use of animals in lead toxicity studies were discussed by
Bornschein et al. (1), with the general conclusion that the practice
of using extremely high exposure levels prohibits interpretation of
the results due to high body burdens and severe growth retardation.
These authors recommend behavioral and biochemical studies at continuous low-level lead exposures using animal models which emphasize the
precise definition of:

1) exposure levels;
2) duration of exposure;
3) lead body burden;
4) nutritional status;

They would use neonatal rats and mice, conducting neurochemical
studies to detect such behavioral defects as learning deficits,
decreased general activity levels and altered habituation rates. To
determine the correct exposure levels, however, it would be necessary
to compare the experimental animals to control animals with natural
body burdens.

Another promising biochemical approach, which may be linked to
the deterioration of the hemopoitic system, involves the subcellular
effects of lead on mitochondria. In experiments described by
Brierley (2), lead at the micromolar level was shown to have a
variety of effects centering around the inhibition of respiration and
breakdown of the regulation of mitochondrial membrane permeability.
The uptake of lead into the mitochondrial matrix is closely related
to the membrane transport mechanism(s) for calcium. Since lead is
known to inhibit several enzymes of the Kreb's cycle and the electron
transport system, heme synthesis may be inhibited at the mitochon-

drial level by the interference of lead on succinyl CoA synthesis
(2). Although the levels of lead used in these experiments are still
orders of magnitude above natural background levels, it is reasonable
that sensitive biochemical tests can be developed to detect abnormal
responses of mitochondria at lead levels slightly above natural
levels. In the absence of adequate neurochemical tests for effects
on the central nervous system, the mitochondrial approach might be
preferable for determining the zero response level for chronic
exposure to neo-natural levels of lead.

Alternative Experimental Models

Studies of the types described above must be re-evaluated to
include control organisms which have body burdens of lead at natural
levels (100-2000 times lower than present levels), which breathe air
at natural levels, and which consume food and water containing lead
at natural levels. Furthermore, since these studies would involve
lead measurements at subnanogram levels, it is essential that they
be carried out under ultra-clean conditions using the best clean-lab
technology to avoid artifact contamination during sample handling
and analysis.

If the mitochondrial approach is used, then unicellular green
algae would seem to be ideal experimental organisms. Their simple
nutrient medium lends itself to chemical purification and the
established procedures for growth are within the limits of clean-lab
technology. The measurement of responses of isolated mitochondria
may be problematic due to the limitations of quantities used and
respiratory control.

The development of a suitable neurochemical test for the zero
response level of lead requires the availability of uncontaminated
small mammals maintained under clean-lab conditions. Although the
technical problems of keeping laboratory animals in a clean room are
great, they are not insurmountable. The question does arise, however,
as to whether currently available experimental animals would be
suitable for this treatment, even after several generations in the
clean room. Since these animals have undergone many generations of
forced reproduction under highly polluted conditions, it is necessary
to assume that natural selection has favored lead resistent individuals of the population. Such organisms would not show the same zero
response level as their native counterparts in rural or remote
ecosystems, nor could they be compared to humans. It may be necessary
to begin again with wild mammals taken from relatively unpolluted
environments.

These two experimental approaches will, at long last, bridge the
gap between biogeochemistry and trace metal toxicology. The initial

response, whether behavioral or physiological, will provide a baseline for further research in the field of neurochemistry and a
guideline for risk assessment to governmental agencies required to
make decisions regarding acceptable levels of lead pollution.   If
this initial response occurs at levels below those found in
laboratory control animals today, a re-evaluation of all research
involving experimental animals will be in order. Likewise, if these
studies show that the zero response level for humans is much lower
than currently accepted values, the present risk assessment determinations are based on the wrong information (18).

References

1. Bornschein, R.L., I.A. Michaelson, D.A. Fox and R. Loch. 1977.
   Evaluation of animal models used to study effects of lead on
   neurochemistry and behavior. *In*: Biochemical Effects of
   Environmental Pollutants. S.D. Lee (ed). Ann Arbor Science,
   Ann Arbor, Michigan.

2. Brierley, G.P. 1977. Effects of heavy metals on isolated mitochondria. *In*: Biochemical Effects of Environmental Pollutants.
   S.D. Lee (ed). Ann Arbor Science, Ann Arbor, Michigan.

3. David, O.J. 1977. Association between lower level lead concentrations and hyperactivity in children. *Environ Health
   Perspectives* 7:17.

4. Elias, Robert W. and Cliff I. Davidson. 1979. Theoretical
   mechanisms for the deposition of aerosol particles on natural
   surfaces. *Manuscript in preparation*.

5. Elias, Robert W., Todd K. Hinkley, Yoshimitsu Hirao and Clair C.
   Patterson. 1976. Improved techniques for studies of mass
   balances and fractionations among families of metals within
   terrestrial ecosystems. *Geochim Cosmochim Acta* 40:583-587.

6. Elias, Robert W., Yoshimitsu Hirao and Clair C. Patterson. 1977.
   Impact of present levels of aerosol lead concentrations on both
   natural ecosystems and humans. International Conference on
   Heavy Metals in the Environment, Toronto, Ontario, Canada.
   T.C. Hutchinson, (ed). Symposium Proceedings Volume 2, Part 1,
   pp 257-271.

7. Elias, Robert W., Yoshimitsu Hirao, and Clair C. Patterson. 1979.
   The magnitude and origin of dry deposition fluxes of lead in a
   remote subalpine ecosystem. *Manuscript in preparation*.

8. Elias, Robert W. and Clair C. Patterson. 1979. The biopurification of calcium in natural ecosystems. *Manuscript in preparation.*

9. Ericson, Jonathan E., Hiroshi Shirahata and Clair C. Patterson. 1979. Skeletal concentrations of lead in ancient Peruvians. *New Engl J Med.* In Press.

10. Fox, J., F. Aldrich and G. Boylen. 1976. Lead in animal foods. *J. Toxic Environ Health* 1:461-467.

11. Hirao, Yoshimitsu, Robert W. Elias, Linda Newbern and Clair C. Patterson. 1979. Anthropogenic lead aerosols circumvent the biopurification of calcium in natural subalpine nutrient pathways. *Manuscript in preparation.*

12. Murazumi, M., T.J. Chow and Clair C. Patterson. 1969. Chemical concentrations of pollutant lead aerosols, terrestrial dusts, and sea salts in greenland and Antarctic snow strata. *Geochim Cosmochim Acta* 33:1247-1294.

13. Nriagu, Jerome O. 1978. Lead in the Atmosphere. *In:* The Biogeochemistry of Lead in the Environment. J.O. Nriagu (ed). Elsevier, North-Holland.

14. Patterson, Clair C. 1971. Native copper, silver, and gold accessible to early metallurgists. *Am Antiq* 36:286-321.

15. Patterson, Clair C. 1972. Silver stocks and losses in ancient and medieval times. *Econ Hist Rev 2nd Series* 25:205-235.

16. Posner, Herbert S., Terri Damstra and Jerome O. Nriagu. 1978. Human health effects of lead. *In:* The Biogeochemistry of Lead in the Environment. J.O. Nriagu (ed). Elsvier, North-Holland.

17. Shirahata, Hiroshi, Robert W. Elias, Minoru Koide and Clair C. Patterson. 1979. Chronological variations in concentrations and isotopic compositions of anthropogenic atmospheric lead in sediments of a remote subalpine ecosystem. *Geochim Cosmochim Acta* In Press.

DISCUSSION

TORIBARA: Are you familiar with our behavioral toxicology group down here? Dr. Weiss and Dr. Laties are two people that are involved in this and I think you people ought to correspond or get together and maybe come up with something.

ELIAS: Well, I certainly would like to talk to these people. We have worked out the techniques for measuring lead at these low levels. We know little about toxicological techniques.

McFEE: How would you propose to produce plants and animals that were uncontaminated.

ELIAS: They would have to be kept in a clean lab situation, a protected sanctuary, free of anthropogenic lead.

McFEE: Where would you get the food, isn't it going to be contaminated?

ELIAS: We would first try preparing algae pellets or similar food that these animals can eat, which is an extremely difficult problem, since it would also have to be prepared under clean lab conditions.

EVANS: On your first table you showed that forest foliage produces some lead. Can you tell us about that?

ELIAS: Yes, there are studies which show small amounts of lead released by foliar emission.

EVANS: They come from the soil and through the plant?

ELIAS: Apparently so. Now, this represents only a tiny fraction of the lead which is on the surface of the needles. And when you calculate the rate of emission, it's approximately 1/10th of one percent of the amount of deposition on needles. So it is a tiny fraction.

ANDERSON: Are you comfortable with the fact that when you wash your leaf surfaces with methanol and acid, that you have not in any way changed the collection efficiency of the vegetation surface by destruction of parts of it or change in its behavior?

ELIAS: Only one time did we actually wash these in the field to determine rates of deposition, and we found the same rate of deposition. The rest has been measured since emergence of the tissue, and we get the same rate of deposition. Now we know, from studies in Virginia, we can measure only between periods

of rain. Our study of rain removal shows that much of this
lead that's removed from the surface of the needle is trapped
at the base of the needle, and is not washed all the way down
to the soil.

McLEAN: In case anyone is thinking of the comments that Arne
Jernelov and I made yesterday about the data on mercury in the
Greenland ice sheet and examining your data with that in mind,
I'd like to say there's absolutely no question that the
methodology that was used for lead in the Greenland ice sheet
was a good methodology, because in fact that's what it was
originally set up to do. And I think one can therefore believe
the kind of data on lead in the Greenland ice sheet whereas I
hope no one's going to believe the data on mercury.

ELIAS: Well, I'm certainly glad you pointed that out. I had
meant to.

McLEAN: The comments on container materials also stand. In
fact, mercury is unique as far as container materials are
concerned.

     One point about your data I would like to raise is I've
seen a recent study in Norway on the examination of lead
fallout on snow pack, measuring the dry deposition by measuring
the concentrations between events. Are you aware of this
work? Because the amount of mercury they were estimating from
dry fallout in remote areas there was considerably less than
the kind of proportions that you were talking about in Yosemite.

ELIAS: I'm not familiar with that study. But in our last
measurement of the snow pack, which we haven't yet analyzed, we
did do this on an event basis. I don't know whether it will be
successful, but we sampled by snowfalls rather than just areas
of snowface. We may be able to confirm or reject that
hypothesis.

McLEAN: You measured the snow as it was coming down?

ELIAS: No, we have only one snowfall collected as it's coming
down, but on the snowface you can easily identify the different
events by the ice crystal structure. And we use this as a
guideline for sampling.

EVANS: I may be mistaken but I was informed that during a
spring melt, or even during the winter when you have snow
packs, that if you have a slight elevation in temperature you
may have differential leaching of parts of the snow pack from
very high in the snow pack through into the ground water before

you actually melt pure water, shall we say. Such that depth
in the snow pack may not represent all the time increments of
snow. Is that correct?

ELIAS: This may be correct but I won't know until we complete
our present study. So far what we've done is to integrate our
values the full depth of the snow pack and subtract what we
measured as deposition on artificial plates from these values
to determine the total input by snow.

JERNELOV: I think it has been demonstrated in many cases that
if you have a sunny day, for example, you will find that some
part of the snow will melt, even if it's minus 10 outside. And
that's the part where you have the highest concentration of
salt because of the lowering of the freezing point. And that
means in this situation that the contaminated part of the water
moves downwards. Then how far down it goes is a question.

SESSION V:
CONTROL PROBLEMS: MONITORING SYSTEMS
AND LEGAL ASPECTS

MONITORING ARIBORNE CONTAMINANTS

G. M. Hidy and P. K. Mueller

Environmental Research & Technology
Westlake Village, California

ABSTRACT

This paper gives a brief survey of air quality monitoring
practice in the United States. The reasons for monitoring are
outlined, and the concept of monitoring systems is introduced.
This "system" includes measurement instrumentation, calibration,
data recording and transmission, data management including validation,
processing and archiving, quality control and quality assurance.
Techniques for measurement of ambient concentrations and surface
deposition are surveyed. Examples of user displays illustrate
management of monitoring data for different purposes. Finally, the
dilemmas inherent in limitations of measurements for monitoring,
data accessibility, and the user identification are discussed in
the light of future directions in monitoring.

INTRODUCTION

The increasing awareness of air pollution has stimulated
widespread efforts to measure the chemical composition of minor
atmospheric constituents in the troposphere. These include both
gases and airborne particles. For purposes of this paper, radioactive material will not be included. The regulation of certain
community pollutants, such as carbon monoxide, sulfur dioxide,
and suspended particulates has a legal requirement to observe
indefinitely pollutant concentrations over much of the United States.
The widespread measurement of designated pollutants in the air or
accumulating at the ground for time periods of longer than a year,
is a conventional definition of air quality monitoring. However,
monitoring could extend to the air-water-soil components and to a
"total" ecosystem as well. This kind of measurement generally is
not done in the United States.

Aside from certain regulatory requirements, why should air
quality monitoring be done? In general, monitoring is needed to
provide facts or deductions about

- the long-term trends in pollution concentrations,

- the geographical distributions in pollution levels,

- the cumulative impact of pollution exposure,

- the establishment of "baseline" conditions in areas where
  change may take place, and

- the relationship of air quality to emissions.

Such information is needed to evaluate the long term evolution
of pollution levels, as well as evaluate the effectiveness of control
strategies. Clearly, monitoring must play an important role in providing information for rational implementation of regulatory policy.
Such data are also of great scientific interest for establishing
the climatology of air chemistry in the lower atmosphere.

In the United States, the evaluation of air quality impact has
focused on relating impacts to concentrations of pollutants. Thus,
virtually all present air monitoring in this country focuses on concentrations, for example, in units of micrograms per cubic meter
of air. There are at least two other useful measures of pollution.
The first concerns the total burden of suspended material over a
distance in the atmosphere. This can be measured vertically, or it
can be measured along a horizontal path in units of (ppm) x (km).
In the case of visible light transmission, this is basically related
to an extinction coefficient, or to impairment in visibility.

The second measure deals with the deposition rate of material
at the earth's surface. Deposition is measured in terms of the rate
of accumulation per unit area per unit time (mg/m^2-day). It may
take place by several processes: dry fallout, absorption, or wet deposition associated with dew or precipitation. Because of the potential
for certain long term ecological effects and degradation of buildings,
deposition has been an important measure of pollution in Europe
(e.g., OECD[1]). In principle, the deposition rate of material at the
ground can be related to the ground level concentration by the formula:

$$\text{Deposition Velocity (m/sec)} = \frac{\text{Deposition Rate in } \mu g/m^2\text{-sec}}{\text{concentration at about 10 meters or less in } \mu g/m^3}$$

Thus, knowing the deposition velocity, one can estimate the deposition rate from atmospheric concentrations of pollutants. Many
investigators have attempted to develop data for the deposition
velocity as a function of surface and aerometric properties to
complete this link since ambient concentrations are easier to measure
than the deposition rate.

To be useful, air quality monitoring requires measurements over
different spatial scales. Generally, high concentrations of
pollutants emitted from sources are confined to zones within tens
of kilometers of the sources. The range of "influence" of pollutants
which are products of atmospheric chemical reactions of emissions
(e.g., ozone and particulate sulfate) may extend hundreds of kilometers from the source of the precursors depending on the balance
between production and loss processes. Generally, the zone of
deposition of pollutants is expected to extend over great distances
from sources for similar reasons. There is evidence for long
distance influences extending several hundred kilometers downwind
in studies from northwestern Europe (e.g., OECD[1]). Thus, monitoring
for pollution impact at ground level can place a range of requirements
on systems. Some of these may take on continental-intercontinental
proportions where deposition processes are involved (e.g., Hidy, et
al.[2]).

Historically, air quality monitoring for nonradioactive constituents in the United States has suffered from lack of long-term
commitment and has emphasized urban scale, near source concentrations.
Presently, the Clean Air Act of 1968 and its amendments require state
and local governments to monitor ambient concentrations for several
regulated pollutants. This nationwide system builds on the older
National Air Surveillance (sampling) Network initiated in the 1950s.
Observations of total deposition were abandoned in the 1960s because
of difficulties of data collection and interpretation for ambient air
concentrations. U.S. monitoring for precipitation chemistry was
undertaken on a very limited basis in the 1950s to early 1960s
(e.g. Junge and Werby[3], Lodge, et al.[4]). Unlike the Scandinavians,

American atmospheric scientists were unable to find support for even
the very limited precipitation network. This is widely recognized
as possible short-sightedness in view of present questions about the
impact of rainwater acidity on ecological processes in the northeastern
United States.

In this report, aspects of the current United States air
monitoring practice are outlined in further detail. Present limitations and needs are considered with recommendations for improvements over the next few years.

CURRENT PRACTICE

The technology of air quality monitoring has emerged as a
measurement "system", which consists of (a) an instrument grouping,
(b) a calibration device, (c) a data recording and transmission
component, (d) data management – including validation, processing
and archiving, (e) quality control, and (f) quality assurance
program. The system may range from nothing more than a high-volume
particle filter sampler and gas bubblers, with intermittent collection
and laboratory analysis and archiving, to expensive, elaborate arrange-
ments of several instruments housed in a station. An example of the
latter type is shown in Figure 1. In these stations, for example,
measurements of sulfur dioxide, nitrogen oxides, ozone and carbon
monoxide may be collected and recorded by automatic, computerized
devices, with on-line transmission to a central data facility. In
addition to gas measurements, particle measurements may be carried
out for total mass concentration and size distribution, and samples
may be collected for laboratory chemical analysis. A listing of typical
measurement for ambient concentration measurements is given in Table 1.
These techniques are adopted for observing both short and long term
population exposures. In Table 2 is a list of methods for multiday
static sampling. Although these methods have the units of a surface
flux, they are not necessarily interpretable as such. With the
exception of particle fallout, these methods should represent an
integrated average concentration in air near a surface over several
days. They are analogous to an integrated exposure of a pollutant.

There is no accepted way of quantitatively monitoring for dry
deposition of either gases or particles. Experimental devices have
ranged from collection plates and dust-fall jars to inventions for
determining instantaneous vertical fluxes at ground level (e.g.,
Platt[6]). Monitoring for wet deposition currently is based on collectors
such as those shown in Figure 2. The device in Figure 2A consists of
two buckets, one of which is open for dry fallout and the other is
opened automatically with onset of precipitation. A precipitation
sample is collected in a chemically inert vessel which is sealed by
the cover after an event. A rain gauge is often used with this device
to record the quantity of precipitation. Samples are removed either
at regular intervals (which should not exceed 24 hours), or on an

Fig. 1A. A modern air quality monitoring station for community
pollutants regualted under the Clean Air Act.

Fig. 1B. Interior of air monitoring station showing gas monitoring
instrumentation, calibration and data recording equipment.

Table 1. Air Quality Measurement Technology Available for Obtaining Population Exposure Data (Source: Mueller et al.[5])

Pollutant	Health Effects Periods		Associated Available Measurement Technology	
	Short-Term	Long-Term	Short-Term	Long-Term
SPM**	24-hr, shorter for very irritating components	Multiday, seasonal, and annual	Flirtation - laboratory chem. continual tape - light absorption/reflection/β-ray attenuation - laboratory chem. nephelometry rotating rod collector - laboratory chem.	Same fallout jars - laboratory chem.
SO$_2$	1-hr to 24-hr	multiday, seasonal, and annual	continuous monitors, various principles bubblers - laboratory chem. impregnated tube or filter sampling - laboratory chem.	bubbler - laboratory chem. impregnated filter sampling
CO	1-hr to 24-hr	none	continuous monitors, various principles bubblers - laboratory chem. impregnated tube or filter sampling - laboratory chem. impregnated plate/diffusion tube exposure - lab. chem.	none needed
NO$_x$	1-hr to 24 hr	multiday, seasonal and annual	continuous monitoring, various principles	impregnated filter sampling - laboratory chem.

Table 1. (Cont'd).

| Pollutant | Health Effects Periods | | Associated Available Measurement Technology | |
	Short-Term	Long-Term	Short-Term	Long-Term
O₃	1-hr to 3-hr	none	bubblers - laboratory chem. impregnated filter sampling - laboratory chemistry continuous monitoring, various BAKI* continual bubblers - laboratory chem. rubber strip exposure	impregnated plate/diffusion tube exposure - laboratory chem. none available except vegetation damage

event basis and shipped in chilled, sealed containers for subsequent laboratory analysis. The device in Figure 2B is refrigerated continuously to maintain sample cooling. It is arranged with a seal to open automatically when precipitation takes place. The purpose of cooling is primarily to preserve sulfite and nitrite in samples.

Laboratory chemical analysis of particulate material, following the collection of samples, typically includes a selection of the listed parameters in Table 3. The reasons for monitoring these constituents are also listed in the table. These reasons are normally related to surveillance of certain materials which are identified with major sources, and are believed to be related to adverse health and ecological impacts. Recent efforts in rain chemistry require analysis for the species listed in Table 4. These parameters reflect interests concerning causes and sources of acidity in precipitation. Certain key trace metal constituents provide information on sources.

Data obtained from monitoring is transmitted and archived, after validation, to banks where it can be accessed by users. The management of such data often represents a major bookkeeping commitment in the monitoring process.

The validation of data also is an important step in the process. Validation begins at the station with standard operating procedures for self-consistency of operations and maintenance and with systematic calibration of instruments. Air quality instruments are not particularly reliable so that frequent, careful calibration is required for useful data acquisition. Validation of data is continued after transmission before archiving. To maximize reliability, validation involves a dedicated scientist and engineer whose experience permits identification of possible errors or inconsistencies in reported observations. Finally, the validation process is cross-checked with station performance audits and laboratory intercomparisons to identify discrepancies in analytical procedures. This chain of activity generally makes air quality a rather expensive commodity, a fact that is not appreciated by many users.

In the data acquisition and management process, the user community and the user specifications need to be kept at the forefront. This is a crucial motivational factor, and the lack of it can often fatally damage monitoring programs. The cost of monitoring is so great, that continuing cost-user benefit tradeoffs for accuracy, precision by parameter days, and timeliness should be considered. When long-term data are available, the user community frequently expands or contracts depending on the questions raised by current problems. Yet experience dictates that there <u>always</u> will be a need for long-term environmental data which is reported using self-consistent methods since most environmental issues have long lifetime impacts which are poorly understood.

Table 2. Available Measurement Technology for Obtaining Population
          Exposure Data for Multi-Day Static Sampling
          (Source: Mueller et al.[5])

Pollutant	Method	Measurement	Applicability
Particles	Fallout	$mg/m^2/day$	Requires 1 to 30 days of sampling. Indicates population exposures to large particles. Provides some correlation with 24-hour filter sampling and population exposure in places where particulate emission controls are poor. Difficult to relate to health effects.
SO_2	Diffusion plate or tube with reactive surface such as $K_2 CO_3$	$mg/m^2/day$	Requires 1 to 30 days of exposure followed by laboratory analysis for sulfates. May respond to other compounds of sulfur. Very cost-effective method. Correlative to volumetric methods, but measurement bias can be significant.
CO	Diffusion plate or tube with a palladosulfite reactive surface	optical density	Requires 1 to 24 hours of exposure followed by light transmission or reflectance measurement at a laboratory. The method is correlative to volumetric measurements for CO.
O_3	Diffusion surface, neoprene rubber strip	crack length length	Requires 1 to 24 hours of exposure followed by observation under a large magnifier. Correlative with volumetric measurements for O_3. The only available noncontinuous method for O_3 that does not require on-site servicing.
NO_x	Diffusion plate or tube with reactive surface such as $K_2 CO_3$	$mg/m^2/day$	Same as for SO_2. Laboratory analysis is for nitrates. Amount of NO adsorbed may depend on NO_2 to NO ratio and is not known. HNO_3 will also be adsorbed. Output will therefore be some function of total NO_x and HNO_3, which may be correlative to volumetric methods.
NO_2	Diffusion plate or tube with arsenite reactive surface	$mg/m^2/day$	Same as for SO_2. Laboratory analysis is for nitrate. Some NO may also be observed. Output should be a predictable function of specific volumetric measurement for NO_2.
HNO_3	Diffusion plate or tube with reactive surface such as nylon or NaCl.	$mg/m^2/day$	Same as for SO_2. Laboratory analysis is for nitrate. NO_x is not a serious interference, and output should be a function of volumetric measurements for HNO_3.

Fig. 2A.   Photograph of DOE/HASL Rain Sampler   Fig. 2B.   The DOE-MAP3S Refrigerated Precipitation
           (Photo Courtesy of C. Hakkarinen)   Collector (Photo Courtesy of C. Hakkarinen)

Table 3. Parameters Commonly Analyzed for in Airborne Particles

Parameter	Method	Reason
Total Mass Concentration	Hi-Vol Filter - Gravimetric	Regulatory Compliance
Total Mass Concentration (<15 μm diameter)	Hi-Vol Filter - Dichotomous Sampler - Gravimetric	Representative of inhaled particle fraction
Total Mass Concentration (<3 μm diameter)	Hi-Vol Filter - Dichotomous Sampler - Gravimetric	Refined particle concentration-characteristic of deep lung penetration; visibility impairment
Sulfate	Filter - Water Extraction and colorimetric analysis	Sulfur gas oxidation product
Nitrate	Filter - Water Extraction and colorimetric analysis	Nitrogen oxide gas oxidation product
Ammonium	Filter - Water Extraction and colorimetric analysis	Principal cation-neutralizes sulfate and nitrate free acid
Lead	Filter - XRFA*	Auto exhaust
Iron	Filter - XRFA	Soil; industrial
Manganese	Filter - XRFA	Industrial
Zinc	Filter - XRFA	Industrial
Chlorine	Filter - XRFA	Related to auto lead; sea salt; industrial
Calcium	Filter - XRFA	Construction; soil
Bromine	Filter - XRFA	Related to auto lead
Vanadium	Filter - XRFA	Fly ash
Sodium	Filter - Plasma Emission Spectr.	Soil; sea salt
Potassium	Filter - Plasma Emission Spectr.	Soil; cation
Magnesium	Filter - Plasma Emission Spectr.	Soil
Silicon	Filter - Neutron Activation Anal.	Soil; fly ash
Nickel	Filter - XRFA	Fly ash; carcinogen
Total Carbon	Filter - Thermometric	Combustion; light absorption
Organic Carbon	Filter - Thermometric	Combustion; secondary aerosol

*X-Ray Fluorescence Analysis

Table 4. Parameters Frequently Specified for Precipitation Chemistry

Parameter	Reason
Acidity, pH	Absorption of Acid Species in Air
Conductivity	Ion Content
Sulfate	Acid Gas Absorption
Sulfite	Acid Gas Absorption
Nitrate	Acid Gas Absorption
Nitrite	Acid Gas Absorption
Organics	Weak Acid Content
Ammonium	A Principal Cation with H^+
Alkali Metals (Ca, Mg, Na, K)	Acidity Neutralization – Soils
Heavy Metals (Pb, Ni, V, Fe, Mn,...)	Potential Toxicity to Soils in Deposition

The needs of users now encourage data processors to provide information in summary forms. These may be in the format of time charts, for example, Figure 3. Here data for the acidity in rain-water in terms of pH are shown for several stations in New York State. Such data indicate the variations or trends in rain acidity over several years at a few sites. But the spatial and temporal coverage in most of the USA is sparse.

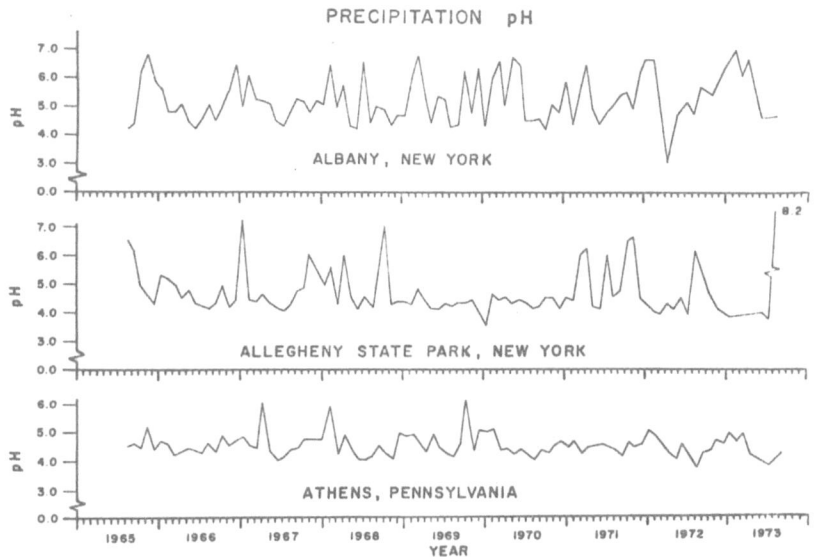

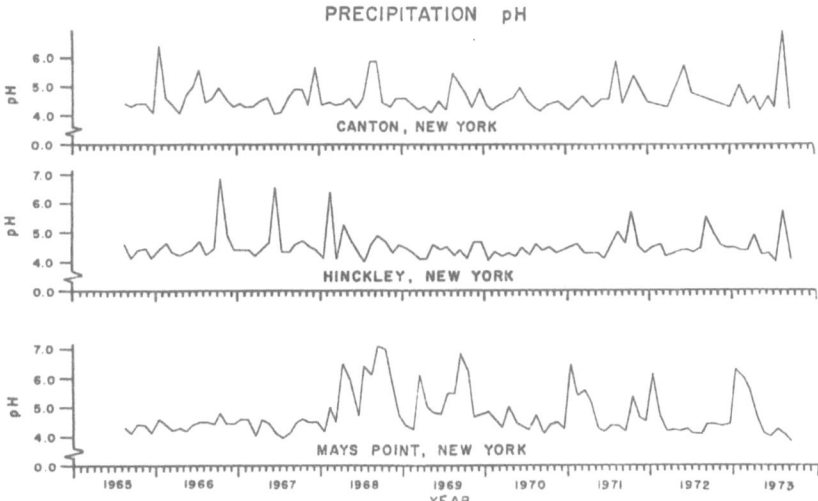

Fig. 3.   The pH of Rainwater Measured at Several Sites
          in New York State over a Period of Years
          (Source: U.S. Geological Survey)

Another useful display format provides cumulative frequency distribution. An example is shown in Figure 4, in which airborne sulfate data taken in the mid-1970s in the New York City area is compared with rural sites in the greater northeastern United States. The data shown in this way suggest similar annual distributions in a major city and at rural stations far to the west of the city. The data also show the occurrence of levels of sulfate higher than, say, 20 μg/m^3 was less than ten percent over the region during 1974-1975. This display gives a useful picture of the frequency of occurrence of pollution events relative to a median condition.

A third type of display may be calculated from network data, to give the spatial distribution of occurrences in pollution levels. As an example, the occurrence of particulate sulfate days higher than 10μg/m^3 is shown in Figure 5 for the month of August 1977 based on the extrapolation of 54 rural stations taking 24-hour samples in the northeastern USA. The extrapolation here is based on an inverse distance squared decrease around each station, summed and averaged for 80 x 80 km squares over the region (e.g., Mueller, et al.,[7]). This display illustrates a geographic scale for exposure levels over the eastern United States for a selected pollutant, which has been identified as having a regional influence (e.g., USEPA,[8]). Other types of displays are available which relate the geographic extent of daily exposures.

A final example is included for the estimate of precipitation chemistry. An estimated annual-average distribution of acidity (pH) in precipitation is drawn in Figure 6 from available stations operating in the region. Interestingly, the deposition of acidity in rain suggests a large-scale zone of minimal pH over much of the northeast. This region is partially coincident with the occurrence of high particulate sulfate concentrations at the ground, but not with high total nitrate levels.

Each of the examples shown represent a user "view" of basic data, which have an environmental application in mind. Data management may be required to provide such "summary" information rather than simple tabulations, which rarely provide an interpretative "message" in themselves. However, such simple displays serve only as a guide for more comprehensive analyses of the many factors and interactions underlying environmental quality data. Powerful numerical analysis schemes include source modeling, receptor modeling and factor analysis.

In the United States, air quality monitoring is underway primarily through the requirements of the Clean Air Act and its amendments. The current picture of air quality monitoring data resources in the United States are summarized in Table 5. As expected, the government agencies are the major underwriters of

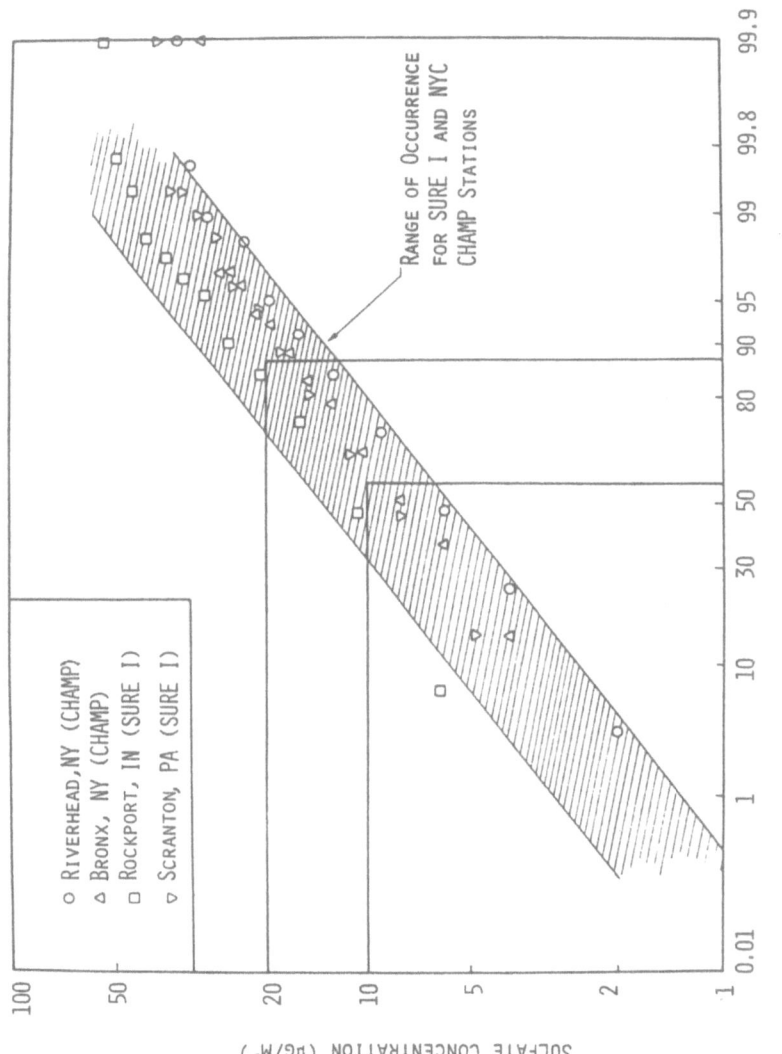

Percent of Daily 24-Hour Values Less than the Stated Concentration

Fig. 4.   Frequency of Occurrence of 24-hour Sulfate Concentrations. Based on 1974-1975 data
from the EPA Community Health Air Monitoring Program (CHAMP) and the Sulfate Regional
Experiment (SURE) (Mueller et al.[7]). The band indicates the range of data from 12 rural
and 2 urban sites. The individual site data is given for four sites to illustrate the
comparison between urban and rural sites in the NE USA.

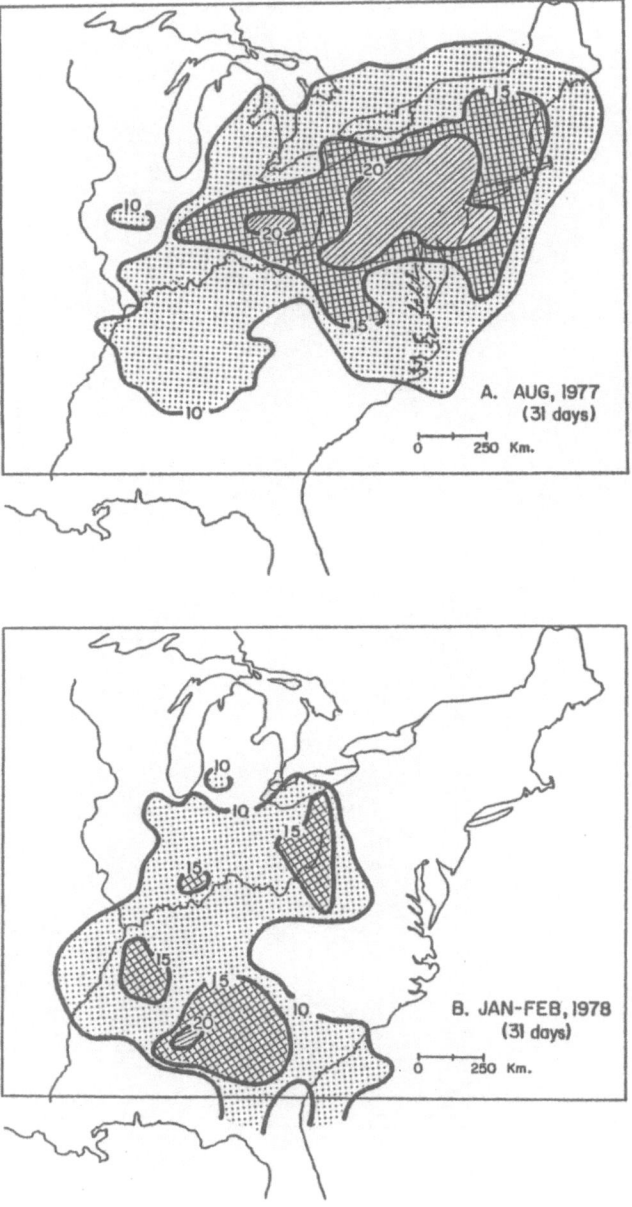

Fig. 5.   Spatial Distribution of Days per Month Sulfates were
           Equal to or Greater than 10 $\mu g/m^3$. This was extrapolated
           from a network of 54 ground stations in the greater north-
           eastern United States. (From Mueller et al.[7]).

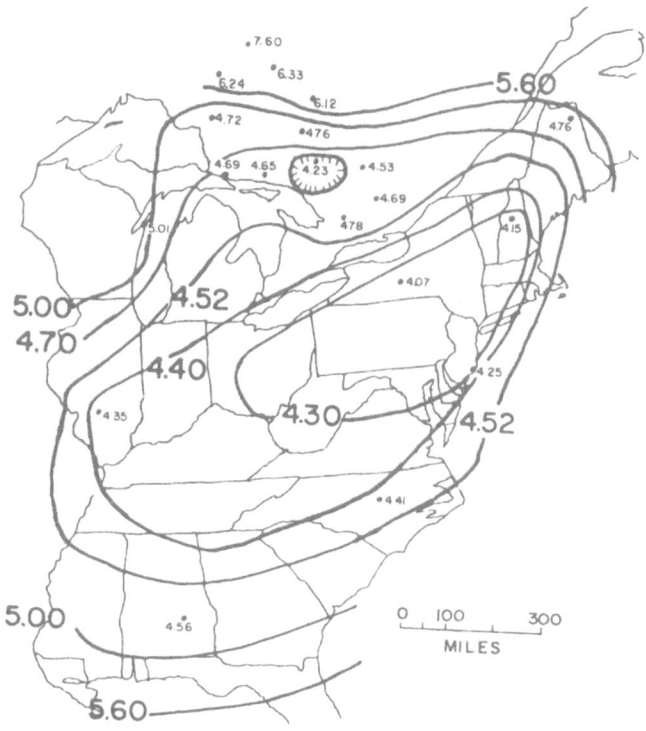

Fig. 6. Calculated Annual Distribution of pH in Rainwater over the
         Eastern United States Based on Available 1972-1973 Data
         for Precipitation Chemistry (from Cogbill[9]).

this activity. The responsibilities for different elements are
distributed among several agencies; however, the Environmental
Protection Agency (EPA) has the principal function to regulate the
pollutants covered by the Clean Air Act. Monitoring is done by
EPA as well as state and local agencies and is collected in the
system called the National Aerometric Data Bank (NADB). For
purposes of surveillance of large source impact, and for special
studies, some monitoring also has been undertaken by the private
sector. Monitoring for dry deposition has not been attempted
recently, but extensive fallout jar data taken prior to the mid-
1960s probably still is accessible. Wet fallout measurements of
a long term nature have been resurrected under sponsorship by
the U.S. Department of Energy, the Department of Agriculture, the
Electric Power Research Institute and Environment Canada. These
programs concern event sampling, while the DOA system is
concerned with monthly averaged, long term trend information.
Since these programs have started within the past 3 to 5 years,
it remains uncertain whether or not they will be maintained for
a truly long term of decades.

Table 5. Air Quality and Related Ground Monitoring Data Resources Operational in the United States

	Program	Description	Reference
A.	Federal, State & Local Air Monitoring - National Aerometric Data Bank (NADB)	Required by CAA--200-300 sites; measureing community pollutants; quality of data spotty, poor hydrocarbon measurements; particulates measured every 6th or 12th day; mostly urban-sub-urban. Reported in SAROAD System.	U.S. EPA[10,11]
B.	Global Environmental Monitoring System (GEMS)	1-2 selected remote-to-urban sites for community pollutants and other pollutants. World-wide network planned by UNEP to be implemented.	Munn[12]
C.	Baseline Studies for Prevention of Significant Deterioration (PSD) Assessment	More than 50 stations--community pollutants; have formal quality assurance requirements.	U.S. EPA[13]
D.	Western Regional Particulate & Visibility Network of EPA	Approximately 10 stations in remote western states; particulate chemistry & visual measurements.	For example, Malm[14]
E.	Industrial Networks for Compliance Monitoring & for Supplementary Control Systems	More than 300 sites near major industrial sources in rural areas; primarily SO_2 & TSP. Often reported to states and to SAROAD.	For example, Newman and Spiegler[15]
F.	Regional Air Quality Study (RAQS) of the Electric Power Research Institute (EPRI)	Follow-on of SURE Network. Nine stations in greater Northeast, 3-6 stations to be added; community pollutants. Western RAWS is planned for added 15 stations by 1980.	For example, Hidy et al.[16]
G.	Regional Precipitation Network of EPRI	Colocated with RAQS stations in eastern U.S.; event sampling and analysis.	--
H.	Regional Precipitation Network of MAP3S	Nine rural precipitation collections in eastern U.S.; 24-hour sampling analysis.	MacCracken[17]
I.	National Precipitation Sampling Network of the Dept. of Agriculture	DOA NC-131 Project, planned stations over entire U.S.; cumulative sampling and analysis for each ecological studies.	Cowling and Gibson[18]

An interesting example of user application of non-air quality monitoring data has recently been reported (e.g., Trijonis, and Yuan [19]; Husar, et al.[20]). Visibility observations have been taken for many years at airports for aviation safety. There is circumstantial evidence that reduced visibility under certain conditions is a measure of pollution associated with suspended particulate matter, in the size range $\leq 3\mu$m in diameter. Airport visibility observations taken over many years have been suggested as a surrogate for pollution. The analysis of such data has provided a provocative indicator of the influence of industrialization in rural areas of both eastern and western United States. Thus, we see that data taken for a completely different application can be applied by innovative investigators in the environmental field when there is motivation to characterize poorly understood processes.

A unique experimental study which may be kept in operation in the future is the National Aeronautics and Space Administration's global air sampling program (GASP), which involves regular observations of selected pollutants by instrumented commercial aircraft flying in the upper troposphere and lower stratosphere. With the possible exception of some Department of Defense sampling, these are the only direct measurements being made routinely aloft sponsored by the United States.

Of course, satellite observations represent regular, routine observations of the Continent. Remote sensing in the visible range of the light spectrum, as well as the infrared range, are available from systems such as the Geosynchronous Observational Earth Satellite (GOES). An example of a regional haze event which is believed to be related to air pollution, associated with finely divided aerosol particles, is shown in Figure 7. The haze is particularly noticeable off the east coast of the United States where the dark ocean surface contrasts it. This event, like others of its kind, appears to take place under certain meteorological conditions which prevail over a very large area. There is evidence that such areas of high-aerosol particle concentration may travel considerable distances before dispersal (e.g., Husar, et al.[20]).

LIMITATIONS AND FUTURE NEEDS

Air quality monitoring suffers from three major problems, which are:

- the acquisition methods are technologically conservative, and may be found to be inaccurate or uncertain with improvements in sampling and analytical methods, as well as station siting;

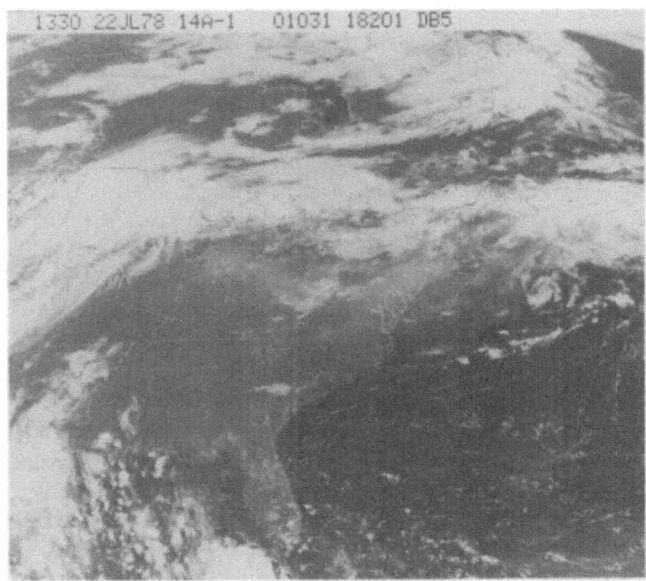

Fig. 7.   Haze on July 22, 1978 observed in the visible range of
          the light spectrum by the camera system of the NASA/
          GOES satellite (photo courtesy of National Aeronautics
          and Space Administration).

*   the data may be inaccessible to a broad user community in
    a timely and efficient manner; and

*   monitoring data often suffer from a lack of user identi-
    fication or "dialog", and consequently active, timely
    feedback for improvement of measurement quality.

     Monitoring data, perhaps by definition, are basically limited
by utilization of basic technology. Since the acquisition of air
quality data is very expensive, the spatial and temporal coverage
has been limited. Furthermore, the methods used over the years
have proven to be subject to interferences and errors, which have
emerged from continued research. The uncertainties in historical
data have been discovered in virtually every measurable parameter,
as our knowledge has expanded (see for example NAS[21]). Notable
in the list are the controversies regarding reactive hydrocarbons,
sulfur dioxide and nitrogen dioxide, not to mention particulate
sulfate, nitrate and solvent soluble organics. Therefore, the
data user has found that considerable sophistication in judgment

is required to interpret "old" information in comparison with new
data reported by data systems. This "traceability" or uncertainty
issue has created endless controversies about the meaning of trend
analyses and long-term changes with emissions of pollution.

   The second limitation of monitoring data is its inaccessibility
to users. Air quality data has been notoriously late in its release
for public use. Even with modern computerized data capacity, neither
the government nor private organizations underwriting monitoring
have succeeded in distributing "fully validated" data in less than
12 to 18 months. In some cases, portions of specialized monitoring
data, such as that of the EPA's Community Health Air Monitoring
Program, have not yet been released even though the observations
were taken over three to four years ago. A major part of the delay
is the verification of the data by data managers. This process can
take a long time if careful a posteriori review of calibrations
and methods is done. For the resources available, the shear effort
of handling the data often has saturated the archiving process.
Without considerably more investment in improved reliability in
instrumentation, data transmittal and management, this situation
is likely to continue for sometime to come. Recently, a federal
interagency committee was organized to review this and other issues
regarding the national monitoring programs, but their recommendations
have not been submitted to Congress as yet.

   Perhaps the Achilles' heel of air quality monitoring programs
is the user identification. With active users, data "comes alive"
with application to advancing knowledge of national processes and
human pertubations on them. In questioning, user's feedback to the
monitoring agency urged improvements and early delivery for observations. Unlike the national weather network, air quality data
has yet to become used routinely for environmental forecasting and
impact evaluation. However, the cases when monitoring, or at least
long term data acquisition, has been coupled with carefully planned
data assessment and evaluation may prove to be invaluable contributions
to environmental science. One such plan was prepared for EPA's
Regional Air Pollution Study (RAPS) (e.g., Burton and Hidy[22]), and
another for the EPRI Sulfate Regional Experiment (SURE) which is
illustrated in Figure 8. Here the data are linked with two goals -
characterization of regional scale pollution, and - the engineering
and refinement of an air quality model to relate cause and effect
for such events. The plan for RAPS was not implemented partly because
of inadequate maintenance of user interest in the data, taken between
1974 and 1977. The SURE project, however, has followed the planned
activity flow organization rather closely, so far. It is too early
to evaluate objectively this approach to the use of long term data
bases, but the SURE and RAPS may serve as useful prototypes for
comparison with future programs.

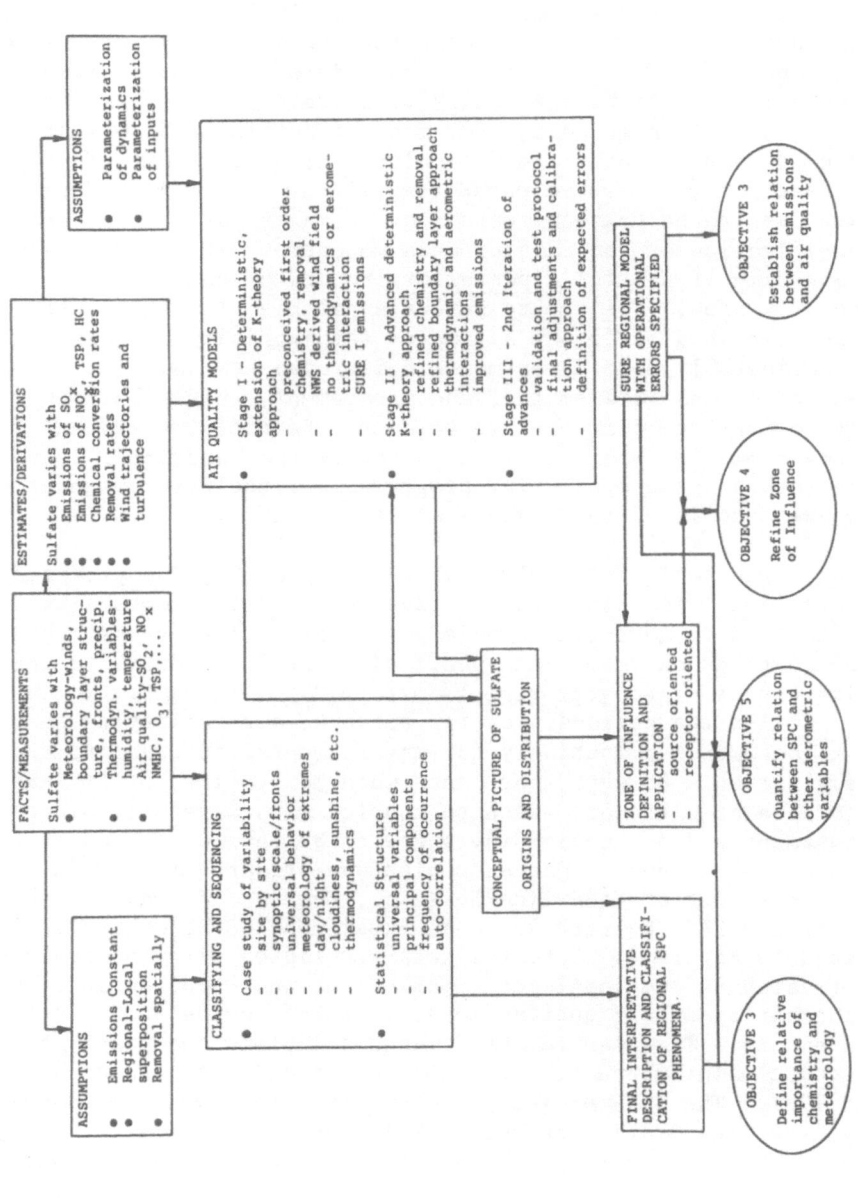

Fig. 8.   Example of a plan for use of monitoring data with other parallel observations for investigation of regional sulfur oxide emissions and ambient sulfur oxide conditions (Muller et al.[23])

With the maturing of environmental monitoring in the United States, the regulatory agencies undoubtedly will find sufficient pressure from users to warrant the long-term investment in monitoring data. Such pressure has not emerged as yet, but almost certainly will be necessary to determine long-term influences on health and ecology of the low levels of community pollutants now under control. Impact evaluation is a very long-term investment which must be recognized and promoted if we are to test the rationality of our decisions for clean air. Despite the pain of criticism of our first efforts (e.g., U.S.HR[24]), a new try at such investments should be made and maintained for the necessary time to reach fruition.

Air quality monitoring will be needed to fulfill the mandates of the Clean Air Act and its amendments for sometime to come. Additional effort probably will be encouraged to improve the basic observations for: (a) reactive hydrocarbon vapors, (b) finely divided particles ($\leq$ 3 μm in diameter), (c) visual effects of air pollution, and (d) certain toxic chemicals released in the air. In addition, there should be more long term surveillance of characteristic "impact" parameters for community health and ecological perturbations. Useful, simple parameters have yet to be well defined, but research should be undertaken to identify them.

ACKNOWLEDGEMENT

This work was sponsored in part by the Electric Power Research Institute under Project 862 to Environmental Research & Technology, Inc.

REFERENCES

1. Organization for Economic Cooperation and Development (OECD), The OECD Programme on "The Long Range Transport of Air Pollutants; Measurements and Findings." OECD, Paris, France (1977).
2. G.M. Hidy, et al., "International Aspects of the Long Range Transport of Air Pollutants." Report to U.S. State Dept., ERT #P-5252, Westlake Village, CA (1978).
3. C.E. Junge and R. T. Werby, The Concentration of Chloride, Sodium, Potassium, Calcium and Sulfate in Rainwater over the United States. J. Meterol. 15: 417-425 (1958).
4. J.P. Lodge, Jr., et al. "Final Report on the National Precipitation Network." National Center for Atmos. Res., Boulder, CO, August (1968).
5. P.K. Mueller, et al. "Background Paper for a Pilot Project on the Assessment of Human Exposure to Air Pollutants." Report WHO/UNEP Proj. No. FP/1301-78-03, Environ. Res. & Technol., Inc., Concord, MA (1979).
6. U. Platt, Dry Deposition of SO_2, in "Sulfur in the Atmosphere" Proc. of the Int'l Symposium, Dubrovnik, Pergamon Press, New York. p. 363. (1978).

7. P.K. Mueller, et al., "Some Early Results from the Sulfate Region-
    al Experiment (SURE)." Proc. 4th Symposium on Turbulence,
    Diffusion and Air Pollution, Amer. Meteorol. Soc., Boston,
    p. 322 (1979).
8. U.S. Environmental Protection Agency, "Position Paper on Regulation
    of Atmospheric Sulfates." Report 450/2-75-007, Research
    Triangle Pk, NC (1975).
9. C.V. Cogbill, "The History and Character of Acid Precipitation in
    Eastern North America," Water, Air and Soil Pollution, 6:
    407-414 (1976).
10. U.S. Environmental Protection Agency, "Position Paper on Regula-
    tion of Atmospheric Sulfates." Report 450/2-75-007, Research
    NC (1971).
11. U.S. Environmental Protection Agency, "Aeros Manual Series, Vol. I:
    Aeros. Overview." Report 450/2-76-00. Office of Air Quality
    Planning and Standards. Research Triangle Pk., NC (1976).
12. R.E. Munn, "Global Environmental Monitoring System (GEMS),
    Action Plan for Phase I." Scientific Committee on Problems
    of the Environment (SCOPE), Report 3, ICSU, Toronto, Canada
    (1973).
13. U.S. Environmental Protection Agency, "Ambient Monitoring Guide-
    lines for Prevention of Significant Deterioration (PSD)."
    Report 450/2-78-019, Research Triangle Pk., NC (1978).
14. W. Malm, "Summary of Visibility Monitoring Workshop " (July), U.S.
    Environmental Protection Agency, Environmental Monitoring
    Systems Laboratory, Las Vegas, NE (1978).
15. E. Newman and D.B. Spiegler, Operational Experience with Air
    Quality Control System – AIRMAP. in "Proc. Symposium," Atmo.
    Diffusing Air Pollution. American Meteorol. Soc., Boston,
    MA (1974).
16. G.M. Hidy, et al. Design and Implementation of the Sulfate Re-
    gional Experiment (SURE). "Proc. 4th Symposium on Turbulence,
    Diffusion and Air Pollution," Amer. Meteorol. Soc., Boston,
    MA p. 315 (1979).
17. M. MacCracken, MAP3S: An investigation of atmospheric, energy
    related pollutants in the northeastern United States, in
    "Sulfur in the Atmosphere," Proc. Int'l Symposium, Dubrovnik,
    Pergamon Press, NY, p. 649 (1978).
18. E.B. Cowling and J.H. Gibson, "The Changing Chemistry of Atmos-
    pheric Deposition and its Effects on Agricultural and Forest
    Lands and Surface Waters in the United States." Unpublished
    manuscript, NC-141, Regional Project on Atmospheric Deposi-
    tion. Colo. State Univ., Fort Collins, CO (1977).
19. J.E. Trijonis and K. Yuan, "Visibility in the Southwest" and
    "Visibility in the Northeast", Reports 600/3-78-039 and 600/3-
    78-075. U.S. Environmental Protection Agency, Research Tri-
    angle Pk., NC (1978).
20. R.B. Husar, et al. Long Range Transport of Pollutants Observed
    Through Visibility Contour Maps, Weather Maps and Trajectory

Analysis. "Proc. 3rd Symposium Atmospheric Turbulence, Diffusion and Air Quality." American Meteorol. Soc., Boston, MA, p. 344-347 (1976).

21. National Academy of Sciences (NAS), "Air Quality and Automobile Emission Control." Vol III. Committee on Public Works, U.S. Senate, 93rd Congress, Report No. 93-24. U.S. Government Printing Office, Washington, DC, p. 37 (1974).

22. C.S. Burton and G.M. Hidy, "Regional Air Pollution Study Program Objectives and Plans." Report 650/3-75-009. U.S. Environmental Protection Agency, Research Triangle Pk., NC (1974).

23. P.K. Mueller, et al. "Implementation and Coordination of the Sulfate Regional Experiment (SURE) and Related Research Programs." Electric Power Research Institute Report, Palo Alto. in press (1978).

24. U.S. House of Representatives (USHR), "Community Health and Environmental Surveillance System (CHESS): An Investigative Report", Committee on Science & Technology, 94th Congress, Serial SS. U.S. Government Printing Office, Washington, DC (1976).

DISCUSSION

HICKS: You observed nitrate in air at much lower
concentrations than sulfate. That's not the case in rain.
Would you consider that to be good evidence of the in-cloud
formation of material?

HIDY: Not necessarily. If you look at the data aloft, the
sulfate to nitrate ratios that we have observed between 500 ft
and 5000 ft, and above were something in the order of 3 or 4,
which is more like what you see in rain. So differences from
ground level conditions may be related to in-cloud processes,
or may just be differences in dry chemistry aloft or in surface
removal processes. Nitrogen oxides may be absorbed in clouds
or may be produced preferentially in air away from ground level
pollution mixtures. One could argue, however, that nitrate
should behave chemically in the same way as sulfate as far as
transport and mixing is concerned. The chemical reactions of
nitrogen oxides in dry air and wet air are considerably more
complex than the sulfate-forming reactions, and probably offer
an explanation for such differences. Your hypothesis that it's
in-cloud is as good as any, I would say.

HICKS: On the basis of the sulfur regional experiments, to
what extent do you suspect that the acidity of precipitation is
indeed regional?

HIDY: We have no historical basis for a judgment on that
subject at present. Using the SURE data alone is not
particularly helpful because we are not obtaining precipitation
data. If we combine the SURE data with the DOE/MAP3S and EPRI
precipitation networks, then there is pretty good evidence that
there is a region-wide connection when there is precipitation
present. An interesting feature in the combined data is that
the maximum precipitation acidity seems to prevail downwind of
where the maximum in the aerosol sulfate is found, by a few
hundred miles. The presence of more nitrate in rain than in
the aerosol particles, relative to sulfate, also is an
interesting feature which makes one wonder about the chemistry
of acid rain compared with dry air processes.

TOMLINSON: That was a most interesting paper and it's given us
a lot of insights we've never had before, I'm sure. Are you
proposing to tie that in with acid rain data that would be
collected at the same time to find out what extent it's washed
out?

HIDY: Yes, we did not take acid rain data in the SURE
experiment. The data were taken in both the EPRI and MAP3S
network, and we are looking at that data in conjunction with
the SURE at the present time.

TOMLINSON: That combination would be most interesting. The other thing I wanted to mention is I happen to live in Canada and I know that we're improving our ability to make such measurements. Are you able to tie this in with Canada now or do you propose to do that?

HIDY: Well, the Canadian network for both precipitation measurements and air monitoring was operating during this time, and so we would certainly like to look at that data at our earliest opportunity.

TOMLINSON: Because of the international nature of the air, it seems to me it could be tied in on the same computer program and this would be extremely useful. Do you know Dr. Welpdale?

HIDY: Clinically it might not be terribly useful for the United States, but it's going to be done.

TOMLINSON: We export a little to the United States too, you know.

VOLCHOK: I'd just like to quibble with you a little bit. In the beginning of the paper you mentioned government data that you can't get. Could you tell me what agency is withholding data from you?

HIDY: Withholding data is a strong term. Let's just say that because of the government bureaucracy, it is extremely difficult at times to obtain certain kinds of aerometric and emissions data for analysis. We have had particular problems in obtaining gas data from the Community Health Air Monitoring Program (CHAMP) and the Regional Air Pollution Study (RAPS).

VOLCHOK: I'd rather you go on a record and say the EPA is the culprit.

HIDY: It's a good point. I will be specific and say the data we have difficulty obtaining comes from the Environmental Protection Agency. I do not believe that there is any effort to suppress or delay information transfer. However, EPA's data retrieval service is cumbersome and complex, making it less efficient than an ideal system.

HICKS: Would you be more specific and say not programs that have been switched to the EPA?

HIDY: You're very sensitive.

STENSLAND: In light of the discussions of the last few months
about problems in measuring nitrate, are you satisfied with the
procedures that were used to collect your data?

HIDY: The filter media used in this study was a teflon-coated
glass fiber substrate, manufactured for us under very strict
specifications. It was used uniformly for all aerosol sampling
in the SURE. The filter medium is not subject to a sulfur
dioxide adsorption artifact, but it is subject to adsorption of
nitric acid. It does not adsorb NO or NO_2, however.
Therefore, the concentrations we are reporting are identified
as total nitrate made up of actual particles and possibly a
component of nitric acid vapor.

STENSLAND: I will present some results next month at the APCA
meeting from experiments where we used Nuclepore filters of
various sizes and in various setups (e.g. stacked filters). We
get very different results with the same filter media, just
having it set up somewhat differently. I assume we will be
hearing much more about this sampling problem in the next few
months.

HIDY: Particulate nitrate in the presence of gaseous nitrogen
oxides is very difficult to sample, as I'm aware our chemist
friends are well aware in the audience here.

STENSLAND: In the ducting situation, could it be that it is
more important that the sources are lined up, as opposed to
having boundaries on two sides? I cannot visualize a plume
from any individual point source spreading from Michigan to the
Appalachians, so it seems like it is more important that the
air flow is such that it goes successively across each of the
different sources and thus gives a buildup in concentration for
this reason.

HIDY: Actually, I didn't have one of my other viewgraphs here
which show the distinction of this ducting event as contrasted
with the stagnation case. In the ducting situation you can
actually see the detachment of the air containing high sulfate
concentrations from the region of high emission density. That
is an important distinction to make and does not have to do
with the apparent accumulation around sources you would get by
just having the light winds blowing along the direction of the
major sources. So there is a distinction there which is
important, and more will be said about the regional scale
processes in the future.

ATMOSPHERIC DEPOSITION OF MAN-MADE RADIOACTIVITY

Herbert L. Volchok

Environmental Measurements Laboratory
U. S. Department of Energy
New York, NY 10014

ABSTRACT

   For the past two decades, the major radioactive component in
rain has been the debris resulting from nuclear tests in the
atmosphere. In studying fallout from these tests, rather extensive
networks of precipitation collections were developed which have
proven to be extremely useful in maintaining an inventory of the
fallout radionuclides on the earth's surface. Additionally, the
models used for estimating current and future human exposure to
these substances are in large part based upon the rainfall data.
A study of total, wet and dry deposition is described, concluding
that dry fallout should not be neglected and that it can successfully be collected on artificial surfaces.

INTRODUCTION AND BACKGROUND

   In addressing the subject of radioactivity in "Polluted Rain",
over the past 20 years, it is clear that the prevalent component
is man-made. Specifically this refers to the material produced
and deposited in the atmosphere by nuclear weapons tests. Other
man-made radioactive releases, such as inadvertent leaks or
ventings from nuclear power plants, nuclear fuel reprocessing
operations, or nuclear weapons production facilities have only
sporadically been observed in precipitation. For the purpose of
monitoring the fallout from nuclear tests, carrying out research
towards understanding the bio/geochemical fate of certain radionuclides, and ultimately the effects on organisms - including man,
several national as well as international deposition networks were
organized and have been in operation for many years. These studies
produced a rather large, comprehensive body of atmospheric

deposition data on nuclear weapons fallout debris, far surpassing in scope any other chemical information on rain. This paper briefly describes the origins and background which gave rise to the precipitation programs plus the scope, the instrumentation employed, and some of the results and implications derived from these studies. Throughout, the word "fallout" is used as a shorthand synonym for "radioactive debris from atmospheric nuclear explosions" whether or not the material has actually as yet been deposited on the earth's surface.

The first nuclear test took place in the United States in July 1945. Since then a number of other nations have tested nuclear devices both above and underground. The total amount of fission which took place in the atmosphere is estimated to be a little more than 200 megatons (MT).[1] Most of the explosions occurred in the Northern Hemisphere, and most of the material produced was injected initially into the stratosphere. Careful and continuous monitoring and inventorying of the atmosphere since 1958 has indicated half-residence times for the fallout particulates of about one year in the stratosphere and about one month in the troposphere. Measurements in the stratosphere and upper troposphere showed that the fission product radioactivity was in or on particles with a diameter of .03 μm or less.[2,3]

The test ban treaty of 1963, signed by the major nuclear powers at that time, virtually ended the large atmospheric inputs, although there has been a more or less continuous series of small and intermediate shots by non-signatories. Hence, using the isotope ^{90}Sr as an example, a total of some 12 megacuries (MCi) was deposited in the stratosphere. The peak inventory was about 6 MCi in early 1963 which decreased to less than .01 in 1977.[4] Except for the fraction which radioactively decayed, all of this material is now on the earth's surface.

Very early in the history of nuclear testing and studies of fallout, it became clear that most of the particulate fallout was brought to the ground in precipitation[5], although a measurable amount deposited dry. In regions of moderate rainfall, the dry fraction probably never exceeds 0.4.[6,7]

THE COLLECTION NETWORK AND THE INSTRUMENTS

In the heyday of atmospheric nuclear testing there were really only two precipitation sample collection networks addressing the questions related to global fallout. The British program, initiated in 1956,[8] increased in size to about 30 stations by the mid-1960s and continues at almost that level to the present.[9] In addition to fairly intensive sampling in the British Isles, the sites include Europe, Iceland, Canada, the West Indies, Hong Kong, South Pacific, Africa, Australia, New Zealand and South America. The sampling period for the most part has been monthly. The rain

collections are made through plastic funnels mounted above plastic
bottles; snow is collected in high walled stainles steel or plastic
pots. The analytical data and interpretation are published
annually (e.g., reference 9).

The U. S. program, initiated in 1954, was designed and operated
by the Health and Safety Laboratory of the U. S. Atomic Energy
Commission. That laboratory is now the Environmental Measurements
Laboratory (EML) part of the U. S. Department of Energy. By 1958,
EML had 32 sites, and the program grew rapidly, peaking at almost
140 monthly stations around the world in 1964. Figure 1 shows
the locations during the period of most intense fallout. It is
obvious in this map that the samplers are neither uniformly nor
optimally distributed. The reasons are logistical (put them where
there are people we can trust to operate them), physical (the only
practical ocean stations are islands), and political (they have to
be in countries to which we have free access). These considerations tended to bunch the sites in the continental mid-latitudes,
with large gaps in the oceans, China, and the U.S.S.R. EML still
runs over 70 collection sites today.

The original sampler was a steep-walled stainless steel pot
from which the water and particulates have to be quantitatively
transferred into a bottle for forwarding back to the Laboratory.
This requires a fairly high degree of specialized competence, not
always available at locations perhaps more interesting and logis-
tically attractive as collection sites. Therefore, a plastic
funnel-ion exchange resin column sampler was developed[10] based on
a Swedish design. In this, as the rainwater (or melting snow)
passes through the resin column, most of the anions and cations in
solution are adsorbed on the resin. Particles are removed at the
top of the column by a filter mat. The data developed in this
program are reported quarterly,[11] and summarized and interpreted
annually.[12]

Later, as we recognized phenomena which could not be treated
and explained using only total or bulk deposition data, we developed
an instrument which quantitatively and automatically collects
separate wet and dry fractions of atmospheric deposition.[13] The
instrument is shown in Fig. 2. The onset of precipitation is
sensed by the sensors mounted on both sides of the roof, causing
it to move off the "wet bucket" and cover the "dry". The wet/dry
deposition collectors have so far been used on a rather limited
basis in radioactivity studies to carry out specific experiments.

All of the analytical procedures used in the precipitation
and dry deposition programs are described in detail in the EML
Procedures Manual.[14]

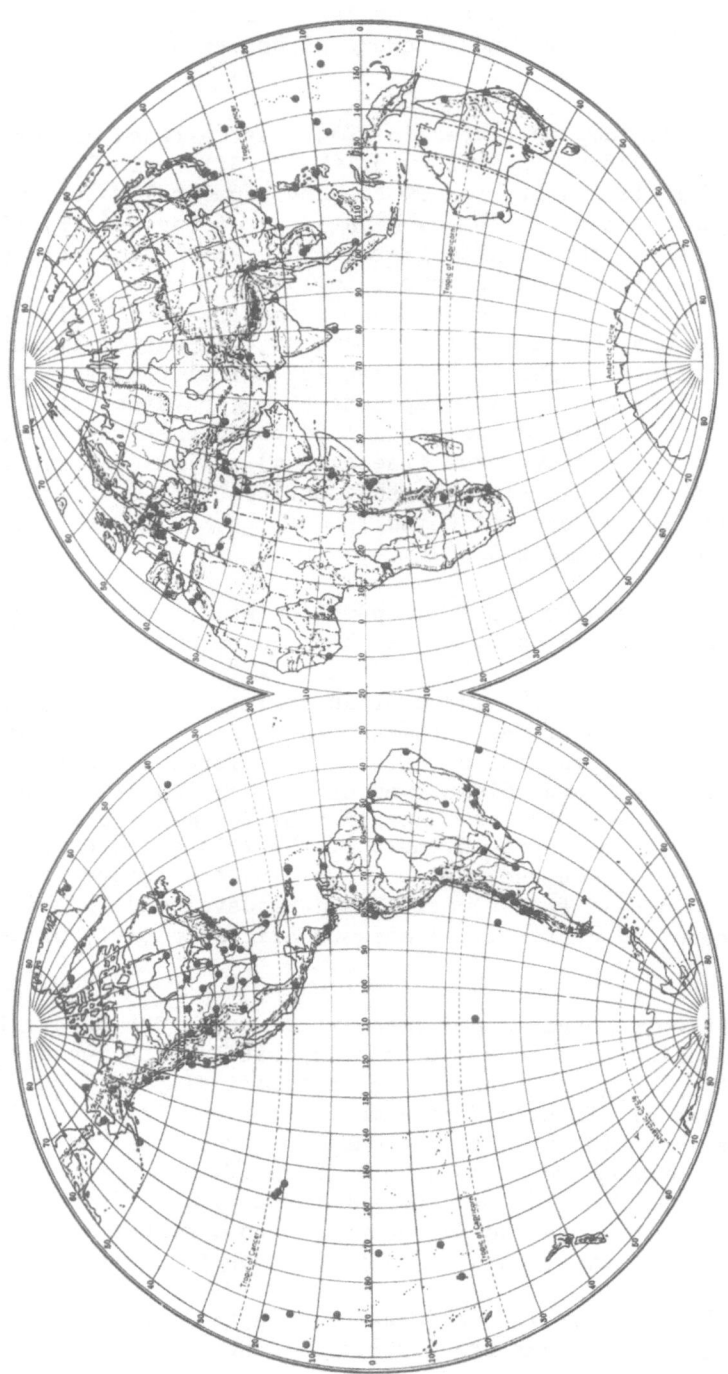

Figure 1.   Sampling Sites in the EML Fallout Collection Network.

Fig. 2. EML wet/dry deposition collector.

RESULTS AND IMPLICATIONS

Worldwide Deposition

     One of the most interesting and challenging aspects of the
precipitation collection program was whether, in fact, the data
produced were applicable to global scale interpretation. One of
the motives in developing this program was to maintain an inventory
and budget of the radioactive debris produced in the atmospheric
tests. For example, the monthly ^{90}Sr information[11] from analyses
of samples collected at the sites shown in Fig. 1 are integrated
to produce the monthly, worldwide ^{90}Sr deposition to the earth's
surface.[12] Another completely independent program produced
periodic inventories of ^{90}Sr in the stratosphere.[4] Clearly,
during periods of no atmospheric weapons tests, that is, when no
additions are being made to the inventory, the stratospheric
depletion should be equal to the deposition. Six such periods
were identified since the test ban treaty went into effect. The
pertinent data are summarized in Table 1. The good agreement adds
credibility to the integrity of the networks, the sampling and
analytical programs and the schemes for integrating the data.

Table 1. Global Inventory Balance of ^{90}Sr (During periods of
          little or no atmospheric nuclear testing)

	kilocuries of ^{90}Sr*		
	(A) Stratospheric Depletion	(B) Global Deposition	(B)/(A)
Jan '64–May '66	2,934	3,366	1.15
Aug '67–Jun '68	173	182	1.05
Oct '69–Apr '70	115	139	1.20
Oct '71–May '73	272	279	1.03
Sep '73–Apr '74	90	61	0.68
Oct '74–Jul '75	59	63	1.08
TOTAL	3,643	4,090	
		WEIGHTED MEAN	1.12

*The data are compiled from earlier citations which can be
 found in references (4) and (12).

Further validation of the worldwide deposition data emerges
by comparison with the United Kingdom network results, mentioned
earlier,[9] and with a program which periodically samples and analyses
soil.[15] The British program involves many fewer sampling sites
than our own, and the method of integrating the data is quite
different. The data obtained from soil samples are obviously quite
different than those from rain in that soil is a time integrator,
accumulating all of the deposited radionuclides. Soil sampling on
a global basis is difficult and expensive, hence this program has
only been carried out six times since 1958. Comparison of the
three programs is illustrated in Fig. 3. Again the agreement is
gratifying, giving further confidence that, in spite of programmatic
differences, each of the network systems provides reasonably good
estimates of the worldwide ^{90}Sr deposition.

These estimates of global deposition of long-lived fallout
radionuclides, based for the most part on the analyses of rain
samples, comprise the most significant inputs to the calculations
of radiation exposure to the world's population from global
fallout.[16] For perspective, this value is often tabulated along
with other sources of human radiation exposure, as in Table 2.

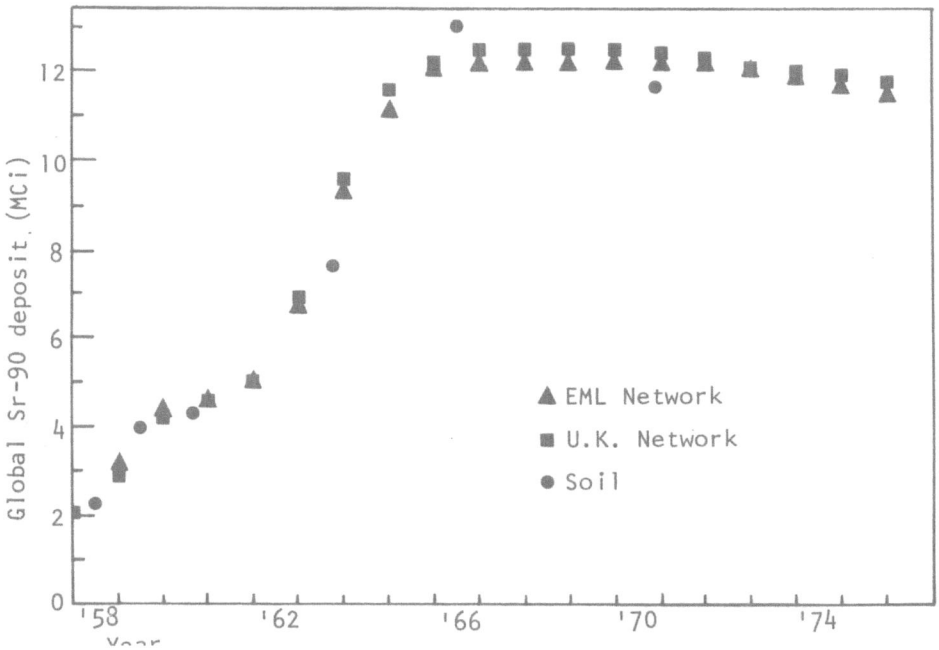

Fig. 3. Comparison of global ^{90}Sr deposit estimates.

Clearly, the annual dose from fallout radiation is less than 10%
of most of our total annual doses and about half what we receive,
on the average, from medical practices.

Table 2.   Average Whole Body Dose Rates to People
in the United States[a]

		mrem/y	%
Natural Background[b]		78	72
Medical		20	18
Fallout		8	7
Nuclear Power Industry		2	2
Consumer Products		1	1
	TOTAL	109	100

[a]Based upon data in reference 16.

[b]Consists of cosmic radiation, terrestrial radiation and radionuclides in the body.

Dry Deposition

    Two of the uncertainties in the fallout collection programs
were the magnitude of the fraction deposited dry and whether or
not the dry fallout was efficiently collected. As early as 1962,
Hardy and Alexander[6] noted that a significant portion of ^{90}Sr was
deposited dry, on soil. Turekian et al.[17] studying the natural
radioisotope, ^{210}Pb, also reported high fractions (as much as 0.6)
of dry deposition.

    Atmospheric chemists and particle physicists generally seem
to agree that for the size particle of interest in these programs
(less than 1 μm) artificial collectors produce only qualitatively
useful data, at best.[18] In fact, in a few of the ongoing precipitation studies for acid rain effects, dry deposition is purposely
neglected in the belief that the resultant data would be useless
and misleading.

At our Laboratory, a series of experiments were carried out to empirically evaluate our dry fallout data. For 13 consecutive months in 1974-1975, wet, dry and bulk collections from New York City were analyzed for ^{90}Sr. The results are shown in Fig. 4; the sum of the wet plus dry data are plotted versus the bulk or total collection. The straight line is a least squares fit. Clearly, the relationship is linear with a high (0.98) correlation coefficient. These results, plus the information derived from the worldwide deposition studies, strongly indicate that the bucket used in our wet/dry instrument quantitatively collects the dry ^{90}Sr deposit. There is an increasing body of evidence that this conclusion holds equally well for aerosols of like and even smaller particle size.

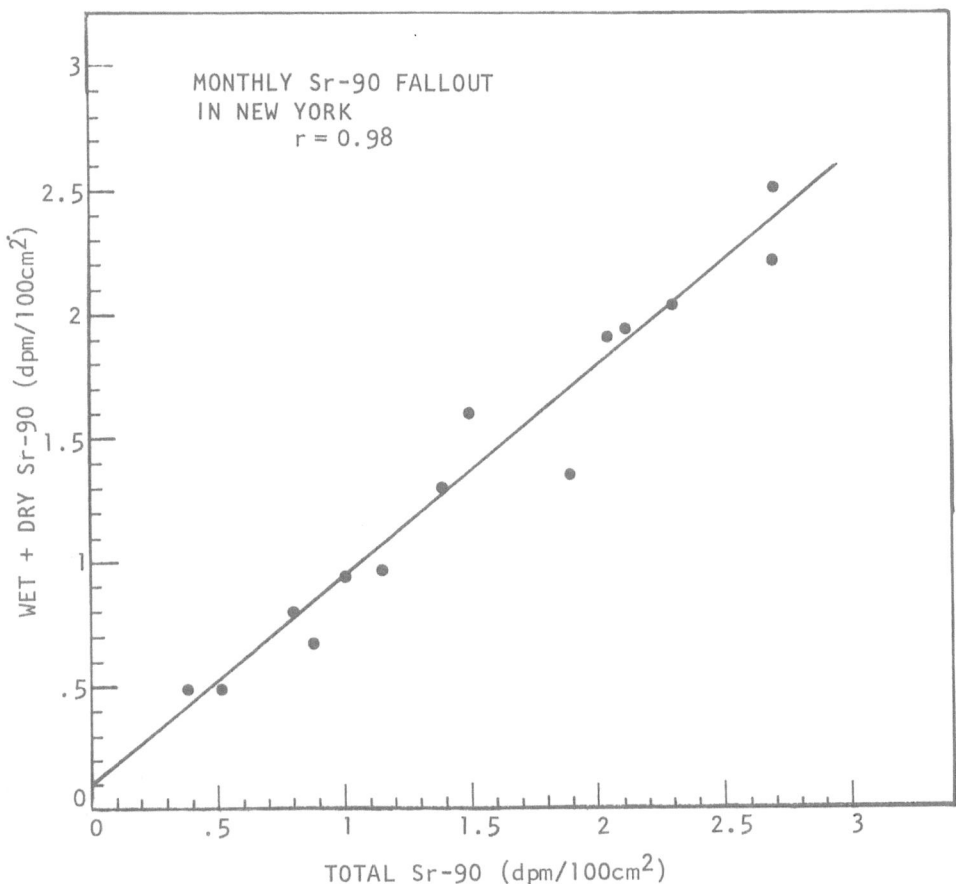

Fig. 4. Wet and dry and total ^{90}Sr fallout in New York City.

Summary

   Numerous interesting and important scientific contributions
have come out of the studies of the atmospheric deposition of
radioactivity. These range in scope from extremely local to global
and cover numerous scientific disciplines. The global networks of
fallout collections have proven invaluable for maintaining a
temporal and a geographical record of the nuclear test fallout and
have provided the data base from which the estimates of human risk
were computed. Separating the ^{90}Sr fallout into wet and dry
components has shown that in many cases the dry fraction is not
insignificant and that an artificial surface successfully and
quantitatively collects dry deposition.

References

1. Announced Nuclear Detonations, 1945-1962, Compiled by the
    Atmospheric Radioactivity Research Project, U. S. Weather
    Bureau, Washington, D. C., HASL-142 (1964).
2. E. A. Martell, The Size Distribution and Interaction of Radio-
    active and Natural Aerosols in the Stratosphere, Tellus
    XVIII:486 (1965).
3. Y. I. Gasiev, S. G. Malakhov, L. G. Nazarov and A. N.
    Silantiev, The Size Distribution of Radioactive Particles
    from Nuclear Weapons Tests and Their Transport in the
    Atmosphere, Tellus 18:474 (1965).
4. R. Leifer and L. Toonkel, Updating Stratospheric Inventories
    to April 1977, U. S. Department of Energy Report EML-334,
    p. I-3, New York (1978).
5. J. H. Harley, A Brief History of Long-Range Fallout, U. S.
    Energy Research and Development Report HASL-306, New York
    (1976).
6. E. P. Hardy and L. T. Alexander, Rainfall and Deposition of
    ^{90}Sr in Clallam County, Washington, Science 136:881 (1962).
7. L. T. Alexander, R. H. Jordan, R. F. Dever, E. P. Hardy,
    G. H. Hamada, L. Machta and R. J. List, Strontium-90 on
    the Earth's Surface, U. S. Office of Technical Information
    TID No. 6567, Washington, D.C. (1961).
8. N. G. Stewart, R. G. D. Osmond, R. N. Crooks and E. M. Fisher,
    The World-Wide Deposition of Long-Lived Fission Products
    from Nuclear Test Explosions, United Kingdom Atomic Energy
    Authority Report AERE HP/R 2354 (1957).
9. R. S. Cambray, E. M. R. Fisher, J. D. Eakins and D. H.
    Peirson, Radioactive Fallout in Air and Rain: Results to
    the End of 1976, United Kingdom Atomic Energy Authority

Report AERE R-8671 (1977).

10. G. A. Welford and W. R. Collins, An Evaluation of Existing Fallout Collection Methods, Science 131:1791 (1960).

11. E. P. Hardy, ed., Environmental Quarterly and Appendix, U. S. Department of Energy Report EML-353, New York (1979).

12. H. W. Feely and L. E. Toonkel, Worldwide Deposition of Sr-90 through 1977, U. S. Department of Energy Report, Environmental Quarterly EML-344, p. I-19, New York (1978).

13. H. L. Volchok and R. T. Graveson, Wet/Dry Fallout Collection, Proceedings of the Second Federal Conference on the Great Lakes, ICMSE - Federal Council for Science and Technology, Published by Great Lakes Commission, pp. 259-264 (1976).

14. J. H. Harley, ed., EML Procedures Manual, U. S. Department of Energy Report HASL-300, New York, revised annually (1979).

15. E. P. Hardy, P. W. Krey and H. L. Volchok, Global Inventory and Distribution of Pu-238 from SNAP-9A, U. S. Atomic Energy Commission Report HASL-250, New York (1972).

16. Sources and Effects of Ionizing Radiation, U. N. Scientific Committee on the Effects of Atomic Radiation, U. N. Publication, 32nd Session, 40 (A/32/40) (1977).

17. K. K. Turekian, J. Nozaki and L. K. Benninger, Geochemistry of Atmospheric Radon and Radon Products, Annu. Rev. Earth. Planet. Sci. 5:227 (1977).

18. Report of the Expert Committee on Dry Deposition R/PXM, Annex B, World Meteorological Organization, Gothenburg, Sweden (1977).

DISCUSSION

HICKS: Regarding the validity of buckets and open pots as
samples we got into this business, as you know, with fallout.
And the hard reality is that it does give a very good
indication of how much material is coming down in dry form.
What we don't understand is precisely why it seems to work. So
there have been, in fact, two groups of people looking at the
problem: one from the sense of: since we don't understand why,
how can we believe it; the other group saying: the hard,
pragmatic reality is that it does agree, and therefore let's
utilize it.

It looks to me like the resolution of the two viewpoints
may eventually be that a natural vegetated surface is going to
eventually remove particles that are impacted upon it. The
buckets seem to do much the same thing. When material gets in
there it's going to be deposited. And it may be as simple as
that, but I really wouldn't recommend that we believe any of
these methods to better than a factor of 2 or 3.

VOLCHOK: I think that's too big a spread, as indicated by the
data in my talk.

JACOBSON: Herb, doesn't it depend on the use to which you want
to put the data? You average out, over a period of months and
years, and come out with a very good correlation coefficient.
However, if you're talking about what happens during an
individual day at an individual site, the variability is
probably so large as to make it unsatisfactory for certain
types of uses, whereas it may be quite satisfactory for others.

VOLCHOK: I tend to agree with you but I have a slightly
cynical attitude because I see a lot of short-term measurements
being made, which are then added together into weeks and months
to analyze the data. I see very little exposition of daily or
hourly data.

WILSON: Herb, I noticed on your chart in which you use soils
in addition to the British & U.S. data, that there was a bit of
a drop in the Seventies. And I get to thinking that your
terrestrial samples were all samples of soils. I didn't know
whether you used that as a generic term or whether you might
have used sediments as well. How do you deal with the notion
that in fact a major amount of deposition will really be in
marine or freshwater sediments--that's part A of the question.
Part B is, what about the end of the food chain?

VOLCHOK: Let me take the first question. We were quite aware that marine deposition was a serious matter and it bothered us for a long time. If you look at the map, it's clear that we're not sampling more than 70% to 80% of the earth's surface. And as I said, with a very poor distribution of samples. In fact I often use a number to elucidate this problem. We're actually sampling about one square foot per million square miles of earth's surface, so it takes a lot of nerve to think that you can make anything out of it; but in fact it worked. And it turns out, after having done a lot of sediment ocean water sampling and analysis, that in any given latitude band the amount deposited is very much proportional to the amount of precipitation that occurred. So neglecting the ocean wasn't as serious as we thought it would be in the beginning.

WILSON: I am concerned about the implications of your findings for concentration, presumably by life systems. We have been led to believe that a number of these compounds, where heavy metals are involved, tend to accumulate in the high end of the food chain. Was your chart of concentration, or was it of flux?

VOLCHOK: I'm sorry, I should have been more specific. That chart was in terms of areal deposition -- I think the units were millicuries per square km. I don't think that showed anything having to do with concentration. But there is a lot of information on the food chain that I couldn't possibly go into.

McLEAN: Could I ask you a question that's been coming up throughout this meeting abou the problems of re-emission. Was this studied?

VOLCHOK: Resuspension has not been extensively studied for strontium 90. The only thing that I've seen was by the English at Harwell. They are doing some studies of resuspension of the surface ocean layer, and seeing some small amount of re-emission in that way. But from land I don't think there have been any studies. We would probably not be able to sense it as it would undoubtedly be a small fraction of the stratospheric fallout.

ELIAS: You made brief reference to the analyses of lead, with respect to your collector, so I really won't hold you to this, but perhaps the materials that are used in this collector which I believe is linear polyethylene or perhaps polypropylene. And these are extremely high in lead. Lead is used as a catalyst to produce these materials. How do you control contamination from that?

VOLCHOK: Well, I really didn't spend any time in my talk on
quality control, which, incidentally, we do a lot of. We run
blanks on the collectors regularly by putting water of the same
pH as rain into clean buckets and letting them sit and rinsing
them out. We don't have a high lead blank.

WILSON: With your present radioactive monitoring system, have
you picked up anything with respect to radioactivity that could
be associated with the combusion of so-called dirty coal?

VOLCHOK: We're not analyzing the network for the emissions
that you get from coal, which is natural radioactivity: mainly
the daughters of radium. We have studied that particular
problem in another way, and so have others in the business.
The conclusion we've come to, which is not part of my talk, and
it's been published, is that the total emission of
radioactivity for the entire coal cycle is a small fraction of
the total emissions of radioactivity from the nuclear fuel
cycle. That's not the power plants but the total cycle.

HICKS: Is the flux to water greatly dissimilar than that to
land?

VOLCHOK: In the early days we thought it might be
substantially higher because we used to think that water would
be a better scrubber of the dry fallout. It turns out that
when you integrate all of the Sr-90 in the oceans, just as you
people integrated the Great Lakes, the flux to land and sea is
the same.

HICKS: How confident about that are you, because you don't
have any soil samples with which to compare these?

VOLCHOK: We have a good deal of sediment.

TORIBARA: Does your natural background include the amount of
potassium that everyone has in his body?

VOLCHOK: Yes, the natural background that I showed in that
chart, consisted of three parts which were about equal -- the
internal material in your body, the external material from the
ground, and cosmic radiation from the sun.

TORIBARA: We made a calculation recently and came out with
something like 35 millirems per year.

VOLCHOK: I think the U.S. average is 26.

TORIBARA: It all depends on your size.

EMISSION QUOTAS FOR MAINTAINING

AIR QUALITY

Daniel R. Mandelker

Stamper Professor of Law, Washington University
School of Law
St. Louis, Missouri

ABSTRACT

In order to control the air pollution, the Environmental
Protection Agency (EPA) has adopted the National Ambient Air
Quality Standards (NAAQS) as required by the Clean Air Act. The
emission quota is applied to major industrial pollutants and
particularly to the sulfur oxides in an attempt to maintain the
standards over an air quality control region. It is a
regionalized and not an interstate control. The scientists at
Argonne have put together a book which tells how to put together
an emission control quota in a local region.

Regions may be categorized as those where the NAAQS are
already violated, where the standards are met and finally where
the air is much cleaner than required. The actions to be taken
in the various categories are discussed. Of special interest is
the situation where additional pollution is allowable--the
problems being those of an equitable and wise apportionment of
the increased pollution. Four emission quota strategies are
considered.

Introduction

Emission quotas, which include emission density zoning as a
quota variant, continue to receive attention as an air pollution
control technique. Their basic function is to relate allowable
increments in pollution to designated land uses. They achieve

compliance with the National Ambient Air Quality Standards while
guiding localized "hot spots" of pollution which exceed the
Standards. Work I have recently completed with the Argonne
National Laboratory, and earlier with Rutgers University, has
explored the potential and legal implications of the emission
quota idea. I want to share the results of that research with
you this morning.

A recent report on the legal issues based on the Argonne
study has been issued by the EPA entitled "Legal Issues of
Emission Density Zoning" (EPA-450/3-78-049). A much thicker
report has been written by the scientists at Argonne and is
called an emission density zoning cookbook which tells how to
put an emission quota together in a local air quality control
region. It is entitled "The Emission Density Zoning Guidebook"
(EPA-450/3-78-048).

Generally, the emission quota as we see it is applied to the
major industrial pollutants and particularly to the sulfur
oxides. We view it presently as a method for controlling new
industrial locations. The reason for that lies in the
difficulty of developing workable models for some of the other
pollutants, especially for the mobile source related pollutants,
although some work has been done on extending the quota idea
beyond the industrial context.

The Statutory Background

Basically, the emission quota is an attempt to maintain the
NAAQS, the National Ambient Air Quality Standards, over an air
quality control region. It is a regionalized control, not an
interstate control. It is based upon the regional control
concept the statute contains and can be effective in controlling
industrial pollutants within the region but does not have an
interstate dimension. Where there is a bistate region, such as
the St. Louis, Missouri-Illinois bistate region, it is possible
to develop an emission quota for that region. In that sense it
will be interstate, but the modeling capability that has been
developed so far at Argonne has not stretched far enough to go
beyond a particular air basin. That is the basic idea behind
the emission quota concept.

The emission quota starts with the NAAQS and modifies them
as a control technique, and I will indicate just exactly how
that happens and why it is done. I thought it might first be
helpful to give you some idea about the quota concept, and
indeed to indicate that the Clean Air Act also imposes a quota

on new industrial development in this country even without an explicit emission quota. This is one of the unspoken, little discussed issues that lies behind the statute that needs a lot more attention.

The NAAQS themselves impose a quota based on pollution increments. Of course, practically every major metropolitan area in the country is a nonattainment area for at least one pollutant. In these areas the NAAQS are violated and no additional pollution is allowable. Let us look instead at the control problem for an area which does not violate the NAAQS for at least some of the pollutants. The law allows increments to existing air pollution in this area provided the NAAQS are not violated. Additional polluting development may occur and the air pollution baseline may be thickened as pollutants are added.

Necessarily then, the Clean Air Act apportions remaining emissions around the country. The Clean Air Act apportions remaining amounts of clean air by setting up the NAAQS as a constraint over additions to the air pollution baseline in areas where the NAAQS are not yet violated. It is necessarily a quota on new polluting industrial development, applied to air quality control regions. The emission quota concept simply makes explicit the air pollution quota assignment implicit in the Clean Air Act.

I have indicated that the emission quota is applied to an air quality control region. As most of you know, the air quality control regions are divided into several categories. In the nonattainment areas the NAAQS are violated and no more growth is allowed until the NAAQS are met. In these areas, the emission quota idea does not apply unless the baseline is improved until the NAAQS are no longer violated.

In the air quality management areas the NAAQS have been met and must be maintained. Here there could be an air pollution control problem if growth and development is not watched very carefully so that the NAAQS are not violated over a period of time after attainment has been achieved. We think the emission quota idea can be most useful in these maintenance areas. I might add that apparently EPA is reconsidering its air quality maintenance strategy. They may be coming to the belief that it is no longer necessary and can be folded into their nonattainment area strategy. But the statute and EPA regulations still carry the maintenance planning requirement, and so far as we know it has not yet been officially dropped.

In No Significant Deterioration (NSD) or nondegradation areas, the air is presently cleaner than what the national

standards require. In those areas, depending upon what class
the area is in, the Clean Air Act allows only modest increments
in pollution. The Act imposes a quota on new development which
is structured by the amount of additional pollution allowed in
the area over the existing baseline.

It is very interesting to speculate on how the additional
pollution increments are to be allocated in areas like the
nondegradation and maintenance areas where some additional
growth is allowable. It is not clear just how that growth is
going to occur. How shall we allocate the additional increments
in pollution that are permitted in these areas?

So far the understanding in EPA has been that additional
pollution increments in the nondegradation areas will be
allocated on a first-come, first-served basis which is simple
enough. I am an industrial developer, and I want to come into a
Class III nondegradation area. An X increment in pollution is
allowable. I use up all of that. If you come along later and
wish to build in that same area, you will not be allowed to do
so because I have used up all of the pollution increment that's
available.

Obviously, this method for allocating pollution increments
may not be the most satisfactory we could put together. In the
legal analysis for our emission quota work, we looked at other
methods for allocating pollution increments, other than the
first-come, first-served method. These are discussed later.
They create so many problems that we more or less decided that
first-come, first-served provides a kind of rough justice. EPA
is still encouraging other methods for allocating emission in
nondegradation areas, including the emission quota. It was
proposed as an alternative for dealing with nondegradation air
quality problems in 1973 but was dropped by EPA at that time
because of technical difficulties, particularly in dealing with
modeling problems. We shall see whether the emission quota can
solve the problems created under the first-come, first-served
allocation method.

The Emission Quota Idea

Well, why have emission quotas? And how do they differ from
the usual controls imposed by the NAAQS?

The emission quota links the amount of air pollution
allowable with land and land use. The NAAQS don't say anything
about land use. That is the next point to make, the next

distinction between the emission quota and the NAAQS as an air pollution control technique. The NAAQS simply say, in an area where the NAAQS will be violated unless additional pollution is controlled, that a given additional pollution increment is allowable. Nothing is said about land use control, in the Act. In fact, the entire NAAQS strategy, does impose land use controls on the country, in spite of Congressional disclaimers. Don't tell your Senator or Representative about this, because they were very burned on land use controls a few years ago when the National Land Use Act was defeated. There is language throughout the Clean Air Act's legislative history stating, I think quite innocently, that the Act does not impose land use controls on the country. It does, of course. As soon as the NAAQS are violated, additional polluting land development must be disallowed.

This kind of indirect, back door, land use planning in the Clean Air Act may result in decisions on locating development consistent with the NAAQS but inconsistent with local land use controls and planning. There may be a real discordance here that impedes the entire program. This issue has not yet been addressed directly. What is the state air quality agency going to do, for example, if a local zoning ordinance permits industrial development and they do not? Well, they have an override power; there is no question about it, they have and will override zoning when that happens.

It would be much better, of course, to develop a system under which the application of air quality controls could be made consistent with local land use planning and controls. That is one of the objectives of the emission quota idea.

The second is to deal with "hot spots". These are violations of the NAAQS that occur in localized areas in an air quality control region due to the location of many polluters next to each other, even though region-wide the NAAQS are achieved. We are required by the Clean Air Act to deal with pollution on a regional basis, and so can be legally indifferent to the "hot spot". This indifference is not justified. The "hot spot" creates an air pollution violation at the point where it occurs. It is dangerous to health at that point and should be subject to a control that will avoid it. The cookbook accompanying our legal report emphasized control of hot spots as the major objective in devising the emission quota technique.

Emission quotas deal with these problems in two ways. First, the NAAQS, which measure pollution in the ambient air, are converted into a pollution- loading standard usually expressed as so much pollution per acre per time unit. Argonne

used seconds in their report but there are other possibilities.
This standard provides a direct link between the land use
expected on the property and the pollution which it is likely to
produce.

Secondly, emission quotas match the pollution-loading factor
to local plans and controls. The cookbook starts with local
zoning, which is on the ground, is mapped, and is easily
available. The cookbook starts with local zoning for industrial
use because we are most concerned with the industrial
pollutants. We could start with the local land use plan as an
alternative, but the land use designations in the plan are not
as detailed and legally binding as the use designations in the
zoning ordinance. For this reason, the cookbook starts with the
local industrial zoning.

The cookbook then takes factors, primarily topography and
weather climatology, to develop a model that indicates the
dispersion of the pollutant around each source and its effect on
receptors. With that as a base and the NAAQS as a control, the
model assigns the pollution determined to be allowable to each
acre of land in the industrial zone. The model is calibrated to
maximize industrial growth consistent with the NAAQS. The model
makes no assumptions about the zoning program; the model simply
takes existing zoning and adopts a growth maximizing objective
for new industrial development consistent with the NAAQS.

This approach does not mean that all industrially zoned
areas can be developed in the muncipality. Part of the reason
is that the relationship between classes of industrial use and
the amount of pollution generated by each use is not very good.
Zoning ordinances are not based on the amount of pollution a
particular industry will emit once it is built. They are based
on other categories--heavy and light, for example--which relate
to the type of manufacturing and not the volume of emissions
produced. Depending upon what kind of industry locates where,
the quota does not guarantee that all of the industrially zoned
areas can be developed without violating the NAAQS. The hope is
that this can occur, and certainly the model attempts that
objective.

I might also point out that the system I have been
discussing so far implies mapped industrial zoning. It implies
that communities have or will want to map industrially zoned
land in advance of its development. That is the key on which
this emission quota is based. If municipalities don't do that,
and it is increasingly common not to do so, they may use a
zoning technique called a "floating zone" under which industrial
development is approved on a case-by-case basis. It will then

be necessary to apply the quota somewhat differently than the way I've described it, although the basic technique for assigning pollution loads to land remains the same.

Emission Quota Variants

Let me next indicate the quota types that have been suggested and the allocation techniques that can be used to implement them.

We developed a four-way classification of quota types, each of which has its advantages and disadvantages. Each of them, however, is based on the underlying assumption that pollution loads are assigned to industrially-zoned land. The difference lies in the way which the administration of the quota is carried out.

The first quota type is known as emission density zoning, which has been the popular term used to describe all of these quota techniques even though in our judgment it describes just one variant. We prefer the phrase "emission quotas" as the generic term. Emission density zoning, or EDZ, is an emission quota which assigns a pollution load directly to units of land. The quota is developed only for industrially-zoned areas, but is then translated into an allocation for each acre of land in those areas.

For example, if I own an acre of land in an industrially-zoned area, I am assigned X amount of pollution per acre per second under EDZ. I then come in and find out that if I build my plant, I will produce 2X units of pollution per acre per second. I must then acquire an additional acre of land so that the air can absorb the rest of the pollution my plant will produce.

That's neat, you think. Well, it is neat. In many ways the EDZ idea is an attractive one. First, it does not require governmental administration as the developer can determine whether he wants to comply. It also has the advantage that it uses the land assembly process as a method of dealing with the pollution problem. There are some legal problems with it, but I think in many ways they are fewer here than with the quota types.

The second type of quota is called the jurisdictional emission quota, or JEQ. What happens here is that the quota is determined regionally and then assigned to local governments

within the region. The local government then decides on its own
how to allocate its emission quota. When there is a discrepancy
between the amount of land an industrial developer holds and the
amount of pollution that will be generated, the EDZ system can
be awkard and unwieldy. The JEQ system has the advantage that a
mediating governmental agency makes this assignment.

The third quota type is the district emission quota, or
DEQ. Under the district emission quota the emission quota is
assigned to zoning districts within communities, rather than to
the community. This quota simply cuts local governments units
out of the picture in the emission quota assignment process and
provides more control at the regional level. Some
administrative and intergovernmental problem are avoided, but
whether regional control will be politically acceptable is
problematic.

A fourth and final type of quota is the floating zone
emission quota, or FZEQ. The FZEQ is determined on a
case-by-case basis. This system has been operating in
Louisville, Kentucky for several years. Under the floating zone
quota, when an industrial developer applies for a source
construction and operation permit, the emission quota is
determined at that time on the basis of the emission quota
assigned to a circle described by a radius of a certain length
which extends out from the plant. A circle is described around
the plant, the emissions available within the circle are
determined and the plant is allowed or not, depending on whether
the emission quota is violated.

This quota can be used in conjunction with the floating
zoning technique for industrial zoning. It has been
successfully used in Louisville. It creates some rather serious
problems if quota areas overlap, because it is difficult to know
who is to get how much of the quota, but it does have the
advantage of flexibility.

Allocation Techniques

How shall we allocate the quotas that are made available?
As I indicated, additional increments of air pollution under the
Clean Air Act have usually been allocated on a first-come,
first-served basis. Once the quota has been exhausted, no more
development is allowed under this approach until the baseline is
improved. Improvement can occur because of plant closings or if
existing plants use improved technology that reduces emissions.

The question is whether this method fairly allocates the emission quota. This problem is important because the Clean Air Act is one of those cross- cutting regulatory programs that affects industry all over the country. It affects industry everywhere, and we ought to be aware of what the effects of these controls are going to be on industrial opportunities.

The alternative to the first-come, first-served approach is to impose limits of some kind. The argument could be made that we do not want the early industrial developer who is greedy and uses up all of the quota because we may feel that industrial development in the community is better served by using up the quota over a period of time. This objective can be met if an annual limit is imposed on the additional pollution to be allowed under the quota in any one year. An annual limit will close out some of the larger, earlier developers who otherwise would use up a substantial share of the quota. Another alternative is to limit the increment in pollution allowable to any one industrial developer. That allocation method has the same effect as an annual quota.

These alternatives are attractive; they certainly seem fair. One argument against them is that the amount of pollution available annually or to any one developer need not be limited. After all, the problem is compliance with the NAAQS and we need not care how compliance is achieved, or when. If we find that the quota is used up quickly in six months or a year rather than stretched out over three or four, that arguably does not make any difference to air pollution control since our objective after all is NAAQS attainment and maintenance.

The contrary argument is that time is always valuable to a pollution control strategy which depends to some extent on baseline improvement through improved technology. With more time perhaps we can develop improved emission control technologies and we may find some plants leaving the area or closing. We may be better able to distribute available emissions over a period of time if we limit the emissions available at any one point in time.

Another problem is that the allocation of the quota may be un- fair if the first-come, first-served method is used. The distri- bution of available emissions may be determined by a race for the quota and the argument can be made that the more heavily financed and successful industries will use up the quota to the exclusion of others. The utility industry, for example, is in a very bad state at the moment in terms of its ability to raise capital. They might be closed out under the first-come, first-served allocation technique by other industries not in that position.

I do not have a definite solution to the allocation problem.
I certainly think the courts would accept any reasonable quota
allocation technique because of the difficulties in formulating
any technique that is clearly superior to any other.

Emission density zoning does not have quota allocation
problems because the developer decides where to develop and how
much land to acquire. All the developer need do is buy enough
land to satisfy the quota and he can then build at any location
to which the quota applies. He may have to acquire expensive
buffer zones. This possibility raises another kind of legal
problem we call "the taking problem". It may be mitigated
because a developer who owns land under EDZ always has attached
to the land a certain number of emission rights and can always
sell those rights to others. He has a marketable commodity which
to some extent mitigates whatever adverse impact the quota has on
his right to develop.

Another final possibility to consider briefly is a technique
we know as transfer of development rights, or TDR, which we did
not consider in our report because we thought it added too
complicated a layer of administration. An emission quota spread
across an air quality control region would be assisted by a
trading system that allowed developers to trade their emission
rights within the region from one to the other. That way we
could achieve the NAAQS and obtain a better distribution of
industrial development. We would not have to be so area-bound in
distributing the available emission rights allowable under the
quota.

TDR is an interesting concept, but we have quite enough
problems to worry about without going that far at the moment. It
is useful primarily as applied to Emission Density Zoning, even
apart from TDR, the exploration of these possibilities for
emission quotas has raised a series of fascinating problems in
the application of legal controls to improve air quality.

DISCUSSION

MORROW: It seems that one of the essential features of this
approach is that somewhere in the beginning you had to have a
model dealing with transport and deposition. You cited the
Argonne model. How does it take into account climatology, let's
say. Does it use a historic average circumstance for that area?
An average wind direction? An average wind velocity?

MANDELKER: It's my understanding they used historic information
to develop that.

MORROW: Then they must also be able to provide how much
fluctuation would be implicit in such a model, probably. Is each
region going to have to develop such a model? You can't take the
Argonne model, if it has topographic information and such
historic information and generalize.

MANDELKER: Argonne developed a method, not an operational model
for one geographic area. That is one of the problems with this
system, since developing a model is fairly expensive and needs to
be updated from time to time. They tell me that once developed
the first time it can be easily updated with new information, but
it does have to be updated.

JACOBSON: I have a basic problem about the philosophy of the
Clean Air Act which leads to the kind of control standards you've
been talking about. As I understand it, the Clean Air Act leaves
the control and implementation of controls to the states and
local agencies within their air quality regions. Now, that
represented our understanding of pollution problems in previous
decades. Our understanding of the situation at the present time
is different, namely that the the most serious and intransigent
air pollution problems are those involving long distance
transport from one state to another, from one region into
another. New York State cannot control its acid rain problem or
its photochemical ozone problem by controlling emissions within
New York State. In fact, I understand there are law suits
referring to this problem. How do you think that the overall
system needs to be changed in order to reflect this long distance
transport problem?

MANDELKER: Mr. Herzberg is going to cover this in his talk; I'll
leave that to him. Let me repeat that the emission quota idea
that I've been talking about is definitely based on the air
quality control region as the area of control, so it does have
that problem.

ELIAS: I think the problem is that the Clean Air Act is a misnomer. It is really a limited Dirty Air Act which says we can only put so much into the dirty air. Now, the problem with your approach--that is buying more land to allow you to put more garbage into the air--is that there's no incentive to reduce the amount of contamination you put into the air. You just simply go out and buy more land and then retain the emission rights to that land. Just like you buy land and retain mineral rights. So what kind of law are we going to get which gives us a Clean Air Act where we have incentives to reduce the amount of contamination that's put in the air?

MANDELKER: Well first, you're right, this technique is based on the threshold concept which underlies the Clean Air Act. The Clean Air Act is not a no-pollution act, it's a limited pollution act, as you point out. I think that the answer to that is, as I understand it, that additional pollution clean up which would bring pollution levels down even further than they are now is simply not cost-benefit justified. We're having enough of a problem with the steel industry and utility industry as it is, getting them to meet existing standards. It is still important to recognize that the Clean Water Act is presently a no-pollution act which has a goal of zero-discharge. The Clean Air Act is not.

Second, with reference to the first part of your question, it's true that EDZ is, in effect, a dispersion technique. You can buy more land to disperse the air pollution. There is some case authority in the courts which indicates that dispersion techniques are not to be allowed. One of the other authors of this report contends that those federal cases would make this system unacceptable under the Clean Air Act. I don't quite agree with him. I don't really think that emission quotas are in the same category as tall stacks, which really do disperse over large areas. We are really only developing another method for implementing the NAAQS.

Your question, in addition to raising the threshold problem, raises the question of what type of control technique we favor. To what extent will we favor dispersion? To what extent will we favor improved technology?

ELIAS: There's another implication though, and that is that the more land you have the more the more you can pollute--the pollution is not dumped on the land you own; it's dumped on the surrounding area. Everybody realizes that, but the law says the more land you have, the more stuff you can dump on everybody else's land.

MANDELKER: First, emission quotas are not law yet. Second, the amount of pollution you're allowed to generate per acre is a sum or an amount which is produced by a model which is going to insure that at least over this air quality region the NAAQS will not be violated. And remember, since the model is based on a source to receptor relationship, the model improves existing air quality controls because it avoids the creation of hot spots. This model is calibrated so that hot spots cannot occur.

WILSON: Is this quota system based solely or almost entirely on industrial emitters?

MANDELKER: At the moment it's been developed for industrial developers only.

WILSON: So it does in fact overlook, for example, what I'm calling heavy residential emissions, where large amounts of fossil fuel are consumed such as in large apartments, buildings and institutions. It seems to me that this is of considerable magnitude when compared to the number of industries, in fact.

MANDELKER: It is, but it's been definitely sliced out of the Clean Air Act. This Act is not being applied to residential development. Some cities control incineration in large apartments and such. The Clean Air Act has a cut-off point and only applies to major sources of pollution. If anything, because of administrative limitations, there is a tendency to cut back even further on the span of the program. So not only is it not a no-pollution statute, it's not even a total control statute because there's lot left out.

WILSON: I have another comment that also relates to Bob Elias' concern. Since you possibly, and wisely so, are going to look at zoning or planning and zoning, are you assuming that implicit in this is that some planning, some day will be effective as a means of getting at these problems?

I feel that my concern goes even further than what Bob has suggested. It isn't simply a matter of incentives to increase control technologies or to give some kind of "brownie points", for that presumably would be another variable that would enter into the picture, but there are surely ways that can be reached through wise planning to reduce the cumulative effect of multiple emitters, for example. That would apply more narrowly to industry. It would also apply in a wider context to transportation systems, or to choices on how you heat and place residential settlements. There is a whole array of planning alternatives which I think we've got to look at with respect to the "Partial Dirty Air Act".

MANDELKER: I agree. I think that emission quotas help with the industrial hot spots, but I think we should understand that the attempt in Congress to inject a land use planning element into this statute has aborted. We don't have it any longer.

THE INTERSTATE CARRIAGE OF POLLUTANTS:

THE LEGAL PROBLEM AND EXISTING SOLUTIONS

Peter J. Herzberg

Sierra Club Legal Defense Fund
1424 K Street, N. W.
Washington, D. C. 20005

I. INTRODUCTION

That air pollution does not confine itself to the state in which it is produced is one of the principal justifications for federal regulation of this environmental hazard. 1/ Yet, to date, the United States Environmental Protection Agency (EPA) has not squarely faced the problem of the wind-blown transport of air pollutants from one state to another. This is especially true with regard to the carriage of sulfates and the formation of oxidants. While some might allege incompetence on the part of the federal agency, it appears that EPA has not dealt with the interstate problem because it knows "which way the wind is blowing," for the control of the interstate carriage of pollutants causes difficult political and economic questions which could raise the ire of local politicians and industrialists already adamantly opposed to assuming the burden of controlling degradation of air quality.

But as the papers given in this seminar demonstrate, the interstate carriage of pollutants is a persistent and serious problem causing documented damage to precious ecosystems as well as damage to real property. Moreover, the interstate transport of pollutants creates unfair economic burdens to be shouldered by the industry of downwind states, perhaps causing losses in the tax base and dislocation for the populations of these downwind

---

1/ See South Terminal Corp. v. EPA, 504 F. 2d 646, 677 (1st Cir. 1974). William V. Luireberg, "The National Quest for Clean Air 1970-1978: Intergovernmental Problems and Some Proposed Solutions," 73 Northwestern Law Review, 432-437 (1978).

states. 2/ Accordingly, the interstate transport problem must be
addressed now and both environmental, as well as economic, damage
from interstate pollution must be mitigated as soon as possible.

This paper is meant to serve as an introduction for the
scientific community to the legal theories presently available to
EPA for confronting the interstate pollution problem. It is hoped
that this paper will set the legal context for current and, hopefully, future scientific research into interstate pollution problems.

II. THE LEGAL MECHANISMS AVAILABLE TO APPROACH THE INTERSTATE
     CARRIAGE OF POLLUTANTS

A. The Federal Common Law of Interstate Nuisance

The common law in the American system is a set of general
principles derived by courts in finding appropriate resolutions
to situations that bring people in our society into conflict. For
instance, monetary recoveries for injuries in our society are,
for the most part, decided on the basis of common law principles.
These common law principles can be overruled by the collective
judgment of society in the form of statutory principles legislated
by Congress or a state legislative body. An example of a legislative rule overruling common law principles is the no-fault statutes
adopted by many states regarding recoveries for property damage
in automobile accidents; these statutory rules override the
common law principle that fault must be proved before a recovery
for damages can be allowed.

In 1972, the United States Supreme Court in Illinois v.
Milwaukee, 3/ confirmed the existence of a federal common law of
interstate nuisance, which grants a state the right to sue a polluting source in another state. The specific case before the Court
involved a claim by the State of Illinois against several municipalities in Wisconsin for dumping sewage in Lake Michigan. However,
the Court described the nature of a nuisance by using an example of
an air pollution situation:

> It is a fair and reasonable demand on the part of a
> sovereign that the air over its territory should not
> be polluted on a great scale by sulfurous acid gas,
> that the forests on its mountains, be they better or

2/ State of New York v. United States Environmental Protection
    Agency, No. 78-1392 (D.C. Cir. 1978).

3/ Illinois v. Milwaukee, 406 U.S. 91, 92 S. Ct. 1385, 31 L.
    Ed. 2d 712 (1972).

worse, and whatever domestic destruction they have
suffered, should not be further destroyed or threat-
ened by the act of persons beyond its control, that
the crops and orchards on its hills should not be
endangered from the same source. 4/

In essence, then, the common law provides a theory for an interstate
air pollution nuisance where pollutants traveling between states
cause a substantial and unreasonable interference with the use
and enjoyment of land in a downwind state. 5/

But while the availability of a common law theory is present
to abate an interstate nuisance, substantive problems exist which
oftentimes render the use of the common law theory virtually
useless. For instance, a state must show that specific actions
of a specific polluter are causing an actual injury. In the
early case of Missouri v. Illinois, 6/ a case which typifies the
problems of isolating causation, the Supreme Court failed to find
fault, thus barring any remedy. In that case, Missouri showed
certain pollution in the Mississippi River around St. Louis and
also showed that the up-river City of Chicago was polluting the
Mississippi watershed (in actuality, Chicago was polluting the
Illinois River which empties into the Mississippi River). But
Missouri could not link specific damages to specific acts of
pollution by Chicago so no recovery of injunctive relief was granted
by the Court.

This failure to be able to establish a link between a specific
polluter's acts and a pollution injury suffered is an especial
problem regarding interstate air pollution. Mathematical models
are the usual way to show a cause-effect relationship as the United
States has no systematic or consistent way of monitoring the changes
that occur in the chemistry of precipitation. 7/ Unfortunately,
these models do not produce accurate results in isolating a
particular source that is causing a downwind problem. 8/

4/ Illinois v. Milwaukee, 406 U.S. at 91, quoting Missouri v.
Illinois, 180 U.S. 208, 238 (1900).

5/ See Restatement, Second, Torts 832, pp. 142-147 (1979).

6/ Missouri v. Illinois, 200 U.S. 496 (1906).

7/ The House Subcommittee on the Environment and the Atmosphere,
Summary of Hearings on Research and Development Related to
Sulfates in the Atmosphere, 94th Cong., 1st Sess. (1975).

Moreover, the federal common law of interstate nuisance can only be used to remedy an existing air pollution problem. It cannot be easily used to prevent a nuisance from occuring, because the essence of the action is to link a specific action of a polluter to an existing injury sustained. As some injuries from pollution can be extremely serious, the inability to prevent injury renders the common law of interstate nuisance meaningless in protecting society from future pollution damages.

Additionally, the common law remedy can only be used by states to abate the interstate nuisance. Individual citizens, even those whose property was being damaged, cannot sue under this particular theory, for the common law theory is based solely on a state's sovereign interest in our government's federal system. 9/

Due to this inability to give definite relief to those injured by interstate pollution, as well as to prevent environmental injuries from occuring, Congress decided to pass legislation to remedy the inadequacies of the common law. The result was the Clean Air Act, which became law in 1970 and which was significantly amended in 1977.

B. The Clean Air Act

While the regulations under the Clean Air Act are quite complicated, the basic structure of this statute is straightforward. In essence, States are required, pursuant to Section 110, to develop State Implementation Plans (SIPs) so that the air quality in the state (or the Air Quality Control Region which might include more than one state) meets National Ambient Air Qualty Standards (NAAQS) and the Prevention of Significant Deterioration (PSD) increments. 10/ The plan must include methods of implementation, maintenance and enforcement so that the state can actually achieve the goals in the SIP. 11/ Moreover, SIPs for areas where air is presently polluted beyond NAAQS (i.e., "non-attainment areas") must contain provisions requiring existing sources to install "reasonably

---

8/ See, e.g. Pierce and Gutfrevno, Evidentiary Aspect of Air Dispersion Modeling and Air Quality Measurements in Environmental Litigation and Administrative Proceedings, 25 Fed'n. of Ins. Counsel Q. 341-347 (1975).

9/ Comm. For Consideration of Jones Falls Sewage System v. Train, 375 F.Supp. 1148 (1974); aff'd en banc 539 F.2d 1006 (4th Cir. 1976).

10/ § 110, 42 U.S.C. §7410; 42 Fed. Reg. 26380 (June 19, 1978).

11/ § 110(a)(1), 42 U.S.C. §7410(a)(1).

available control technology" (RACT) and requiring demonstration
that the non-attainment area will make "reasonable further progress"
toward meeting NAAQS by December 31, 1982. 12/ It must also contain
provisions requiring new facilities with the potential to emit 100
tons or more of pollutants per year to undergo preconstruction review after which a permit may be issued only if: (1) it is shown
that total emissions from all area sources, plus the proposed new/
modified source, will be "sufficiently less than" the total prior
to submission of the permit application; (2) the new/modified
source complies with the stringent "lowest achievable emission rate"
(LAER) standard; and (3) all other major sources owned by the
owner of the proposed new/modified source within the State are also
in compliance with all emission standards under the Act. 13/

By contrast, States which have achieved air quality which is
cleaner than national standards (i.e., "attainment" states) need to
incorporate in their SIPs Part C of the 1977 amendments, known as
the "Prevention of Significant Deterioration" (PSD) regulations.
14/ Because such areas already have "clean" air, the PSD section
does not require any controls at all on existing sources. All new
or modified sources which have the potential to emit less than 100
tons of pollutants per year, or which actually emit 50 tons of
pollutants per year or less, are also exempt from PSD review. 15/
PSD requirements are applicable to 28 categories of new or modified
sources which have the potential to emit more than 100 tons of
pollutants per year, and all other new or modified sources which
have the potential to emit more than 250 tons of pollutants per
year. 16/ While new or modified sources which are subject to PSD
review may not be constructed without a permit, PSD permit requirements are somewhat less stringent than Part D permit requirements.
For example, facilities in PSD areas must adopt "best available
control technology," (BACT) 17/ an easier standard to meet than
LAER required in Part D. For BACT takes into account factors such
as "energy, environmental, and economic impact and other costs" on a
case-by-case basis, none of which are considered in LAER. 18/

12/ § 172, 42 U.S.C. § 7502.

13/ § 173, 42 U.S.C. § 7503.

14/ 43 Fed. Reg. 26385 (June 19, 1978). These regulations are under
     judicial review in Alabama Power Co. v. Costle, No. 78-1006.

15/ § 165, § 169(1), 42 U.S.C. §§ 7475, 7469(1).

16/ Id.

17/ § 169(3), 42 U.S.C. § 7479(3).

18/ § 171(3), 42 U.S.C. § 7501(3).

In essence, the Clean Air Act requires a minimum federal level of control technology on all sources as well as a preconstruction permitting review program, but otherwise grants states the discretion to develop necessary additional standards in the SIP beyond the federal technology standards as long as the additional provisions of the SIP allow the state to meet NAAQS standards and applicable PSD increments. This scheme obviously corrects many of the weaknesses of the common law system -- e.g., a preventive approach can be taken under the permitting review process so that pollution injuries are less likely to occur; no causation between an injury and an act need be shown, as a SIP sets pollution emission levels on the basis of the technological standards as well as standards set to meet NAAQS or applicable PSD increments; also, the Clean Air Act provides for citizen suits so that individuals may sue for emission violations.

But this scheme in itself still has weaknesses which inherently do not enable it to handle the interstate abatement problem. First, the SIP system will only control pollution in the receptor state, which state might bear an undue burden in having to control the interstate problem; i.e., an upwind state will not need to install control equipment meeting stricter standards than the technology standards required by the Clean Air Act as the upwind state itself is likely to meet NAAQS and the applicable PSD increment even though its emissions interfere with NAAQS or attainment or PSD increments in the downwind state. [19] Moreover, this outcome is also encouraged by the fact that the impact of interstate pollution will be in a different state, the citizens to which the upwind state's politicians are not accountable. Second, the minimal power accorded the federal government in this scheme, as opposed to the significant discretion given to the states, leaves the federal government without the ability to force a SIP to control to better than NAAQS levels so as to protect citizens of another state from interstate pollution. In fact, present EPA regulations do not require upwind states to control their contributions to the pollution of other states at all, but put the whole burden of control on the receptor state: "Pollutants entering a state from sources in neighboring states...and contributing to the violation of a NAAQS in a nonattainment area must be included in the demonstration of reasonable further progress and attainment." [20]

---

[19] Even if an upwind state did not meet NAAQS or the applicable PSD increment, it would only need to devise a SIP meeting these standards even though pollution from the state would still contribute to a problem in another state after meeting NAAQS and PSD.

[20] 44 Fed. Reg. 20372, 20378 (April 4, 1979). See also, 43 Fed. Reg. 21673, 21674 (May 19, 1978).

Realizing that the basic Clean Air Act scheme would not remedy existing interstate pollution problems, Congress, while amending the Clean Air Act in 1977, passed two provisions to deal directly with the interstate pollution problem. Unfortunately, EPA has not implemented either provision and, consequently, has not met the interstate pollution problem pursuant to Congress' mandate.

Section 110(a)(2)(E) attacks the interstate problem directly:

...The Administrator shall...approve...[a SIP]...if he determines that it was adopted after reasonable notice and hearing and that...(E) it contains adequate provisions (i) prohibiting any stationary source within the State from emitting any air pollutant in amounts which will (I) prevent attainment or maintenance by any other State of any such national primary or secondary ambient air quality standard, or (II) interfere with measures required to be included in the applicable implementation plan for any other State under Part C to prevent significant deterioration of air quality or to protect visibility, and (ii) insuring compliance with the requirements of Section 126, relating to interstate pollution abatement;

The Senate Report 21/ explains the Congressional intent in legislating this provision:

The]1970] act did not specify any abatement procedure in the event that a stationary source in one State did emit air pollutants which adversely affected the air quality control efforts of another State. As a result, no interstate enforcement actions have taken place, resulting in serious inequities among several States, where one State may have more stringent implementation plan requirements than another State. For example, an implementation plan for the State of Ohio was not even proposed until 1976. It has now been challenged in court and has not yet been effectively implemented. As a result, there are no enforceable control requirements applicable to most of the significant major

---

21/ At the conference on the Clean Air Act, the Senate version of this provision was concurred in by the House with only minor amendments. House Conference Report, No. 95-564, 95th Con., 1st Sess., pp. 145-146 (August 3, 1977).

stationary sources of sulfur oxides in Ohio. The emission
from plants in Ohio are transported across the Ohio River
to West Virginia, which must then cope with pollution not
generated by a source under its own control; and must
require more stringent control of West Virginia sources
to attain the ambient air quality standards.

In the absence of interstate abatement procedures,
those plants in States with more stringent control requirements are at a distinct economic and competitive
disadvantage. This new provision is intended to equalize
the positions of States with respect to interstate pollution
by making a source at least as responsible for polluting
another State as it would be for polluting its own State.

Additionally, Congress provided an administrative procedure
for downwind states to enforce an upwind state's SIP in Section
126. This administrative avenue of enforcement was provided as
additional remedy to a state in addition to the state's ability
to bring a citizen's suit:

This petition process is intended to expedite, not
delay, resolution of interstate pollution conflicts. Thus,
it should not be viewed as an administrative remedy which
must be exhausted prior to bringing suit under Section 304
of the Act. Rather, the committee intends to create a
second and entirely alternative method and basis for
preventing and abating pollution. 22/

But while Congress has attempted to solve the interstate
problem by requiring SIPs to adopt provisions so that interference
with NAAQS or PSD increments in downwind states does not occur, as
well as by providing an administrative remedy to enforce such SIP
provisions, EPA has done nothing to implement this scheme. State
SIPs need not take into account interference with air quality in
a downwind state. 23/ Nor have regulations been drafted to set
up a Section 126 process. In fact, EPA has simply not dealt with
the interstate problem at all.

Accordingly, some law suits have been filed. For instance,
the State of New Jersey has appealed EPA's designation of those
areas in the country that EPA considered in attainment for oxidants.
24/ The theory behind New Jersey's suit is that EPA's attainment

---

22/ H. Rep. 95-294, 95th Cong., 1st Sess., p. 331 (May 12, 1977).

23/ 44 Fed. Reg. 20372, 20378 (April 4, 1979).

24/ State of New Jersey v. United States Environmental Protection
     Agency, No. 78-1392 (D.C. Cir. 1978).

designations were improper, as emission from upwind states as far away as Minnesota and Texas were contributing to New Jersey's difficulty in meeting NAAQS for oxidants. While New Jersey's theory of regional nonattainment is appealing, it has unfortunately raised the interstate pollution problem on an appeal of an EPA action that is probably legally justifiable and an EPA action to which Section 110(a)(2)(E) does not appear applicable. In other words, as often happens in our legal system, New Jersey's theory is sustainable but the procedure by which it was raised is inappropriate. Therefore, I do not have much hope for the New Jersey suit.

On the other hand, an appeal regarding interstate pollution has been taken in the United States Court of Appeals for the Third Circuit. 25/ In this case, West Virginia -- with EPA approval -- has granted a variance from meeting air pollution standards to two West Virginia coal-fired electric generating stations, thereby allowing those sources to burn coal with a higher sulfur content than previously allowed. Plaintiffs allege that this higher sulfur coal will cause interference with NAAQS standards and PSD increments in neighboring Pennsylvania as well as with Pennsylvania's own sulfate standard. 26/ Section 110(2)(E) is directly applicable in this case. Accordingly, it appears that this case has a greater chance, from a procedural viewpoint, for success in placing the interstate pollution problem squarely before a court. Moreover, in my mind, the legal issues regarding the applicability of Section 110(a)(2)(E) to this situation is clear; the only significant question in this litigation is whether plaintiffs can factually show interference with NAAQS standards, applicable PSD increments and Pennsylvania air quality standards. A court decision on this matter should be made within a year.

But a win in the above case will not solve for all times the interstate pollution problem. This can be seen from the result of the West Virginia case; at best, the variance will be voided and the original SIP provision, which does not cause interference with Pennsylvania air quality, will be made applicable to the two West Virginia electric generating stations. Yet a difficult policy problem lurks in the wings with regard to existing SIPs that, unlike West Virginia's original SIP, do not meet the Section 110(2)(E) requirements of not interfering with NAAQS attainment and PSD increments in downwind states -- i.e., what requirements under Section 110(a)(2)(E) are these inadequate SIPs to meet? For

---

25/ Council of Senior West Virginians et al. v. Costle, No. 79-1026 and consolidated cases (3rd Cir. 1979).

26/ Union Electric Co. v. EPA, 427 U.S. 246, 96 S.Ct. 2518, 49 L. Ed. 2d 474 (1976).

if a state's sources do interfere with NAAQS attainment or PSD
increments in another state, how are the two states to allocate
the burden of control on the interfering sources in the upwind
state and the polluting sources in the downwind state? As an
example, should sources in the receptor state control its sources
at a higher standard and cost than the sources of the interfering
source? Or should they be the same? Or should each state control
its emissions, including its interfering emissions, at a level
necessary to assure that NAAQS and PSD is not interfered with
in either state? Or should a regional implementation plan be pro-
mulgated (but who should undertake this task -- EPA or the states)?
Or should the burden be placed on those sources that will be able
to reduce pollution for the problem pollutants with the least
marginal cost to its operations while being financed for its
pollution reductions by equal contribution from all regional sources
contributing to the pollution problem?

   Accordingly, the interstate problem is far away from a solution.
I expect many more suits to be filed in the near future forcing EPA
to face up to the interstate problem. I hope that these will
eventually lead EPA to forthrightly consider that interstate problem.
When it does, however, I also hope that the scientific community
will have assembled the necessary data for EPA to make a wise and
thorough decision.

DISCUSSION

KRUPA: I'm from Minnesota, and I think there may be a mixup in somebody's data because we just finished doing a reverse trajectory model and we can acount for 50% roughly of the oxidant found in Minnesota from the Ohio Valley. The point I'm making is that you're stating that the transport of oxidants is from the Midwest to New Jersey; I think it's the other way particularly during June-September. In the summertime the trajectory goes from east to west.

HERZBERG: The eastern states have been yelling about the interstate problem; the western states have not. I too have seen recent studies that do show the sloping back, that interstate pollution is not necessarily a west-to-east problem; but that it can also be an east-to-west problem. However, I understand that there are less days where there is an east-to-west problem than there is an west-to-east problem. The point is, though, that EPA has not attempted to fairly allocate the interstate pollutants.

JACOBSON: What about pollutants that do not have an ambient air quality standard. For example, one of the areas that have been discussed at this meeting is the effects of acidic precipitation in the Adirondacks. What kind of proof, what magnitude of information would be necessary to actually influence this question of interstate transport?

HERZBERG: An example of that is sulfates, for which there is no NAAQS.

        The Sierra Club in fact brought a suit to get an NAAQS for sulfates promulgated, I believe a year or two ago. By the time they got to court, though, the new administration had come into office and said that they were going to look at the sulfate problem. It is hoped by the Sierra Club that the Carter Administration's inquiries would lead to a sulfate standard. The Sierra Club dropped its suit voluntarily on the basis of that promise by the agency, by the new personnel. I understand that EPA did begin to look at the problem, but they've dropped looking at the problem, and I understand that there's not really much research going on at Research Triangle Park. Now, this information is second hand so I can only report to you what I have second hand. My understanding is right now there's really not much action as far as the sulfate standard is concerned.

JACOBSON: Suppose there is no sulfate standard. There's never going to be an acid rain standard, certainly. But there is a problem. How do you approach that?

HERZBERG: I think you have to force the promulgation of an
NAAQS for sulfates or acid rain, or must use the total
suspended particulate NAAQS, or one of the other existing
standards like ozone. That's the way the agency has been
approaching it, we think, up to this point. At least that's
true as far as my understanding of what's going on in
Pennsylvania. But this approach is inadequate. It just
doesn't give you a handle. A lawyer feels very uncomfortable
going into court with total suspended TSP standard on an
interstate problem; at least I do. I think we might be forced
to do it, but that's not to say we wouldn't be uncomfortable
because the logical connections and easy flak that an expert
witness can get while testifying on a TSP standard having an
impact in another state could be really devastating. But I'm
afraid it might be the situation we are forced to unless we
have a sulfate standard, or some better oxidant standard, or
some acid rain standard.

By the way, the Sierra Club and a lot of
environmental groups, as well as the oil industry, have
appealed the recent change in the oxidant standard. The oil
industry is saying that the change was too low; the
environmental groups saying it was much too high. I suspect
that the standard will stay the way EPA has changed it, except
for maybe the secondary standard where I think that EPA has a
lot of problems.

STENSLAND: Could you comment on the water quality laws. That
is, have they solved the interstate transportation pollutants
simply by saying zero discharge everywhere? And then if so,
and they change this, are they going to have to come back and
address the same problem? And then part B is, don't you have
to be considering that problem at the same time you're
considering the air quality problem?

HERZBERG: I think the water quality problem is under better
control than the air pollution problem. Let's put it this
way: The facade is out there that it's under better control.
I don't know in fact whether it is.

Under the water pollution laws, there are regional
planning agencies and there are basin plans for rivers. In
these basin plans for rivers, you can take into account the
interstate impacts from various pollution sources. So
theoretically the interstate problem should be under better
control. Whether in fact they are, I don't know. I haven't
really looked in depth at that.

I'm not sure that EPA has better models, but at least I am saying that theoretically they are looking at the problem, which is a substantial advance over what EPA is doing in the air situation. In the water area, I think they're saying even if they don't have better models, they're using the models anyhow. They're not saying that in the air area. EPA is just basically throwing up their hands and saying "we don't have the models at least right now so we're not going to do anything about it".

Finally, I do think there is an interrelationship between the air and water pollution problems, especially in those situations where acid rain caused by long-range transport of air pollutants impacts a water body by effecting its pH. However, I think this problem can be approached from an air pollution perspective-at least, from a legal standpoint, that approach gives an adequate handle.

POLLUTION PROBLEMS ACROSS INTERNATIONAL BOUNDARIES

Erik Lykke, Director General

Ministry of Environment

Oslo, Norway

ABSTRACT

On the basis of available data which indicated that airborne
pollutants were transported distances of hundreds of kilometers,
the Nordic countries decided in 1970 to propose the OECD conduct
a study of the transport of air pollutants. The OECD in 1972
adopted a "Co-operative Technical Programme to Measure the Long
Range Transport of Air Pollutants." This programme covering
Northwest Europe focussed primarily on sulphur dioxide and
sulphur compounds and was concluded in 1977 after confirming
that these compounds do travel hundreds of kilometers across
international borders.

Because any effective control must include both Eastern and
Western European countries, Norway in 1975 proposed the
establishment of a European-wide monitoring programme for air
pollutants. Such a programme was adopted in 1977 in
coordination with the World Meteorological Organization, and the
first phase was started in January of 1978. The programme will
continue for three years. Ten countries initially participated
and were joined by Belgium in 1974.

An international conference on the effects of acid precipitation was organized in Norway in 1976 with another planned for
1980. In April of 1979, 35 governments participating in the
United Nations Economic Commission for Europe adopted a draft
Convention on Transboundary Air Pollution. This Convention
requires signatory states to do all in their powers to limit and
reduce, as far as possible, air pollution. Thus, a first
international cooperation to combat air pollution has begun.

INTRODUCTION*

During the last decade, it has become increasingly evident that air pollution is not only a problem of densely populated and heavy industrialized areas, but also affects large geographical regions far beyond the major sources of pollution.

The degree of concentration of some pollutants at "black spots" has often been reduced or stabilized, but some pollutants, such as those sulphur compounds that are transported long distances by air, now affect much wider areas. It is now generally recognized that sulphur dioxide and particulate matter, in addition to being pollutants in themselves, react chemically and physically in the atmosphere. In particular they combine and then form aerosols or minute quantities of solid or liquid matter dispersed throughout the air. The resulting pollutants affect health, reduce visibility, and cause acid rain.

A number of substances may have long term effects even in small quantities. While it may be true that acute ill-health and death caused by short term exposure to intense loads of pollution have decreased, risks to humans of long term effects from exposure to substances that may give rise to genetic changes, cancer and deformities at birth, are at present causing considerable concern. Several countries have monitored trace elements in the air and in rainfalls and have generally found them contained in fine particulates. They are also known to travel over long distances in considerable quantities. Analyses of Greenland ice layers confirm this for tetraethyl lead. Many halogenated organic compounds also travel long distances through the atmosphere.

It has also been established that pollutants not only are transported from one country to another and across the oceans, but even, as in the case of chloro-fluro-methanes influence the delicate processes in the stratosphere upon which life on earth depends.

Photochemical pollution may also have an international dimension since studies conducted in Europe and Eastern North America have established that photochemical pollutants and their precursors can be transported up to several hundred kilometers.

---

*See in particular OECD: "The state of the environment", a report presented to the Environment Committee meeting at Ministerial level, May 7-8, 1979.

The potential long-term effect of air pollution on climate is at present causing considerable concern. Increased concentrations of carbon dioxide and other pollutants may cause a rise in global air temperature leading to changes in precipitation affecting croplands and deserts. Sulphates and other fine particulates that affect visibility could also reduce solar radiation.

For all pollutants where long range transport has been established and may have adverse effects, the implication is that emission control on a local scale is insufficient and that regional or multi-national approaches to emission control must be developed.

THE OECD STUDY ON LONG RANGE TRANSPORT OF SULPHUR COMPOUNDS*

In 1968 a first analysis of European data showed that a central area of acid precipitation in Europe was expanding from year to year. Emissions of sulphur oxides for the region as a whole had risen by a factor of 2 since 1950. The emissions of nitrogen oxides, which are also important in relation to atmospheric chemistry and precipitation acidity**, had increased even more rapidly than those of sulphur dioxide. On the basis of the available data the Nordic countries decided in 1970 to propose that the OECD should conduct a study of the transport of air pollutants and acidification of precipitation in Europe.

After further preparations and discussions the OECD in 1972 adopted a "Co-operative Technical Program to Measure the Long Range Transport of Air Pollutants". Initially, ten countries decided to participate - Austria, Denmark, Germany, Finland, France, the Netherlands, Norway, Sweden, Switzerland and the United Kingdom - and in 1974 Belgium also joined.

The objective of the programme was to determine the relative importance of local and distant sources of sulphur compounds in terms of their contribution to the air pollution over a region, special attention being paid to the question of acidity in

---

*"The OECD Programme on Long Range Transport of Air Pollutants. Measurement and Findings." Paris, 1977.

**At the present time, it seems to be widely assumed that in broad terms, sulphur dioxide is responsible for about two-thirds of the acidity in precipitation and nitrogen oxides are responsible for about one-third.

precipitation. The co-ordination of measurements, data
analysis, modelling, and preparation of the final report was
the responsibility of a Central Co-ordinating Unit at the
Norwegian Institute for Air Research. The programme was
supervised by a Steering Committee composed of representatives
of the participating countries.

The research programme which was developed covered Northwest
Europe and comprised three essential elements:

1.   a survey of sulphur dioxide emissions all over the
     region.

2.   the measurement of sulphur compounds, essentially the
     concentration in the air and deposition through
     atmospheric precipitation, at a network of ground
     sampling stations and by aircraft.

3.   the development and testing of mathematical air
     dispersion models to relate emission data with
     concentration and deposition data.

The programme was concluded in 1977. It confirmed that sulphur
compounds do travel several hundred kilometers or more in the
atmosphere and that the air quality in any European country is
measurably affected by emissions from other European countries.

In general the concentrations and the resulting total
deposition of sulphur compounds are at a maximum in the major
emission areas and decline with increasing distance from them.
However, certain localized areas, e.g. southern Scandinavia and
Switzerland, have higher total deposition figures than would be
expected by their distance from the major sources. This
situation is brought about by high amounts of wet deposition
due to a greater incidence of precipitation than in neighboring
areas because of orographic effects.

Due to uncertainties in the emission data and in the model
estimates, the true figure for the contributions from emissions
in one country to depositions in another country for any period
of time is only approximately known. Moreover, because of
meteorological variations, these contributions may differ
considerably from one year to another. This does not, however,
change the overall picture which shows a considerable transport
of sulphur pollutants over the national borders in Europe. The
total deposition in a given country, which is due to foreign
emissions of sulphur dioxide, depends on the size and
geographical position of the country in relation to the major
emission sources.

From the model estimates, countries can be grouped roughly into
net receivers or net donators of sulphur pollution. For 1974
the estimates showed that countries ranged from those with
depositions almost three times their emissions, to others with
emissions three times their depositions. In about half of the
countries in the OECD study, the total deposition estimated for
1974 exceeded the indigenous emissions.

Further, the study showed that in half the countries, the major
part of total estimated deposition in 1974 originated from
foreign emissions. Some countries probably received three
quarters of their deposition from abroad. At the other end of
the range were countries where indigenous sources accounted for
three quarters of the estimated deposition.

The results of the study show clearly that if some countries
find it desirable to reduce substantially the total deposition
of sulphur within their borders, individual national control
programs can achieve only a limited improvement.

The programme provided a first comprehensive insight into the
long distance transport of air pollutants on a continental
scale. In the course of the OECD study several unknowns were
identified which should be taken into account in other research
projects. They relate to emission data, precipitation data,
atmospheric dispersion and deposition processes, atmospheric
chemistry, vertical processes, vertical profile measurements by
aircraft sampling and the improvement of sampling equipment
used in aircrafts.

THE EUROPEAN MONITORING AND EVALUATION PROGRAM

The OECD LRTAP study established that any effective control of
air pollution in Europe must include both Eastern and Western
European countries. When the question of environmental
protection was taken up at the Conference on Security and
Cooperation in Europe in 1975, Norway therefore proposed the
establishment of a European-wide monitoring programme for air
pollutants. The proposal was adopted, and preparatory work was
started within the United Nations Economic Commission for
Europe (ECE) in Geneva. In 1976 a task force was established
to develop a detailed plan for the future program.

The experience from the OECD LRTAP programme had shown that
more specific and sensitive methods were needed for some of the
measurements, particularly for background levels, and that the
measurements program, in addition to sulphur dioxide and

sulfates, should include other chemical components of
importance for the interpretation of the results. To meet
these requirements extensive studies were initiated and
laboratories in the Nordic countries are cooperating in
developing improved methods.

In January 1977, the Task Force presented a detailed plan to
the ECE. In the meantime the World Meteorological Organization
had agreed to coordinate the meteorological part of the work,
and the United Nations Environment Programme had decided to
contribute financially to the international coordination of the
programme, as this would constitute a valuable contribution to
their Global Environmental Monitoring System (GEMS). In the
course of 1977, the ECE countries agreed to adopt the programme
along the recommended lines, and the first meeting of the
Steering Body for the European Monitoring and Evaluation
Programme (EMEP) for the long-range transmission of air
pollutants took place in September 1977.

The chemical part of the programme is being coordinated by a
Chemical Coordination Center established at the Norwegian
Institute for Air Research. The chemical measurements were
initiated in October 1977 and are now regularly reported from
more than 40 stations in about 14 countries, and extensive
intercalibrations are being carried out. The meteorological
part of the work is being coordinated by two Meteorological
Centers, one at the Hydro-Meteorological Institute in Moscow
and the other at the Norwegian Meteorological Institute.

The first regular phase of the programme started in January,
1978 and will last for three years. During this period, work
will concentrate on sulphur dioxide and related compounds.
Using annual emission surveys and trajectory calculations, the
concentration, deposition, and transboundary flux of sulphur
compounds will be continuously estimated in order to evaluate
the atmospheric transport and fate of sulphur dioxide. The
model estimates will be compared with the daily measurements at
the monitoring stations.

In the implementation of this first part of the programme,
advantage will be taken of the experience gained by the OECD
study. Further research and development will be carried out,
when possible, to improve atmospheric modelling and data
evaluation techniques. In connection with the measurement
programme the use of aircraft and remote sensing techniques
will be considered, and the Steering Body of the programme has
recommended that participating countries, when appropriate,
establish pilot stations for the testing and calibration of
methods.

An extension of the EMEP programme to include regular
monitoring of other important air pollutants than sulphur
compounds is envisaged. The current evaluations of the
dispersion of sulphur compounds give a good methodological
basis for examining the transfer of pollutants for which
emissions are difficult to estimate, such as trace elements.
It will also be of interest to find out more about the fate of
that fraction of the pollutants which is transported out of the
European region, and the input into Europe of pollution from
other continents, particularly North America.

EFFECTS OF ACID PRECIPITATION

A special research project "Acid precipitation - Effects on
Forest and Fish" (the SNSF-project) was established in Norway
in 1972 with participation from a number of research
institutions. Work under this project will continue until 1980
and form the basis for a future monitoring programme.

Extensive surveys of fish population and chemistry in Norwegian
lakes south of 63°N have revealed that the majority of the
most acidic lakes have lost their trout population in recent
years. Small lakes at high altitudes are generally first
affected. These reports are similar to the reports of regional
fish extinction in acidified areas in Sweden, Canada, and USA.
Present indications are that inland areas affected by fish
decline are increasing in southeast Norway.

No other environmental factor than water acidity seems to
explain the gradual regional loss of fish population.
Indications are that egg and fry mortality is the main cause
for failing reproduction, but during serious pH-drops,
particularly in spring snowmelt and during heavy autumn rain,
even larger fish are killed. The tolerance to acid stress in
fish depends on many factors, such as content of salts in the
water, fish age, size and genetic background.

Also the dry deposition of acidifying compounds is important.
During episodes of high precipitation in the autumn after
comparatively dry summer periods, there is often a drop in
river water pH, which may become lower than the pH of the
precipitation causing the runoff event. In these cases there
is regularly an increase in the sulphate concentration of the
runoff water, which can be accounted for by deposition, mainly
dry deposition, in preceding periods.

There is a coincidence in regional occurrence both in Europe
and in North America of

1.   acid lakes and rivers, where the acidity has no
      obvious sources in the drainage basins, and

2.   areas with both high deposits of acid precipitation
      and geological conditions which do not favour
      neutralization of acid - quartz-rich bedrock, sparse
      overburden.

This does not mean that only the H⁺ content in the
precipitation is of importance. Especially the increase in
recent decades in the content of mobile sulphate anions in
precipitations may have had great consequences.

Changes in agricultural practices and forestry may contribute
to a soil acidification and probably to acidification of
surface waters in some areas. However, available statistics on
land use in some of the most affected regions in Southern
Norway show that such changes cannot be the most important
reason for the regional acidification in the studied area.

It is feared that acid precipitation also will have an effect
on soil and vegetation. It has been shown in field and
laboratory experiments that acid precipitation reduces
microbiological activity in the soil and affects the
mineralization of nitrogen and the leaching of mineral
components. In spite of this, it has so far not been
established that this reduces forest growth in Scandinavia. It
is possible that the adverse effects of the pollutants are
balanced by the advantageous contribution of nitrogen
compounds. It is also possible that the changes which take
place in the soil are not large enough at present to affect
plant growth.

Under the auspicies of the Norwegian research project an
international conference on the effects of acid precipitation
was organized in Norway in 1976. A new scientific conference
is now being planned in Norway in 1980. The meeting will be
concerned with the ecological impact and fate of acid
precipitation, and also other pollutants found in association
with acid precipitation, e.g. heavy metals and low
concentrations of sulphur dioxide, and oxides of nitrogen. The
effects of these substances on:

a)   terresterial plants (natural assemblages, forest and
      agricultural crops) and soils,

b)   aquatic plants and animals, and water quality will be
     debated.

Among the topics will no doubt also be the reported
relationship between acidification and mercury contamination of
fish.

In addition to the SNSF project we have recently launched a
special study of the chemical, biological, economic and
transportation aspects of liming in the affected lakes and
rivers. We expect that it will prove possible to retain or
re-establish pH values which will make it possible for
indigenous fish species to survive in selected smaller lakes
and streams. It may also be possible to create a number of
"pockets" in rivers where fish may survive during brief
pH-drops. We do not expect, however, that it will prove
feasible to apply lime on a large scale within the whole area
of Norway affected by acidification.

INTERNATIONAL HARMONIZATION OF EMISSION POLICIES

At the end of April this year 35 governments participating in
the United Nations Economic Commission for Europe adopted a
draft Convention on Transboundary Air Pollution. The
Convention will be signed by governments at an ECE environment
meeting at ministerial level which will be held in Geneva in
November this year (1979).

This Convention represents the end result of a long and arduous
process which started with a proposal, put forward in the first
instance by Norway, for a convention to combat acid
precipitation and other international air pollution problems.
The parties to the ECE Convention are the governments of
Eastern and Western Europe and North America.

The Convention requires signatory states to do all in their
power to limit, and as far as possible gradually reduce and
prevent air pollution, including transboundary air pollution.
It provides for exchange of information, consultation,
cooperation in research and development, and on the basis of
these, for the development of measures to combat the discharge
of atmospheric pollutants. Such measures will include the use
of best available and economically feasible technology
particularly in new installations. The Convention also
provides for specific consultative procedures when major
projects are planned which may adversely affect the environment
in other countries.

I quote the article of the Convention which deals with
monitoring and related research:

"The Contracting Parties stress the need for the
implementation of the existing co-operative programme for
the monitoring and evaluation of the long-range
transmission of air pollutants in Europe (EMEP) and, with
regard to the further development of this programme,
emphasize:

(a)   the desirability of Contracting Parties joining in and
      fully implementing EMEP, which as a first step, is
      based on the monitoring of sulphur dioxide and related
      substances,

(b)   the need to use comparable or standardized procedures
      for monitoring whenever possible,

(c)   the desirability of basing the monitoring programme on
      the framework of both national and international
      programmes. The establishment of monitoring stations
      and the collection of data shall be carried out under
      the national jurisdiction of the country in which the
      monitoring stations are located,

(d)   the desirability to establish a framework for a
      co-operative environmental monitoring programme, based
      on, and taking into account, present and future
      national, regional, subregional and international
      programmes,

(e)   the need to exchange data on emissions at periods of
      time to be agreed upon, of agreed air pollutants,
      starting with sulphur dioxide, coming from grid-units
      of agreed size, or on the fluxes of agreed air
      pollutants, starting with sulphur dioxide, across the
      national borders, at distances and at periods of time
      to be agreed upon. The method, including the model,
      used to determine the fluxes, as well as the method,
      including the model, used to determine the
      transmission of air pollutants based on the emissions
      per grid-unit, will be made available and periodically
      reviewed, in order to improve the methods and the
      models.

(f)   the willingness to continue the exchange and
      periodical updating of national data on total
      emissions of agreed air pollutants, starting with
      sulphur dioxide,

    (g)   the need to provide meteorological and physico-
           chemical data relating to processes during
           transmission,

    (h)   the need to monitor chemical components in other media
           such as water, soil and vegetation, as well as, a
           similar monitoring programme to record effects on
           health and environment,

    (i)   the desirability of extending the national EMEP
           networks to make them operational for control and
           surveillance purposes."

According to a resolution which has already been accepted in
principle, and which will be adopted together with the
Convention, implementation of the main provisions of the
Convention will begin as soon as possible after the Convention
has been signed, and before the Convention is ratified in
accordance with international legal procedure. The sulphur
problem will in particular be the subject of priority attention
by the Contracting Parties. An executive organ, composed of
representatives of the participating countries, will review
progress and supervise the implementation of the Convention,
assisted by a special secretariat unit which will be
established within the ECE.

The Convention represents an important step forward although it
does not contain the precise and binding commitment to reduce
sulphur pollution which Norway and the other Nordic countries
argued for during the negotiations.

Combatting air pollution will, for the first time, become a
matter of international cooperation. The industrialized
countries of Europe and North America will now have an
instrument which will facilitate the development of more
rational and mutually beneficial sulphur control policies.

A reduction of sulphur pollution will result in less damage to
vulnerable ecosystems both in areas close to emission sources
and in distant areas. In local areas there will also be health
benefits as well as a reduction in damage due to corrosion.

The polluted air masses which spread over Europe and North
America come from many areas and many different emission
sources. For this reason alone it will no doubt take time
before we can expect any improvement with respect to acid
precipitation. The Convention will, however, by its very
existence create pressure on governments to take the acid

precipitation problem into account when determining their
pollution control policies.

The present expanded use of coal in energy supplies will have
negative impacts on the air pollution situation. This can be
resisted only if the relevant authorities exert efficient
efforts to require the best technical solutions - and most
stringent measures - for limiting and reducing pollution
discharges. The goal should, in my view, be, as a minimum, to
prevent a deterioration of the overall air pollution situation
in the next decade, even with an increased dependence upon coal.

The Convention will also provide a machinery to tackle other
emerging international air pollution problems.

Bearing in mind that we still have a considerable way to go in
coping with long range air pollution problems, we should see to
it that the objectives and the potentials of the Convention are
utilized to the maximum extent possible to prevent further
deterioration and bring about a reduction of emissions.

SESSIONS CHAIRMEN

SESSION I: THE CHEMISTRY OF POLLUTED RAIN

                Taft Y. Toribara, Ph.D.
                The University of Rochester
                Rochester, New York

SESSION II: THE MERCURY PROBLEM

                George H. Tomlinson, Ph.D.
                Domtar Inc./ Research Centre
                Senneville, Quebec

SESSION III: EFFECTS ON PLANTS

                Morton W. Miller, Ph.D.
                The University of Rochester
                Rochester, New York

SESSION IV: ANTICIPATED PROBLEMS AS YET NOT QUANTITATED

                Paul E. Morrow, Ph.D.
                The University of Rochester
                Rochester, New York

SESSION V: CONTROL PROBLEMS (MONITORING SYSTEMS AND LEGAL ASPECTS)

                Dennis J. Sugumele
                Rochester Gas and Electric Corporation
                Rochester, New York

SPEAKERS

Jay A. Bloomfield, Ph. D.
Dept. Environ. Conservation
Albany, New York 12233

Robert W. Elias, Ph. D.
Polytech. Inst. of Virginia
Blacksburg, Virginia 24061

Lance S. Evans, Ph. D.
Brookhaven National Laboratory
Upton, New York 11973

K. Lal Gauri, Ph. D.
Univ. of Louisville
Louisville, Kentucky 40208

Peter Herzberg, J. D.
Sierra Club Legal Defense
    Fund, Inc.
Washington, D.C. 20005

Bruce B. Hicks, M. Sc.
Argonne National Laboratory
Argonne, Illinois 60439

George M. Hidy, D. Eng.
Environ. Research and Tech., Inc.
Westlake Village, Calif. 91361

Jay S. Jacobson, Ph. D.
Boyce Thompson Institute
Cornell University
Ithaca, New York 14853

Arne Jernelöv, Ph. D.
Swedish Water and Air Pollution
    Research Institute
Göteborg, Sweden

Sagar Krupa, Ph. D.
Univ. of Minnesota
St. Paul, Minnesota 55108

Howard M. Liljestrand, Ph. D.
Calif. Inst. of Technology
Pasadena, California 91125

Erik Lykke, Director General
Ministry of Environment
Oslo 1, Norway

Daniel R. Mandelker, J. D.
Washington Univ. School of Law
St. Louis, Missouri 63130

Michael J. Matteson, D. Eng.
Georgia Inst. of Technology
Atlanta, Georgia 30332

William W. McFee, Ph. D
Purdue University
West Lafayette, Indiana 47907

Ronald A. N. McLean, Ph. D.
Domtar Inc.
Montreal, Canada H3C 3M1

Edward L. Mills, Ph. D.
Cornell Biological Field Station
Bridgeport, New York 13030

Carl L. Schofield, Ph. D.
Dept. of Natural Resources
Cornell University
Ithaca, New York 14853

D. S. Shriner, Ph. D.
Oak Ridge National Laboratory
Oak Ridge, Tennessee 37830

Gary J. Stensland, Ph. D.
Illinois State Water Survey
Urbana, Illinois 61801

Arend J. Vermeulen, Chem. Engr.
Dept. of Environmental Control
Overveen, The Netherlands

Herbert L. Volchok, Ph. D.
Environ. Measurements Laboratory
New York, New York 10014

Stephen O. Wilson, M. A.
New York State Energy Research
    and Development Authority
Albany, New York 12223

John M. Wood, Ph. D.
Univ. of Minnesota
Navarre, Minnesota 55392

PARTICIPANTS

Kurt Anderson
Empire State Electric Energy
Research Corporation
Schenectady, New York

Joseph Gorsuch
Eastman Kodak Company
Rochester, New York

Kenneth Juris, Ph.D.
Empire State Electric Energy
Research Corporation
Schenectady, New York

John Kadlecek, Ph.D.
State Univ. of New York
Albany, New York

Daniel Matias
New York State Gas and Electric
Company
Binghamton, New York

Jean Muhlbaier, Ph.D.
Gneral Motors Research Lab.
Warren, Michigan

Cornelius Murphy, Ph.D.
O'Brien & Gere Engineers, Inc.
Syracuse, New York

Scott O. Quinn, Ph.D.
New York State Department of
Environmental Conservation
Albany, New York

Dudley Raynal, Ph.D.
Syracuse University
Syracuse, New York

John Schott, Ph.D.
Calspan Corporation
Buffalo, New York

Ronald Scrudato, Ph.D.
State University College
Oswego, New York

Ralph Tedesco
New York State Gas and Electric
Company
Binghamton, New York